The Crisis of Food Brands

To mormor and morfar, in loving memory – AL
To all of those who have given their help over my 25-year association with food and farming – MKH
To Victoria – JV

The Crisis of Food Brands

Sustaining Safe, Innovative and Competitive Food Supply

ADAM LINDGREEN,
MARTIN K. HINGLEY
and
JOËLLE VANHAMME

Routledge
Taylor & Francis Group

LONDON AND NEW YORK

First published 2009 by Gower Publishing

2 Park Square, Milton Park, Abingdon, Oxfordshire OX14 4RN
711 Third Avenue, New York, NY 10017

Routledge is an imprint of the Taylor & Francis Group, an informa business

First issued in paperback 2018

Gower Applied Business Research
Our programme provides leaders, practitioners, scholars and researchers with thought provoking, cutting edge books that combine conceptual insights, interdisciplinary rigour and practical relevance in key areas of business and management.

Adam Lindgreen, Martin K. Hingley and Joëlle Vanhamme have asserted their moral right under the Copyright, Designs and Patents Act, 1988, to be identified as the editors of this work.

British Library Cataloguing in Publication Data
The crisis of food brands : sustaining safe, innovative and
  competitive food supply. – (Food and agricultural
  marketing)
1. Food industry and trade – Management 2. Food industry
and trade – Quality control 3. Consumers – Attitudes
I. Lindgreen, Adam II. Hingley, Martin III. Vanhamme,
Joelle
381.4'5641

Library of Congress Control Number: 2008944177

ISBN 978-0-566-08812-4 (hbk)
ISBN 978-1-138-38103-2 (pbk)

Contents

List of Figures

List of Tables

About the Editors

Adam Lindgreen

After graduating with degrees in Chemistry, Engineering and Physics, Professor Adam Lindgreen completed a M.Sc. in Food Science and Technology at the Technical University of Denmark. He also finished a MBA at the University of Leicester, as well as a One-Year Postgraduate Program at the Hebrew University of Jerusalem. Professor Lindgreen received his Ph.D. in marketing from Cranfield University. Since May 2007, he has served as a Professor of Strategic Marketing at Hull University Business School.

Professor Lindgreen has been a Visiting Professor with various institutions, including Georgia State University, Groupe HEC in France and Melbourne University; in 2006, he was made an honorary Visiting Professor at Harper Adams University College. His publications include more than 65 scientific journal articles, 6 books, more than 30 book chapters and more than 75 conference papers. His recent publications have appeared in *Business Horizons, Industrial Marketing Management*, the *Journal of Advertising*, the *Journal of Business Ethics*, the *Journal of the Academy of Marketing Science*, the *Journal of Product Innovation Management, Psychology & Marketing* and *Supply Chain Management*; his most recent book is *Managing Market Relationships* (Gower Publishing, 2008). The recipient of the 'Outstanding Article 2005' award from *Industrial Marketing Management*, Professor Lindgreen also serves on the board of several scientific journals; he is the Editor of the *Journal of Business Ethics* for the section on corporate responsibility and sustainability. His research interests include business and industrial marketing management, consumer behaviour, experiential marketing and corporate social responsibility.

Professor Lindgreen has discovered and excavated settlements from the Stone Age in Denmark, including the only major kitchen midden – Sparregård – in the south-east of Denmark; because of its importance, the kitchen midden was later excavated by the National Museum and then protected as a historical monument for future generations. He is also an avid genealogist, having traced his family back to 1390 and publishing widely in scientific journals related to methodological issues in genealogy, accounts of population development and particular family lineages.

Martin K. Hingley

Dr Martin K. Hingley has degrees from three UK universities: He first graduated (Agricultural and Food Marketing, B.Sc Honours) from the University of Newcastle upon Tyne; he has an M.Phil in Marketing from Cranfield University; and he received his Ph.D. in Marketing from the Open University. Dr Hingley is a Reader in Marketing and Supply Chain Management at Harper Adams University College, the leading university in the United Kingdom specializing in agri-food business. He is also a Visiting Fellow to Hull

University Business School and has previously held a Fellowship endowed by Tesco plc. Dr Hingley has wide-ranging business experience in the international food industry and has spent some time in the provision of market and business analysis with the Institute of Grocery Distribution, a leading UK research and training organization.

Dr Hingley's research interests are in applied food industry marketing and supply chain relationship management. He has presented and published widely in these areas, including in the *British Food Journal, Industrial Marketing Management* and the *Journal of Marketing Management.* He serves on the board of several scientific journals and also regularly guest edits such journals.

Joëlle Vanhamme

Dr Joëlle Vanhamme is Associate Professor at the IESEG School of Management at the Catholic University of Lille and Assistant Professor at the Rotterdam School of Management at Erasmus University. Dr Vanhamme has a degree in Business Administration and two master degrees (Psychology and Business Administration), all from the Catholic University of Louvain. She also has been awarded the CEMS master degree from the Community of European Management Schools. Subsequently, Dr Vanhamme received her Ph.D. from the Catholic University of Louvain; the Ph.D. thesis examined the emotion of surprise and its influence on consumer satisfaction. She has been a Visiting Scholar with Delft University of Technology, Eindhoven University of Technology and the University of Auckland; she is currently a Visiting Professor with Hull University Business School. Dr Vanhamme's research has appeared in journals including *Business Horizons, Industrial Marketing Management,* the *Journal of Business Ethics,* the *Journal of Consumer Satisfaction, Dissatisfaction and Complaining Behavior,* the *Journal of Customer Behaviour,* the *Journal of Economic Psychology,* the *Journal of Marketing Management,* the *Journal of Retailing, Psychology & Marketing* and *Recherche et Applications en Marketing.*

About the Contributors

Luís Kluwe Aguiar

Mr Luís Kluwe Aguiar is Senior Lecturer of Marketing and International Business at the Royal Agricultural College and an Associate Lecturer at Open University Business School. Mr Aguiar received his M.Sc. degree in Agricultural Economics from the University of London-Wye College. Prior to becoming a full-time academic, he worked for many years in both the private and public sectors and lectured on a part-time basis. His extensive international research and consultancy work focuses mainly on marketing and consumer studies, especially in relation to ethical consumerism.

Kristian Alm

Dr Kristian Alm is Associate Professor of Ethics at the Institute for Strategy and Logistics, BI Norwegian School of Management, Oslo. Dr Alm received a Ph.D. from the University of Oslo in 1999 with a doctoral thesis entitled: 'Visions of freedom from guilt.' He is also the coordinator of SRI-EBEN, a cross-national network of ethical investments. Research interests include the ethical management of the Norwegian Government Pension Fund Global.

Dao The Anh

Dr Dao The Anh is an Agricultural Scientist with a Ph.D. in Economics. He is the director of the Centre for Agrarian Systems Research and Development of the Food Crop Research Institute in Vietnam. His major interests include the strategies of small-scale family farmer households, farmer organizations and participatory policy design. He has published a number of papers on Vietnamese agriculture in local and international journals (for example, *Acta Horticulturae*, *Agriculture and Rural Development*, and *Cahiers Agricultures*).

John Byrom

Dr John Byrom is Deputy Head of the School of Management at the University of Tasmania. He is also Associate Dean (Launceston and Cradle Coast Campuses) within the Faculty of Business. His research interests centre mainly on the retail sector, and his research has been published in the *International Review of Retail, Distribution and Consumer Research* and the *Service Industries Journal*, amongst others. Dr Byrom is also Academic Editor of the *Journal of Place Management and Development*.

Conor Carroll

Mr Conor Carroll is a Lecturer in Marketing at the Kemmy Business School, University of Limerick, Ireland. Mr Carroll has written more than 55 published marketing case studies, many of which are used extensively in leading European and International Universities. He has won numerous accolades for his case study writing at both national and international competitions. Furthermore, Mr Carroll has delivered seminars on case study writing at Harvard Business School. He is the winner of the 2004 Teaching Excellence Award at the University of Limerick for his teaching activities within the University; he also has won other international teaching awards.

Luca Cesaro

Dr Luca Cesaro is Senior Researcher and Head of the Research Unit at the National Institute of Agricultural Economics in Rome. He received his Ph.D. at the University of Padova. He is also Professor of Economics and Management at the University of Padova. He has coordinated several European and national research projects in the field of agricultural economics and farm management, published more than 150 articles in various journals and edited 6 books. His research interests include rural and regional policies analysis and evaluation, forest economics and policy.

Sylvain Charlebois

Dr Sylvain Charlebois has been an Associate Professor in Marketing in the Faculty of Business Administration of the University of Regina, Canada, since 2004. His current research interest lies in the broad area of food distribution and safety, and he has published many articles in international academic journals. He has given lectures on food safety in many countries around the world. He has authored two books related to agriculture issues and marketing and is a co-author of the *Food Safety Performance World Ranking 2008*. Dr Charlebois is currently working with the C.D. Howe Institute, is a Faculty Research Fellow for Viterra, and serves as an Associate Researcher for the Montreal Economic Institute.

Tove Christensen

Dr Tove Christensen is a Senior Researcher in the Consumption, Health and Ethics Division, Food and Resource Economics Institute. She received her Ph.D. from the Institute of Economics, University of Copenhagen, and recently has had papers accepted in *European Journal of Integration* and *Journal of Food Distribution Research*. Other publication activities, directed toward Danish media, include agricultural and economic journals and magazines, as well as a book, conference proceedings, research reports and working papers. Her research interests include the optimal regulation of externalities related to agricultural activities, governmental intervention, and consumer perceptions and behaviour in relation to food safety and (other) non-market goods.

Federica Cisilino

Dr Federica Cisilino is a Researcher at the National Institute of Agricultural Economics. She received her Ph.D. from the University of Siena. She is Professor of Agro-environmental Economics and Management at the University of Udine. She has participated in several European, national and regional projects in the field of agricultural economics and sustainable farming, as well as published in several journals and books at the national and international level. Her research interests include agri-food economics, statistical economics applied to the agricultural sector, rural policies analysis, innovation in small and medium-sized enterprises, and sustainable and organic farming.

John W. Cone

Dr John W. Cone is Senior Scientist in Animal Nutrition at the Animal Nutrition Department of Wageningen University. Dr Cone received his Ph.D. from Wageningen University. He has published in *Animal Feed Science and Technology, Journal of Animal Physiology and Animal Nutrition, Journal of the Science of Food and Agriculture and Plant Physiology, Livestock Science,* and *Netherlands Journal for Agricultural Sciences,* among several others. His research interests are animal nutrition and animal rearing, plant sciences, rumen physiology, feed evaluation and in-vitro analytical techniques. He is a member of the British Society of Animal Science and serves on the editorial advisory board of *Animal Feed Science and Technology.*

Ana Isabel Costa

Dr Ana Isabel Costa is Research Fellow in Economics and Management of Innovation, Technological Change and New Product Development at the School of Economics and Management, Portuguese Catholic University. Dr Costa received her Ph.D. from Wageningen University. She has published in several journals, such as *Appetite, Food Quality and Preference, Journal of Food Products Marketing, Netherlands Journal for Agricultural Sciences,* and *Trends in Food Science and Technology.* Her research interests include innovation management, food marketing and consumption behaviour, research methodology in behavioural sciences, economic psychology, applied and social psychology, food science and technology, and human nutrition. She serves on the editorial advisory board of *Trends in Food Science and Technology* and *Scientific Journals International.*

Paul Custance

Dr Paul Custance is Principal Lecturer in Marketing at Harper Adams University College and received his Ph.D. from the University of Nottingham. He was Director of ruralconsultancy.com, the reach-out arm of the Business Management and Marketing Group at Harper Adams University College, for over 20 years. During that time, he helped prepare a wide range of reports for government bodies, regional development agencies, multinational companies and local and regional enterprises. He also has published in

Business Strategy and the Environment, Journal of Business and Industrial Marketing, Journal of Rural Enterprise and Management and other journals. His current research interests include work-based learning, rural enterprise, branding and supply chain management.

Hans Dagevos

Dr Hans Dagevos is a Sociologist of Consumption and Senior Researcher/Project Coordinator at the Agricultural Economics Research Institute (LEI) of Wageningen University and Research Centre in The Hague, the Netherlands. Dr Dagevos received his Ph.D. with honours from Utrecht University in 1994. During the last decade, his research interests have been in the field of contemporary consumer culture, with a special focus on food consumption. He is co-editor of several books, including one about citizen–consumers and more recently one about the so-called obesogenic society, as well as the (co-)author of numerous articles in Dutch and scientific journals. Recently, he has published in *Food Policy, Appetite, British Food Journal* and *Risk Analysis*.

Per Engelseth

Dr Oecon. Per Engelseth is Associate Professor in International Marketing at the Department of International Marketing, Ålesund University College. Dr Oecon. Engelseth received his Dr Oecon from BI Norwegian School of Management in 2007. His research interests include the use of packaging to support efficient information exchange and goods identification, product traceability in supply networks, safety and quality issues in fish supply networks and developing knowledge, culture and integration to support purchasing and product supply. These issues are covered in a series of conference papers from the *Agrichain and Network* (University of Wageningen), *NOFOMA, ISL, RM-Summit* and *IMP* conferences, among others.

Roberto Esposti

Professor Roberto Esposti is Senior Researcher and Aggregate Professor of Agricultural and Environmental Economics at the Department of Economics, Università Politecnica delle Marche at Ancona (Italy). He received his Ph.D. from Trento University (Italy) and has been a Visiting Scholar at the Department of Agricultural and Applied Economics, University of Wisconsin at Madison (USA). He has published in several international scientific journals, including *Agricultural Economics, Agricultural Economics Review, American Journal of Agricultural Economics, Cahiers d'Economie et Sociologie Rurales, Empirical Economics, European Review of Agricultural Economics, Growth and Change, Journal of Productivity Analysis,* and *Regional Studies, Research Policy* among others. His research interests include agricultural technological change and productivity, agricultural development and economic growth, agricultural biotechnologies and environmental implications, rural development and agricultural, and rural development policies.

Muriel Figuié

Dr Muriel Figuié is a Sociologist at CIRAD, specializing in consumer studies and food risk perception analysis. She has conducted extensive food consumption surveys in Vietnam, with a focus on the demand for quality and health concerns. Her interests lie in the variety of perceptions and behaviour among 'experts' and 'laypeople'. Her publications include papers in *Agricultures, Risk Analysis,* and *Vietnam Social Sciences.*

Stephan Hubertus Gay

Dr Stephen Hubertus Gay is Scientific Officer at the European Commission's Joint Research Center, Institute for Prospective Technological Studies. Dr Gay received his Ph.D. from University of Natal. He has published in journals, including *Agrekon, European Review of Agricultural Economics, South African Journal of Economics,* and *Management Sciences,* among others. His research interests include analyses of the Common Agricultural Policy, trade modelling and non-tariff barriers in food trade.

Fatma Handan Giray

Dr Fatma Handan Giray has been a Research Fellow at the European Commission's Joint Research Center, Institute for Prospective Technological Studies, since 2005. She received her Ph.D. in Food Marketing and Marketing Research from Ankara University. Before joining IPTS, Dr Giray worked at the university (1991–1998) and for the Prime Ministry of Turkish Republic Southeastern Anatolia Project Regional Development Administration (1998–2005), where she participated in many research and implementation projects. Her research interests are rural development, agriculture and food marketing and food security and quality.

Berit Hasler

Dr Berit Hasler is a Senior Researcher and Head of the Social Science Section at the Department of Policy Analysis, National Environmental Research Institute, University of Aarhus. She received her Ph.D. from Institute of Economics, Royal Veterinary and Agricultural University, Copenhagen (now Food and Resource Economics Institute, Faculty of Life Science, University of Copenhagen). She has published in journals such as *Ecological Economics, Environmental and Resource Economics, Environmental Pollution, Journal of Environmental Management,* and *Nordic Hydrology,* as well as in books and conference proceedings. Her research interests include environmental and agricultural economics, valuation of non-market goods and services, integrated environmental and economic modelling and analysis, regulation and economic incentives, cost-minimization and modelling of nature protection and restoration, and nitrogen and pesticide pollution abatement.

Meike Henseleit

Dr Meike Henseleit is Research Associate at the University of Giessen, Germany. Dr Henseleit received her Ph.D. from the University of Bonn. She has published in several journals, including *Berichte über die Landwirtschaft, Ernährung im Fokus* and *BfN-Skripten*. Her research interests include consumer behaviour, externalities, and especially impacts on the climate, attitudes and corporate social responsibility.

Athanasios Krystallis

Dr Athanasios Krystallis is a Researcher at the Institute of Agricultural Economics and Policy (IGEKE) at the National Agricultural Research Foundation of Greece. He received his Ph.D. from the University of Newcastle upon Tyne. His scientific interests focus on consumer behaviour with emphasis on perceived intrinsic and extrinsic food quality and food safety and their implications on marketing strategy of various food products including, for example, organic foods and genetically modified products. Also, his expertise includes the statistical modeling of food consumer behavior in the above-described issues (quality and safety management). Dr Krystallis has participated in several research projects at the national and the EU level. His research comprises more than 35 peer reviewed publications in, among others, *Appetite*, the *British Food Journal, Risk Analysis*, the *European Journal of Marketing, Food Quality and Preference, Health Risk and Society* and the *Journal of Consumer Behaviour.*

Rainer Kühl

Dr Rainer Kühl is a full Professor at Justus-Liebig University of Giessen, Germany, and holds the Chair of Food Economics and Marketing Management. After graduating and receiving his Ph.D. in agricultural economics from the University of Kiel in 1984, his career path led him from the university to the agribusiness industry, where he worked for 3 years at the national centre of agricultural cooperatives. In 1987, he became an Assistant Professor for Agricultural Economics at Kiel University. In 1992, Kühl became Professor at the University of Bonn, until 1999, when he moved to his current position. His main research interests are food industry analysis and business strategies, vertical integration and contracting in the food supply chain, strategic marketing and management of innovations, quality management and consumer behaviour.

Monique Libeau-Dulos

Ms Monique Libeau-Dulos is Scientific Officer at the European Commission's Joint Research Center, Institute for Prospective Technological Studies. Ms Libeau-Dulos coordinated the project 'Quality assurance and certification schemes managed within an integrated supply-chain.'

Michalis Linardakis

Dr Michalis Linardakis is an adjunct Assistant Professor of Statistics in the Department of Regional Economic Development at the University of Central Greece. He received his Ph.D. in Applied Statistics from Athens University of Economics and Business. Dr Linardakis has published in *Acta Radiologia*, the *Journal of the Royal Statistical Society-Series C (Applied Statistics)* and *Sociologia Ruralis*, among others. His research interests are in applied statistics in economics, business, marketing, education and medicine.

Nguyen Thi Tan Loc

Ms Nguyen Thi Tan Loc is a Researcher in marketing at the Fruit and Vegetable Research Institute of Vietnam. She holds a M.Sc. in Marketing from a French training institute specializing in tropical agriculture. Her research interests include the food distribution chains to supermarkets and consumer food preferences. She has published papers in local Vietnamese journals, including the *Journal of Consumer Association* and *Agriculture and Rural Development*, and has also contributed to papers in *Acta Horticulturae*.

Nguyen Ngoc Luan

Mr Nguyen Ngoc Luan is Head of Department for Sustainable Rural Development at Rural Development Centre–Institute of Policy and Strategy for Agriculture and Rural Development (RUDEC–IPSARD) in Vietnam. He has published in *Journal of Development and Integration* and on the websites of *Rudec* and *Ipsard*. His research interests include consumer habits and satisfaction, willingness to pay, food safety, the value chain for agricultural products and new models of sustainable rural development, among others.

Spyridon Mamalis

Dr Spyridon Mamalis is Assistant Professor in Marketing in the Department of Business Administration at the Technological Educational Institute of Kavala. He has published in, among others, *Agricultural Economics Review*, *Food Quality and Preferences*, the *Journal of International Food and Agribusiness Marketing* and *Tourism Economics*. His research interests include consumer behaviour, travel and leisure marketing, marketing research and food marketing. He serves on the board of many journals.

Paule Moustier

Dr Paule Moustier is a Food Market Specialist at CIRAD, the French research centre specializing in tropical agriculture. She has been based in Vietnam for the last 5 years and is presently in charge of a project concerning small farmers' access to supermarkets and other quality distribution chains. In her 15 years of experience in Africa and Asia, she has been involved in research and training in the areas of food marketing, peri-urban agriculture and institutional economics applied to commodity chain analysis. She has published in various French journals on agriculture and social science (*Autrepart, Economie Rurale, Cahiers Agricultures, Fruits, Les cahiers de la multifonctionnalité*), as well as in *Acta Horticulturae, Development Policy Review* and *Supply Chain Management*. She has also coordinated books published by CIRAD, IDRC and ADB.

Morten Raun Mørkbak

Mr Morten Raun Mørkbak is a Ph.D. student at University of Copenhagen, Institute of Food and Resource Economics, Consumption, Health and Ethics division. He received his Masters degree in Environmental and Natural Resource Economics from the faculty of Life Sciences (formerly Royal Veterinary and Agricultural University), University of Copenhagen. His publication activities have been directed toward Danish media, including magazines, research reports, working papers and conference proceedings. His research interests centre on the economic valuation of both market and non-market goods, risks and different perceptions of risk and cost–benefit analyses, with special emphasis on food safety issues and environmental restoration projects.

Melina Parker

Dr Melina Parker is Manager of Milton Farm Pty Ltd, a sixth-generation family firm based on Tasmania's north-west coast. Milton Farm grows vegetables for fresh and processed markets across Australia and is also engaged in new product development, principally wasabi. Dr Parker has a Ph.D. from the University of Melbourne in the fields of strategic management and relationship marketing.

Stephen Parsons

Mr Stephen Parsons is an Economist with longstanding interests in government policy as it relates to farming, the countryside and food supply. From 1997–2002, he was responsible for the direction of the Marches Farm Enterprise Programme, which gave practical support and guidance for the development of non-farm but on-farm enterprises in those parts of rural England bordering on Wales. Since then, he has been involved in further encouraging entrepreneurship in the countryside, particularly with a focus on projects that take advantage of the benefits to health and rehabilitation that the rural environment provides. He is currently Vice-Chair of the National Care Farming Initiative (UK).

Virginie Diaz Pedregal

Dr Virginie Diaz Pedregal is a post-doctorate candidate in Sociology at the French Agricultural Research Centre for International Development (CIRAD), MOISA joint research unit. She received her Ph.D. from University Paris 5-Sorbonne. She has published in *Consommations et Sociétés, Flux, Développement Durable et Territoires* and *Journal of Macromarketing*, among others. She is also the author of two books on fair trade and fairness of distribution, published at *L'Harmattan*. Her research interests include fair trade, food consumption, consumer behaviour toward food risks and producer organizations for sustainable development.

Marco Platania

Dr Marco Platania is an Agricultural Economics Researcher at Mediterranea University, Reggio Calabria (Italy), where he teaches models and theories of food consumption, agricultural economics and agroindustrial economics. He received his Ph.D. from University of Catania. A member of the Society Italian of Agricultural Economists (SIDEA) and European Association of Agricultural Economists (EAAE), Dr Platania has authored several papers in Italian and international publications. His research interests include information and communication technologies applied to the agricultural sector, quality food product markets, agricultural marketing, consumer behaviour, and local and rural development.

Donatella Privitera

Ms Donatella Privitera, an Agricultural Economy Researcher with a Masters degree from the Catholic University of Milan, teaches marketing of agricultural products, the economy of landscape and the economy of food product markets for the Faculty of Agriculture at University Mediterranea, Reggio Calabria. A member of Society Italian of Agricultural Economists (SIDEA) and European Association of Agricultural Economists (EAAE), she has authored more than 45 scientific publications, mainly associated with her research in the following areas: information and communication technologies in agriculture; agribusiness in Italy; quality food product markets and strategies; consumer behaviour and trends in agro-food marketing; multifunctional agriculture and its implications for agro-tourism; and landscape marketing. At present, she is participating in cooperative research projects developed between the Calabria Regional Authority and University of Reggio Calabria.

Jofi Puspa

Dr Jofi Puspa is a Researcher (Post Doctoral fellow) at the Justus-Liebig University of Giessen, Germany, Chair of Food Economics and Marketing Management (Professor Kühl). After finishing graduating from the Environmental Biology Faculty (1990) at Satya Wacana University, Indonesia, she worked at PT Bayer Indonesia and PT Glaxo Indonesia in their pharmaceutical departments. She received a full scholarship from PT Bayer Indonesia to take her Masters degree in Business Administration. In 2000, she moved to Germany and finished her Ph.D. in Nutritional Science at Justus-Liebig University of Giessen, Germany in 2005. Since then, she has been working at her current position. Her research interests are psychological and economics aspects of consumer behaviour, especially food consumption, nutrition and disease prevention; personal communication; managing innovation; and the strategic marketing of food products

Takeo Takeno

Dr Takeno is Associate Professor of Information Systems at Iwate Prefectural University. Takeno received his Ph.D. of Engineering from Tokyo Metropolitan Institute of Technology. He has published in *Journal of Computers and Industrial Engineering, Journal of Engineering Design and Automation, Journal of Japan Industrial Management Association, Journal of Japan Information-Culturology Society, Journal of Logistics and SCM Systems,* and *Journal of Production Economics,* among others. His research interests include information system applications in production and logistics systems and mathematical programming.

Luciana Marques Vieira

Dr Luciana Marques Vieira is Lecturer in Management for the Postgraduate Program (PPG) at Unisinos Business School, Brazil. She received her Ph.D. from the University of Reading, England. Luciana has published in international journals, such as *British Food Journal* and *Brazilian Business Review.* Her research interests include global agri-food chains management and international operations strategy.

Pénélope Vlandas

Ms Pénélope Vlandas was Scientific Officer at the European Commission's Joint Research Center, Institute for Prospective Technological Studies. Ms Vlandas coordinated the stakeholder consultation process in the framework of the project 'Quality assurance and certification schemes managed within an integrated supply-chain.'

Tim Voigt

Mr Voigt is a Research Fellow at Justus-Liebig University of Giessen, Germany, Chair of Food Economics and Marketing Management (Professor Kühl). After graduating with a degree in Business Economics from Philipps-University Marburg in 2005, he started his Ph.D. thesis on 'Cooperation and Innovation in the Context of Strategic Management.' His research interests include strategic paths in the food industry, innovation management, evolutionary economics and industry simulation models. Since 2005, he has been Manager of the Institute for Cooperative Science at Justus-Liebig University of Giessen.

Keith Walley

Dr Keith Walley is Head of the Business Management and Marketing Group at Harper Adams University College. He received his Ph.D. from Bradford University. He has extensive commercial research experience in sectors ranging from retailing to agricultural inputs and auto-components. He has published in a variety of management journals such as the *British Food Journal, Industrial Marketing Management, Business Strategy and the Environment, International Small Business Journal, Journal of Business and Industrial Marketing* and *International Studies of Management & Organization.* His research interests include competitiveness, competitive advantage, coopetition and consumer behaviour.

Lisa Watson

Dr Lisa Watson is Assistant Professor of Marketing at the University of Regina Faculty of Business Administration. She has previously published in such journals as *European Journal of Marketing* and *Social Marketing Quarterly.* Her research interests lie in the areas of consumer welfare, social marketing and entrepreneurial marketing.

Reviews for *The Crisis of Food Brands*

"The Crisis of Food Brands *brilliantly spotlights the challenges facing global agri-business. Professors Lindgreen, Hingley, and Vanhamme have collected together a range of fascinating studies demonstrating that global agriculture requires a professional marketing approach that takes into account the needs and concerns of all stakeholders. This is especially so given we are dealing with that most precious of commodities – food."*

Prof. Michael B. Beverland, Royal Institute of Technology, Australia

"The food industry is faced with unprecedented issues in production and marketing, the adoption of new technologies, and available labour-issues that are both considerable and fundamental. Against a rich background of quantitative and qualitative research sourced from the United Kingdom and overseas, The Crisis of Food Brands *offers a timely, fresh, and wider perspective on these critical issues. The book offers both theoretical and practical insights, and is a must-have for anyone with responsibility for marketing food, communicating about the food industry, or connecting with consumers."*

Sir Don Curry, Chairman of Defra, UK

"The editors are to be congratulated in assembling a range of contributions which as an ensemble present a holistic and whole-chain overview of key current controversies within global food and agricultural marketing systems. There is a strong focus on food safety scares and on the management of risk and perceived risk at both the product and brand levels in the international marketplace. But equally consideration of the wider emerging underlying ethical issues and responses in supply chains are addressed, including those of organics, animal welfare, the roles of quality assurance systems, the measurement of consumer's responses to ethical products and competitive behaviour within food manufacturing and retailing. It is a very well crafted body of evidence, which highlights tensions and harmonies in the implementation of theory into practice. It will be relevant for a whole range of readers."*

Prof. Wynne E. Jones, OBE, Principal, Harper Adams University College, UK

"Food is essential for survival and health of man but, at the same time, sensitive to fraud, mistakes, superstition, protectionism and much more. The potential for controversies is plentiful. This interesting book gives a lot of relevant examples like functional food versus organic, versus conventional, versus genetically modified. There are detailed descriptions of very problematic marketing of contaminated products with subsequent recalls. Chapters from Vietnam, Australia and Europe give widespread aspects. Forty four contributors bring statistically based documentation to the hopefully many readers. The book is highly recommended."*

K. Porsdal Poulsen, Prof. h.c., Technical University of Denmark

Foreword and Acknowledgments

As the global market for and movement of food grows apace, it brings to the fore the considerable and fundamental issues of the cost of choice and the worldwide impact of freer markets in production, marketing, technologies and labour. Specifically, there are governance, market structural and ethical consequences that must be considered. Developments in agri-food may not easily cross national boundaries in terms of market adaption, regulation and, most important, consumer acceptability. Issues that have emerged from developed, western food and agricultural economies (quality, health, environmentalism, animal welfare) are becoming global issues, as is the case for the development of innovative production and food science and systems, centralized supermarket channel power and the brand strength of global companies. From all of the undoubted benefits of the modern agri-food economy, there are also many problem areas to be addressed if we are to realize the best and fairest systems in the delivery of good food choices for all.

The overall objective of this book is to provide a wide-ranging collection of cutting-edge research on controversies in food and agricultural marketing, especially in terms of the consequences for businesses and appropriate marketing strategy plans. This book's 19 chapters are organized into four sections: (1) Food crisis and responsibility, (2) Agri-food systems, product innovation and assurance, (3) The consumer view and (4) Fair engagement? The section themes are introduced and the chapters briefly outlined here.

Food Crisis and Responsibility

The agri-food industry has been rocked by notable and, at times, momentous crises of disease and food quality (for example, contamination, poisoning). Sometimes, problems are localized to specific product or geographic areas; however, in this age of global sourcing and markets, such localization is seldom the case for long. In many cases, the response has been support for isolationism based on nationalistic market protection (for example, the responses of European neighbours to the UK BSE outbreak in the 1990s). However, this approach has become increasingly unrealistic. What globalization has taught us is that protectionism is unworkable when world trade and interlinked, cross-border production, value adding and consumption is commonplace. Another lesson we note is that the developing world is fast catching up with the developed world in terms of not only food products, systems and product branding but, perhaps more significantly, in the development of consumer power and lobby strength.

In this first section of the book, we consider crises and critical incidents for food. One of the major lessons to be learned is that global brand power is no insulation against errors of judgment (and worse) – matters critical to food quality and safety. Similarly, it matters increasingly less whether organizations operate in developed or emerging economies, because consumer expectations of brand owners the world over are converging in their demands for moral and ethical leadership on issues of quality, safety and assurance, whether domestic or international. The chapters in this section demonstrate the need for a coordinated and interrelated stakeholder approach to food safety and quality and in response to crisis incidents. Such an approach would further help head off the possibility of future calamity. Therefore, the ideal is sound quality and safety systems, as well as the proactive network integration of all key stakeholders, including governance, principal channel actors, consumers and media.

Chapter 1 (Carroll) presents the failure of the launch of Dasani® bottled water by Coca-Cola® in Europe. This chapter encapsulates the key issues associated with the risks and dangers in international product development, even for global brand giants. The chapter also provides a significant lesson in crisis management when faced with a controversy concerning product contamination. In particular, it offers a key consideration of the importance of stakeholder engagement and role of the media in handling crisis and responsibility.

In a second offering by the same author, Chapter 2 (Carroll) explores how another global household name, Cadbury, handled a major and costly contamination crisis. In this case, stakeholders' perception of the crisis provides the main focus of attention, including how the interpretation and handling of the problems differed among those involved. Corporate organizations, to their detriment, underplayed the perception and reality of the risk to consumers; the media was the catalyst in bringing the controversy into focus. The chapter thus identifies a significant lesson with regard to organizational learning.

Chapter 3 (Charlebois and Watson) also provides a case study of contamination risk, this time regarding an outbreak of *E. coli* in fresh spinach. This chapter conceptualizes the process of risk and crisis management by creating a model that encapsulates a response framework for all of those concerned and that makes critical recommendations for systematic handing of future crises.

By taking a supply chain view of food quality and safety issues in a continuous channel, Chapter 4 (Engelseth, Takeno and Alm) draws on supply chain management and integrates this knowledge stream with issues of ethics, food safety and channel information transparency. This approach highlights the importance of a whole chain and network perspective in dealing with matters of risk, responsibility and assurance.

In Chapter 5, Pedregal and Luan investigate the controversies surrounding marketing a food product in an emerging economy whose growing demand outstrips supply and the consequent dangers of 'cutting corners'. This case involves milk, specifically, passing off powdered milk as fresh in Vietnam. The issues identified in previous cases regarding the risks and controversies surrounding product and brand development in western economies are being fast followed in developing countries. Domestic and exporter organizations ignore consumer sensitivity at their peril; this case highlights the importance of organizational and government accountability and the need to implement processes for consumer protection and engagement.

Agri-food Systems, Product Innovation and Assurance

Controversy often surrounds the development of new agri-food production systems and methods of production. In some cases, the contention can loom for a considerable length of time, whereas in others, it may be simply part of the natural process for consumer acceptability in a world of increasing choice. Consumers often lead busy lives and assimilating detailed choices about complex food production systems and scientific data can be difficult. As a result, consumers rely on the short-cut guidance provided by product promotional campaigns to make their purchase decisions.

This section of the book therefore details some controversial food systems, ingredients and processes that consumers now face. Genetically modified production systems, foods and ingredients, for example, create not only emotive and often polarized responses from concerned parties but also consumers who are left bewildered by the partisan views for and against them. Similarly, functional foods may offer some tremendous benefits but also open up a world of confusion surrounding the multiplicity of ingredients, product offers, combinations and benefits. The chapters in this section consider controversial food systems and processes to rationalize and create frameworks for consumer choice. Furthermore, this section explores the important theme of the approach to market governance and assurance, along with appropriate ways to communicate them to consumers in a world of conflicting information and expanding choices.

The opening contribution, Chapter 6 (Gay, Giray, Vlandas and Libeau-Dulos), sets the scene by outlining and defining the different types of and differences between quality assurance schemes. Their impact on the food chain is notable and empirical support identifies the impact of assurance on stakeholders. The authors draw some conclusions regarding the present and future role of assurance schemes.

Chapter 7 (Cisilino and Cesaro) considers the challenges and development of a particular agricultural system, namely, organic production of wine and the subsequent issues of marketing this system. The authors adopt an Italian perspective. Furthermore, this chapter follows the theme established in the prior chapter regarding assurance and market regulation and makes some recommendations for system and market developments.

The Italian context again appears in Chapter 8 (Platania and Privitera), though this chapter refers to a production system diametrically opposed to the organic method. That is, it focuses on the contentious market for genetically modified products and outlines consumer attitudes toward such products. These attitudes are complex, but they do not necessarily reflect poor label information. Controversy and confusion surround the issue, the role of media is significant in generating and driving opinion, and great general mistrust of the food industry exists. Some future strategies for developing genetically modified products are proposed.

Chapter 9 (Esposti) is the first of three chapters that considers different issues related to the development of functional foods. This first chapter considers the issues and controversies surrounding the challenges of meeting consumer expectations and demands through innovation and niche marketing, as well as how expectation and technological development can come together to combine what might have been seen as divergent ideas, such as functional and 'natural' foods.

The theme of functional foods continues in Chapter 10 (Puspa, Voigt and Kühl), which draws on different sector disciplines (pharmacy and food) to raise to the challenge of managing organizational competences and capabilities in an area still open to debate, that is, scientific evidence of positive health links and diverse consumer attitudes. The authors use network marketing theory to provide a framework for managing competences in new market areas, centred on health and well-being, where matters of consumer trust are vital.

The final chapter on functional foods, Chapter 11 (Krystallis, Linardakis and Mamalis) focuses on the controversial area of children's diet and health. This empirical study considers the role of parents and their aspirations for their children regarding diet, as well as how buyers manage the difficult equation of managing health, value and taste aspirations.

The Consumer View

In the increasingly demand-driven world of food choice, it seems easy to make assumptions about how consumers see the world, and then to fall into the trap of identifying neat, standard viewpoints of consumer needs and wants. Yet consumer attitudes are diverse, and opinions constantly evolve. Consumers (especially in developed economies) may offer, when asked, received wisdom or attitudes that they believe are expected of them, or perhaps even what they think the questioner wants to hear; but these responses do not necessarily translate into their buying behaviour. In the real world of harsh economic choices, consumers can rationalize contradictory personal actions – believing and stating one thing but behaving in an entirely opposite way when it comes to actual purchases – as is all too apparent in the controversies surrounding ethical questions of production methods and their consequences on the planet, people and other species.

This third section attempts to plumb consumer attitudes toward controversial food market systems, products and developments. Although consumers appear concerned and responsive to the world around them, they are not supine, nor do they compliantly buy in to associated product offers developed on the back of ethical agendas.

Thus, Chapter 12 (Walley, Custance and Parsons) starts off by providing an overview of empirical work that evaluates UK consumer views regarding food, farming and the impact of economic activity on the environment, animal welfare and production practices. This longitudinal study tracks the crises, twists and turns in the food and farming controversy that has provided signposts on recent history. The catalogue of consumer responses outlines levels of concern, as well as the responses and compromises made by consumers who make their food choices against a backdrop of constant change.

Chapter 13 (Mørkbak, Christensen and Hasler) continues the theme of consumer attitudes toward animal welfare and food safety, this time using case material related to chickens and consumers' willingness to pay for various welfare attributes. The authors identify several differences between the stated willingness to pay and actual behaviour and emphasize the role of industry (for example, retailers) in purchase decision making.

To focus on ethics with regard to animal welfare, Chapter 14 (Costa and Cone) uses a detailed empirical investigation. The authors emphasize that consumers are driven by personal and self-interests; it is habit and hedonic preference, not ethical considerations, that tend to drive purchase behaviour.

Chapter 15's (Dagevos) study of consumer behaviour centres on organic foods, in an attempt to determine whether these foods are 'real' goods or 'feel' goods. That is, do consumers choose them for perceived or emotive reasons, or do they have a more grounded dimension? The proposed model encompasses an expanded marketing mix, comprised of eight Ps, to reflect what the author sees as a more rounded approach to decision making.

The last chapter in this section, Chapter 16 (Henseleit), investigates the issue of green consumerism and, as in previous chapters, consumer preparedness to pay for ethical choices. The author outlines the gap between consumer interest in environmentalism and actual consumer behaviour. This investigation also suggests some directions for business organizations that want to market green products more effectively and engage more constructively with their consumers.

Fair Engagement?

Power, influence and control all are highly significant to the global food economy. Supply chains can be long, even intercontinental, and the food sector, like all business chains and networks, contains certain power brokers. In demand-led chains, power often migrates to large, significant, global buyers, such as multinational brand manufacturers or retailers. In this often power-imbalanced food market, it becomes important to identify how smaller-scale or weaker players (though the range is large, most suppliers of big global branding and retailing organizations are smaller) manage their relationships in this context.

In this last section, Chapter 17 (Parker and Byrom) investigates the role of retailers as powerful and controversial figures in food and agriculture. The authors draw on relationship marketing theory and chronicle buyer-supply relationships in the Australian food supply. The controversy surrounds the often adversarial and power-skewed environment that delivers control to retailers over food suppliers. Case studies of buyer-seller relationships chart the issues of disaffection and offer some suggestions for directing and resolving such conflict.

The theme of supply channel equity continues in Chapter 18 (Moustier, Figuié, Anh and Loc), in the context of supermarkets in Vietnam; this chapter highlights the challenges and controversies of the global impact of foreign investment and the evolution of trading formats in emerging economies. Although the poor are excluded from the supermarket sector, the range and scale of street trading provides large-scale employment and supplier market access. The authors make recommendations for balancing the needs of diverse stakeholders in the supply, retailing and consumption of food.

Finally, Chapter 19 (Vieira and Aguiar) considers the issue of fair trade and its role in the global economy, as well as how small-scale specialist producers can compete in the complex and harsh world of global trade. The context, honey producers in Brazil, effectively represents the issues of trade barriers and market facilitators, which enables the authors to make recommendations for overcoming the disadvantages and deprivations of origin.

Closing Remarks

The double-blind process for selecting entries for this volume required the assistance of many reviewers who dedicated time and effort to provide helpful feedback to the authors. We greatly appreciate their work, which helped improved the chapters herein. We extend a special thanks to Gower Publishing and its staff, which has been most helpful throughout the entire process. Equally, we warmly thank all of the authors who submitted their manuscripts for consideration for this book. They have exhibited the desire to share their knowledge and experience with the book's readers – and a willingness to put forward their views for possible challenge by their peers. Finally, we thank Erasmus University Rotterdam, Harper Adams University College and Hull University Business School for their support in this venture. Special thanks go to Elisabeth Nevins Caswell and Jon Reast.

We are hopeful that the chapters in this book fill knowledge gaps for readers but also that they stimulate further thoughts and actions regarding issues of contention in the agri-food environment. Controversy can be both fast emerging and long lasting, and stakeholder understanding may be fragmented and contradictory. One clear message from the chapters of this book, however, is the ongoing need for an integrated, coordinated, open and transparent approach to handling and managing present, current and potential controversy.

Professor, Dr Adam Lindgreen, Hull University Business School.
Reader, Dr Martin K. Hingley, Harper Adams University College.
Associate Professor, Dr Joëlle Vanhamme, IESEG School of Management.

Food Crisis and Responsibility

1

The Dasani Controversy: A Case Study of How the Launch of a New Brand Jeopardized the Entire Reputation of Coca-Cola®

BY CONOR CARROLL*

Keywords

crises, contamination, crisis management, Dasani®.

Abstract

This chapter discusses the failed launch of Dasani® by Coca-Cola® into the European market. The Dasani case highlights the importance of crisis management and the implications of getting a new product launch fatally wrong, along with the dangers of a contamination scare and the ensuing implications for the parent brand. The following topics are discussed: the need for effective crisis communications during a food scare and the need for effective scenario planning. Also discussed is the role of the media during a crisis, the importance of stakeholder support during a crisis and, finally, the need for cultural awareness and sensitivity in international marketing.

Introduction

Dasani, a 'pure' still water brand, was launched in the UK in February 2004, with a huge promotional budget. The Dasani brand is well-established in the US, launched as a purified water product several years prior to tremendous success. The parent brand Coca-Cola had hoped to emulate that success by launching it in the UK and Europe, using the same

* Mr Conor Carroll, Kemmy Business School, University of Limerick, Department of Management and Marketing, Plassey, Co. Limerick, Ireland. E-mail: conor.carroll@ul.ie. Telephone: + 353 61 202 984.

marketing formula. But things went badly wrong. Coca-Cola could never have envisaged the level of negative publicity that would ensue, turning the launch of Dasani into one of the worst marketing debacles ever witnessed in Europe that damaged the company's European reputation once again after a recent contamination scare in Belgium.

Dasani initially sold for 95p per 500ml bottle, but the source of contents was ordinary tap water from Thames Water, a water utility company in the southeast of England, which charges only 0.03p for the same amount of water. The Dasani beverage thus was 3000 times more expensive than its key ingredient. A media frenzy ensued when consumers discovered that the expensive bottled water was in fact just processed tap water. The newly launched brand faced a barrage of negative publicity, but the company persevered with the launch. When it seemed that things could not get any worse, they did; the entire range of Dasani products had to be recalled in the UK when a known cancer-causing chemical was found in the water. This event signalled the death knell of the new brand in the UK, the postponement of the pan-European rollout of Dasani and the loss of millions of pounds in investments. The debacle highlights the importance of effective crisis management for reputation management. By tracing the UK launch of the maligned Dasani brand, this case illustrates the valuable lessons to be learned from this epic calamity. The case of Dasani will join the pantheon of other infamous UK marketing gaffes, such as 'The Hoover® Free Flight Promotion', Ratners and the crap comment, and the Post Office®'s confusing Consignia brand.

Dasani Launch

Dasani had the support of a £7 million promotional campaign for its launch. Coca-Cola initially developed the Dasani brand in the US, in an effort to capture the lucrative bottled water market. In just more than 20 years, this market had undergone a remarkable transformation. Previously, people drank bottled water only if they feared a contaminated tap water supply. Today, people consume huge volumes of bottled water as the drive toward healthier lifestyles continues. The success of bottled water brands largely rests on their claims of absolute purity and association with active and healthy lifestyles. Since its inception, the Dasani brand had achieved tremendous sales growth in its domestic US market, fully utilizing Coca-Cola's powerful distribution system. Based on this success, Coca-Cola decided to launch the brand in continental Europe, first in Britain, and then in other countries. Coca-Cola executives thought that the brand's American success would be quickly replicated in Europe. But things went wrong – badly wrong! They never envisaged the level of negative publicity that would ensue, turning the launch of Dasani into one of the worst marketing debacles ever witnessed in Europe.

Coca-Cola, throughout is history, has had it fair share of crises and controversy. In the mid-1980s, the company suffered its New Coke fiasco, in which the company replaced the original Coke brand with a newer version in an effort to defeat its archrival PepsiCo. The Pepsi® brand was gradually eroding Coke's market share with its Pepsi Taste Challenge campaign, and Coca-Cola executives thought a new formulation and rebranding of Coke was required. After extensive product and taste testing, New Coke was launched with huge fanfare. However, a cult following of diehard Coke fans were outraged with the demise of their beloved brand. National boycotts and protests were organized, forcing Coke to rescind its new strategy. Coke Classic reappeared on shelves.

Then in the 1990s in Belgium, a carcinogenic scare erupted, when a number of people became sick from drinking contaminated Coke. The contamination was later traced to a chemical used in the cleaning transportation pallets. However, the company was widely criticized for its slow reaction and handling of the crisis, which severely damaged its reputation in Europe.

Yet Coca-Cola remains one of the world's most ubiquitous brands, with a massive global presence and double-digit growth throughout the late twentieth century, achieved particularly through expansion to international markets. As this spectacular expansion levelled out, the company was striving to achieve new growth opportunities, and bottled water, the next big thing in the beverage industry, provided an avenue for future growth. The water sector achieved substantial growth rates compared with those for carbonated soft drinks (see Table 1.1), and all the large beverage companies continue to vie for a larger slice of the bottled water market. These companies pour massive investments into promoting their brands, including PepsiCo's Aquafina®, launched in 1995 as a non-carbonated, purified drinking water. Coca-Cola offered Dasani in response in 1999; it has grown to be the second most popular drink in the US. Coca-Cola previously distributed other water brands through its distribution networks, including a brand called Naya that enjoyed stellar success in the 1990s, with 30 per cent annual growth. However, after introducing the Dasani brand, Coca-Cola halted Naya's lucrative distribution agreement, seeking a bigger share of the bottled market in terms of both manufacturing and distribution.

Dasani achieved enormous success for a relatively new brand, ensured largely by Coca-Cola's huge marketing muscle and extensive distribution network. The brand of purified 'tap' water with added minerals sat beside natural spring waters from mountain peaks on shelves, yet it still won over customers. The American public appeared not particularly concerned or were simply apathetic about the origin of brands, as long the bottled water was safe. With this first foray into the bottled water market a barnstorming success, Coca-Cola viewed Europe as the next target. The company had long craved a successful new product launch in new product categories, rather than more similar line extensions, such

Table 1.1 Water market at a glance

- In the US, bottled water earns an estimated $8.3 billion in revenue (2003), compared with $1.1 billion in 1984.
- The UK bottled water market is estimated to be worth £1.1 billion, compared with an estimated worth of £360 million in 1998.
- Fastest growing sector of the drink market.
- The volume and value of UK bottled water marketing is expected to double by 2011.
- The sector includes still, sparkling, sport bottles, kids packs, flavoured and even light sparkling.
- 80 per cent of bottled water is sold as still water.
- Water with added flavours and minerals is a huge growth area. Some such waters are classified as health waters, while others aim at the fitness sector (for example, Reebok Fitness Water).
- Volvic is the UK market leader, followed by Evian, Highland Spring and Vittel.

---

as Vanilla Coke that achieved only modest growth targets. In the US, Coca-Cola even segmented the bottled water market by price, dividing it into three tiers: Dannon as the low-tier brand, Dasani in the middle and Evian as their high-price distribution., using strategic alliances to boost its portfolio with Danone, the French group.

Coca-Cola chose the UK as the launch site, because it appeared ripe for exploitation. British consumers drink far less bottled water than their continental neighbours, at 34 litres per capita, compared with Germany's 116 litres, France's 149 litres, Spain's 126 litres and Italy's 203 litres per capita. Predictions suggested the market for bottled water would grow rapidly in response to environmental factors, such the drive toward a healthier lifestyle and growing concerns about the safety and quality of local water supplies.

Aquafina, the key Dasani competitor in the US market, had not launched yet in the UK because of contractual obligations with PepsiCo's European distributor, which gave Coca-Cola an extra impetus to launch there. Dasani would gain an important head start in the quest to establish a sustainable market presence in the purified bottled water market. Other major players already had well-established market positions in the UK, including the European behemoths Danone and Nestlé, both of which owned multiple water brands (see Table 1.2). These traditional water brands faced the prospect of two huge beverage companies, with massive marketing resources, aggressively trying to enter their market. In the US, Coca-Cola and PepsiCo outspend their rivals by up to three times on advertising. Already the Dasani brand was on sale in almost 20 different countries; now Europe was next.

The company had existing brand names for purified bottled water products in some European countries, including the Irish Deep River Rock® and Portuguese Bonaqua® brands. Coca-Cola decided to retain the Dasani brand name for the major European drive, a made-up name that tried to evoke the brand's core values of 'relaxation, pureness and replenishment'. A series of advertisements featured the catch phrase, 'Prepare to get wet!' The company planned to dedicate an initial marketing budget of more than £7 million and position the new brand as 'urban water for the fast-living generation'. The target market for Dasani was 20 to 35 year olds who would see Dasani as a lifestyle brand. Prior to the general launch, Coca-Cola placed a series of advertisements in retail trade journals,

Table 1.2   Dasani's key competitors

Aquafina	**Evian**	**Perrier**
Purified still water Source: various Owned by PepsiCo US market leader	Natural mineral water Source: France Owned by Danone	Sparkling mineral water Source: Vergeze, France. Owned by Nestlé
Vittel	**Volvic**	**Highland Spring**
Natural mineral water Source: Vosges, France Available in over 80 countries Owned by Nestlé	Natural mineral water Source: Auvergne, France Owned by Danone Produces over 2 million bottles a day In the UK it sells over £98 million and is the market leader	Natural spring water Source: Scotland Leading UK supplier

publicizing the imminent arrival of the brand to interested retailers. The company's dominance in retail shelves greatly increased the likelihood of a successful launch. In some cases, the company simply leveraged its power with retailers by allocating space within Coca-Cola refrigerators to the Dasani brand, supplanting existing water brands, and forcing retailers to stock the Dasani brand as their only water brand. This strategy antagonized some small retailers who wanted to stock local bottled water brands, leading some of these retailers to remove Coca-Cola refrigerators.

Dasani Media Firestorm

The brand was sold in distinctive blue bottles, with a label that described 'pure, still water' which simply mimicked the format used so successfully in the US in the UK context. However, prior to the release, trade journalists noted that the water was just purified tap water and published articles about the source of this new Coca-Cola product, namely, Thames Water. A national syndicated news organization published these stories, which spread like wildfire in tabloid, broadsheet and other news sources. A media frenzy developed surrounding Dasani and Coca-Cola, perhaps because of the cultural sensitivity of the topic and the price mark up, which consumers viewed as virtual extortion. Coca-Cola had a major incident on its hands that threatened the very survival of its nascent UK brand.

The media compared Dasani with a classic BBC comedy, *Only Fools and Horses*, in which the two lead characters, Del Boy and Rodney, sell tap water as 'Peckham Spring' for ridiculously high prices. Numerous media showed images from that well-known episode, in which the characters fill bottles with a hose in their council flat, with the comment that the water comes from 'a natural centuries old source – the Thames!' The irony was not lost on the media or the public; the Dasani plant was only a few miles from Peckham in London. Commentators argued that Coca-Cola had showed barefaced cheek in selling ridiculously overpriced tap water to the unwitting general public. The Dasani brand was appearing in headlines for all the wrong reasons. Consumers' confidence in the brand was decimated as a result, and retailers grew nervous about stocking this brand, though most continued to do so in the hope that the initial furore would dissipate.

Coca-Cola's incident management team swung into action to assess and respond to the barrage of negative publicity. The crisis management team pressed ahead with the launch of the brand, reassuring retailers, releasing press releases and responding to media queries. Company spokespeople tried to reiterate that the product was entirely safe and that the water had undergone a 'highly sophisticated filtration process' developed by NASA engineers. Yet journalists and headline writers retained the opinion that the general public was being duped by a large multinational. Table 1.3 lists examples of the types of headlines the Dasani launch garnered from the UK and international press, across the various forms of media – press, radio, Internet and television. Eventually Coca-Cola acknowledged that the bottled water was in fact processed tap water, taken from the regular mains of Thames Water, in Sidcup, Kent. A spokesperson for the brand commented, 'We would never say tap water isn't drinkable. It's just that Dasani is as pure as water can get – there are different levels of purity.'[1] Some press commentators openly criticized Coca-Cola's handling of the crisis, particularly a radio interview on BBC Radio 4 in which the spokesperson's media skills were panned as the 'most embarrassing and excruciating' they had ever heard. A journalist commented that the spokesperson failed to give a straight answer, fudging the issue, and

Table 1.3   Typical newspaper coverage of Dasani® launch

'Real thing or rip off' – *The Evening Standard*
'Eau de Sidcup' – *The Daily Mail*
'Eau bother' – *The Financial Times*
'Eau dear' – *The Guardian*
'Coke puts bottled water plant on ice' – *The Independent*
'Should i really despise Coca-Cola?' – *The Independent*
'Junk medicine' – *The Times*
'For Coke, it's water down the drain' – *The International Herald Tribune*
'Has Coca-Cola's bubble burst?' – *Sunday Telegraph*
'How Coca-Cola conned the world?' – *Daily Mail*
'How Coca-Cola is selling water from the tap at 95p a bottle' – *The Daily Mail*
'Coke's 95p tap water versus the real thing' – *The Sun*
'Coke's pure water claim hard to swallow' – *The Times*
'Water waste; after 7m launch, Coke drops 'pure' water over cancer fears'
– *The Mirror*
'Coke takes a bitter gulp of "the realty thing". The drink icon's humbling over Dasani
is not the first humiliation for a company that seems to be taking too many wrong
turnings' – *Sunday Telegraph*
'Remember that dodgy Coke water they won't be bringing it back now'
– *The Express*
'Soft drinks giant copies del boy's crazy scheme; Coke sells tap water...for 95p'
– *The Express*

---

even quoted Oscar Wilde, saying that any publicity was good publicity for a firm. In the wake of this media storm, PepsiCo sought to launch its Aquafina brand, to capitalize on the failure of Dasani, but the firm's UK Britvic® Soft Drinks distributor vetoed the effort.

Furthermore, Coca-Cola had incurred the wrath of the Natural Mineral Water Association by using the words 'pure, still water' on Dasani's product label. The Association considered the label misleading and referred it to the Food Standards Agency, arguing Dasani could not use the word 'pure' because it added calcium and magnesium during the filtration processes. Coca-Cola noted that its lawyers confirmed compliance with labelling regulations, but the Food Standards Agency initialled an investigation. The Natural Mineral Water Association then created a generic packaging logo, featuring a leaf and a drop of water, to certify water from an accredited natural mineral water supplier. To be classified as mineral water, the water must naturally contain certain specific quantities of minerals. The bottled water industry continues to protect the usage of terms such as 'mineral' and 'spring' which represent key differentiating factors (Table 1.4).

Product Recall

During this problem-laden product launch, another shock came in the form of a massive product recall, triggered when samples of Dasani revealed excess levels of

Table 1.4   Classification of water types

Natural Mineral Water	Spring Water	Table Water	Tap Water
The water must be free from any pollution, have a stable composition, originate from a protected source and have no treatments. The addition of carbon dioxide is allowed to make it sparkle. Must comply with strict EU guidelines. Examples include Buxton, Badoit, Perrier, Vittel and Volvic.	Spring water must originate from an underground source and must be bottled at source. Companies are allowed to treat the water to improve the taste or remove undesirable elements, under EU guidelines.	Typically bottled filtered water, used in the restaurant trade. It can be filtered and treated.	Water companies treat this water, making it safe for domestic use

bromate, a known carcinogen, in the water. The company had to withdraw close to 500 000 bottles of Dasani from the market, because samples showed that the traces of bromate were double the EU limit.[2] Coca-Cola had added calcium chloride, which contains bromide, to change the taste of the water and match consumer preferences for 'designer water'. The bromide oxidized into bromate, and long-term exposure to this chemical can increase the risk of cancer. The company immediately decided to pull the product off the shelves (see Table 1.5), just a few weeks after the launch. The recall signalled Dasani's death knell. All Dasani products were destroyed by Coca Cola, and within 24 hours, 85 per cent of the recall was complete.[3] However, Coca-Cola continued to maintain that the situation was an isolated incident and that the risk was small.[4]

In the wake of the decision to axe Dasani, on 24 March, Coca-Cola formally announced it was delaying the proposed rollout of the brand in France and Germany, stating that the 'timing is no longer considered optimal'. The brand was scheduled to launch in France a month behind Dasani's launch in the UK. Coca-Cola instead released a French press statement in the wake of the recall in the UK:

Although the incident was isolated, specific to Great Britain and rectified, Coca-Cola has decided to suspend the launching of the Dasani brand in France and Germany. Indeed, the timing for launching the brand in these two countries is not regarded any more as optimal.... In France, we have not yet created customer interest, the production of Dasani in France had not started and we wished to limit as much as possible the impact for our distributors. This is why it was essential to make a fast decision. Our determination and the motivation of our teams to develop the company in the bottled water category in Europe and France remains intact.

Table 1.5   Coca-Cola's statement issuing a recall of Dasani®

Voluntary Withdrawal of Dasani in UK – 03/19/04

To ensure that only products of the highest quality are provided to our consumers, the Coca-Cola system in Great Britain is voluntarily withdrawing all Dasani products currently in the marketplace in UK. The withdrawal began on Friday, March 19, 2004 and will be 80-85% completed within 24 hours.

Calcium is a legal requirement in all bottled water products in the UK, including Dasani. To deliver the required calcium, the company adds back Calcium Chloride into the product. Through detailed analysis, the company discovered that its product did not meet its quality standards. Because of the high level of bromide contained in the Calcium Chloride, a derivate of bromide, bromate, was formed at a level that exceeded UK legal standards. This occurred during the ozonization process the company employs in manufacturing.

Immediately after the company identified this issue, it consulted with the Food Standards Agency. The FSA has confirmed that there is no immediate health or safety issue. The withdrawal is a precautionary measure. The company welcomes consumers to return purchased product by contacting the free phone consumer care line 0800 ------- or to call if they have other concerns. The care line number appears on all Dasani packs.

The company is working closely with all its stockists to remove the product from the market place. This withdrawal only affects Dasani in the UK market. Consumers rightly expect that products of The Coca-Cola Company meet only the highest possible standards for quality as well as all UK regulations.

Ironically, the company planned to launch Dasani using a bona fide spring water source from Belgium. The failure of Dasani in Britain wrecked all plans for a pan-European water brand.

The Dasani brand still enjoys success in the US, untouched by the commotion in Europe. However, the Dasani brand name in Europe will probably remain in the graveyard of failed brands. The Dasani disaster may be one of the biggest faux pas in marketing history – industry professionals voted it the worst case of marketing mismanagement ever – yet Coca-Cola remains committed to entering the fast growing water market in Europe in the future. In looking for potential sites for a bottled water brand, a Dasani Version 2.0 could be hitting supermarket shelves in the not too distant future. The company is considering establishing a plant in Derbyshire, with a name yet to be decided. It is estimated that Coca-Cola lost approximately £40 million on the failed launch.

Affected Stakeholders

A crisis may affect just one stakeholder group, a combination or all of the company's stakeholders. The Dasani debacle had far-reaching consequences for all of Coca-Cola's stakeholders. The series of mishaps during the launch eroded any consumer confidence in the new brand and also damaged their confidence and trust in the parent brand. Coca-Cola was also derided and lampooned in the media for its inept mismanagement of the launch and for failing to send out forthright spokespeople.

Thames Water, the supplier, was enraged at the suggestion that its water was inferior and needed further purification. Furthermore, the contamination led Thames Water customers to become worried about their water supply. Retailer confidence was also extremely shaken by the failed launch, which could have jeopardized Coca-Cola's future trade relations with key retailers. Retailers may not stock future Coca-Cola water brand launches, for fear of a repeat contamination scare, and instead just stock trusted, well-established water brands. Competitors initially worried about the launch of Dasani, which would take up sought-after shelf space, but its failed launch became an opportunity for them to educate customers about the differences among the different types of water, strengthen their individual brand propositions and launch their own new water brands.

Government organizations such as the Food Standards Agency may have directly affected the marketing strategy of the firm, in terms of both labelling and manufacturing. The contamination scare placed the spotlight firmly on Coca-Cola UK. Powerful lobby groups, such as The Natural Mineral Water Association, highlighted to other stakeholders the differences in product offerings. For Coca-Cola shareholders, this calamity likely shook investors' confidence in the management team, which lost millions of pounds in the investment.[5] The company desperately wanted a share of the burgeoning bottled water market to achieve growth targets. The crisis influenced investment decisions by both existing and prospective shareholders. Even US media organizations reported on the contamination scare, affecting the brand internationally. Coca-Cola's foreign subsidiaries in Europe also had to cancel their launch plans, severely curtailing their bottled water strategy and giving competitors an edge.

Lessons Learned

In essence, the launch was severely curtailed and fatally damaged by Coca-Cola's poor appreciation of cultural dynamics and sensitivity. Coca-Cola failed to understand that the European customer is more discerning about buying bottled water, because the market has been exposed to many spring water and mineral water brands. The source of water matters in a European context. The brand launch may have been successful if it used purified tap water from a rural source, rather than water from the heart of London. Bottled water is typically associated with clean, natural mountain springs; Dasani was associated only with a classic BBC comedy. Just because one format is successful in one country does not mean it will be successful in another. It is surprising that Coca-Cola's UK executives did not appreciate these important cultural sensitivities. The goal of effective crisis management may be ending the external crisis as reported in the media in as few news cycles as possible.[6] In contrast, Coca-Cola executives focused on using their plant resources, failing to look at the bigger picture.

Their decision making was also flawed from the start, beginning with the unsuitable location for sourcing the product. Prospective customers also saw the price as exorbitant, which the media considered extremely newsworthy. The company failed to communicate why the product was worth such a hefty price premium. In the wake of the crisis, Coca-Cola needed stakeholder support,[7] and therefore, it should have highlighted to stakeholders why the product was a good value proposition and why customers were not being duped. Instead, the UK market sees the Dasani brand as expensive tap water in a blue bottle. A better choice would have been positioning the product correctly compared with existing mineral water brands.

Dasani spokespeople exhibited poor crisis communication skills, fudging their responses to questions and replying in marketing babble about the brand. The bad press they received further exacerbated the problem, whereas positive media relations are vital in a crisis. The media processes a company's message and disseminates this information to various stakeholders, which means that the interpretation they offer of an event or the company's response has enormous repercussions for perceptions of the company. Ulmer advocates the use of a credible and well-trained spokesperson in a crisis situation.[8] Spokespeople must have significant media training and a well-crafted, well-rehearsed message to send to concerned stakeholders if the company hopes to be seen as trustworthy and credible. Effective crisis communications is just one part of a holistic crisis management recovery plan.

Another error pertains to the need for greater product testing prior to the launch. The product recall may have been handled well, but the contamination should not have happened in the first place. Consumers demand the utmost confidence in the safety of their food or beverages. In today's business environment, quality and safety are paramount, and threats to these aspects can have serious consequences for the company's share price, sales, relationships with key trading partners, reputation and scrutiny from the media and advocacy groups. If a crisis creates a public panic, the government may even intervene, which again places greater scrutiny on the organization and its behaviour, from which investigations, litigation, greater governmental controls and legislation may emerge. As Tsang observes, firms must strike a delicate balance between full disclosure and silence, based on stakeholder reaction.[9]

The use of the word 'pure' in the product's labelling seemed misleading to modern consumers, suggesting the water came from a natural source in the countryside. Prior to the recall, Dasani was already being investigated by the Food Standards Agency, which could have resulted in a damaging reprimand. The launch provoked powerful lobby groups; in particular, the Natural Mineral Water Association was enraged by the use of the word 'pure'. Dasani represented a significant threat to members, because of Coca-Cola's power. Furthermore, Thames Water vehemently denied Coca-Cola's suggestion that its water product was not sufficiently pure and required further purification to make it consumable. The failure to disclose uncertain information may negate public confidence in that entity, if the uncertainty is disclosed at a later date.[10] As Johnson and Slovic note, if an organization conveys uncertainty, the audience can perceive as either a signal for incompetence or honesty.[11]

Finally, this incident was the second major contamination scare faced by Coca-Cola in Europe in recent years. In 1999, a dioxin contamination occurred in Benelux countries, and the company faced widespread criticism for its handling of the crisis, which made it seem unresponsive and uninterested about stakeholder concerns. The company issued a recall only after the Belgian Government issued a ban on Coke products. In the UK case, it appears Coca-Cola learned its lesson and voluntarily recalled the entire inventory. The experience of a crisis thus can provide an organization with positive effects, such as change and learning.[12] The company successfully managed a national recall in a 24-hour period. For an effective crisis management plan to work, it must be well-prepared, supported by the right team and given sufficient resources. The recall might even have been serendipitous in the wake of all the bad press coverage surrounding the Dasani launch; it provided Coca-Cola with an exit strategy.

Conclusions

In the aftermath of the crisis, Coca-Cola has several options available. The UK bottled water market still has great growth potential, and if the company wants to realize its growth targets, it needs to have a brand in this product category. The company could have relaunched the Dasani brand using the same water source, but consumers, distributors, retailers and shareholders lacked confidence in the brand in the wake of the contamination and negative media coverage. Frewer and colleagues find that trust in information links to the accuracy, knowledge and concern for the public welfare demonstrated by a food company, whereas distrust relates to perceptions of a history of providing erroneous information and deliberate distortions of information.[13] The company thus could relaunch the Dasani brand using a different water source and clearly communicate that the product is safe and sourced from an entirely natural source, not the Thames. Similarly, it could launch a new water brand in Europe using a different source, which would enable Coca-Cola to use its strong marketing muscle and leverage its extensive distribution network. The company must develop a clear, believable positioning strategy for the new brand; it might use its existing brand names in Europe, such as River Rock or Bonaqua. However, stakeholders may be extremely cynical about the idea of Coca-Cola getting involved again in the water business. Finally, the company could develop links with existing water brands and act as a distributor, though this option would mean it would lose out on valuable revenue streams. An expensive alternative would involve buying an existing water brand with an established presence and develop it further using Coca-Cola's distribution and marketing clout.

The Dasani fiasco highlights various key issues surrounding crisis management and reputation management, including the importance of considering different stakeholder needs, media training, fast responses to a crisis, effective strategic planning and the need for risk assessment and scenario planning. If companies fail to conduct sufficient stakeholder research, they may risk losing their hard-won reputations in an instant, along with cash, careers and brand value. Stakeholders during a crisis event want communication to gain a better understanding of events, determine causation, associate blame and determine the impact of the crisis.[14] People constantly use cues and filters to assess various entities, whether they be individuals, groups, products or companies. Coombs argues that crises create two main threats to an organization: reputational threat, which jeopardizes how a firm's stakeholders perceive that organization, and operational threat, such that the firm cannot function properly.[15] This episode placed the reputation of Coca-Cola in the UK at risk. Managers placed their new brand as well as the parent brand in jeopardy by sourcing tap water from the heart of London. Surely they should have realized how consumers, shareholders and the media would react upon discovering that fact, regardless of the level of public relation spin.

References

1.   Poulter, S. (2004), 'Coke pulls the plug on its £70 million Dasani', *Daily Mail*, 25 March, p. 32.

2.   Vickers, A. (2004), 'All 500 000 bottles of Dasani are recalled because of high levels of chemical linked to tumours; cancer alert over Coke's "Tap water"', *The Express*, 20 March, p. 4.

3.   Cowell, A. (2004), 'Coke recalls bottled water newly introduced to Britain', *The New York Times*, 19 March, p. 3.

4.   Jones, A., & Liu, B. (2004), 'Dasani recall leaves Coke dangling over murky water', *Financial Times*, 20 March, p. 3.

5.   Johnson, J., Jones, A., & Liu, B, (2004), 'Coca Cola hit by new debacle at Dasani', *Financial Times*, 20 March, p. 1.

6.   Carmichael, B. & Rubin, J. (2002), 'Dow Corning – a new model for crisis management', *Strategic Communication Management*, Vol. 6, No. 1, pp. 22–25.

7.   Heath, R. L. (1997), *Strategic Issues Management: Organizations and Public Policy Challenges*. Thousand Oaks, CA: Sage.

8.   Ulmer, R. R. (2001), 'Effective crisis management through established stakeholder relationships. Malden Mills as a case study', *Management Communication Quarterly*, Vol. 14, No. 4, pp. 590–611.

9.   Tsang, A. S. L. (2000), 'Military doctrine in crisis management: Three beverage contamination cases', *Business Horizons*, Vol. 43, No. 5, pp. 65–73.

10. Miles, S. & Frewer, L. J. (2003), 'Public perception of scientific uncertainty in relation to food hazards', *Journal of Risk Research*, Vol. 6, No. 3, pp. 267–284.

11. Johnson, B. B. & Slovic, P. (1995), 'Presenting uncertainty in health risk assessment: Initial studies of its effects on risk perception and trust', *Risk Analysis: An International Journal*, Vol. 15, No. 4, pp. 485–494.

12. Elliott, D., Smith, D., & McGuinness, M. (2000), 'Exploring the failure to learn: Crises and the barriers to learning', *Review of Business*, Vol. 21, No. 3/4, pp. 17–24.

13. Frewer, L. J., Howard, C., Hedderley, D., & Shepherd, R. (1996), 'What determines trust in information about food related risks?', *Risk Analysis*, Vol. 16, No. 4, pp. 473–486.

14. Sellnow, T. L. & Seeger, M. W. (2001), 'Exploring the boundaries of crisis communication', *Communication Studies*, Vol. 52 No.2, pp. 153–167.

15. Coombs, T. W. (2002), 'Deep and surface threats: Conceptual and practical implications for "crisis" vs. "problem"', *Public Relations Review*, Vol. 28, No. 4, pp. 339–345.

2 Cadbury's Salmonella Scare: Good or Bad Crisis Management?

BY CONOR CARROLL*

Keywords

crises, food scare, crisis management, Cadbury.

Abstract

In this chapter, the author discusses the *Salmonella* scare that enveloped the iconic Cadbury brand in late June 2006, when the company had to recall seven of its leading branded products in the UK and Ireland due to possible contamination. This chapter details the following topics: the difference between issues and crises; the different lenses stakeholders use to observe a crisis; crisis lifecycles; and crisis communications response strategies.

Introduction

Cadbury Schweppes, the world's largest confectionary company, had to recall seven of its branded products in the UK and Ireland due to *Salmonella Montevideo* contamination in June 2006. *Salmonella* causes severe food poisoning and leads to extensive diarrhoea and vomiting. Yet the chocolate company knew of the possible contamination as early as January 2006 and did not inform regulatory agencies. The company was castigated in the media by the Food Standards Agency and lambasted for its negligence. Finally, the company decided to issue a recall of more than a million affected products, though in January, it stated that its chocolate was safe for public consumption despite the contamination and did not pose a risk to the public. The Food Standards Agency instead declared that Cadbury products posed an 'unacceptable risk to the public'.[1] Cadbury's crisis management strategy thus was counterintuitive compared with the traditional

* Mr Conor Carroll, Kemmy Business School, University of Limerick, Department of Management and Marketing, Plassey, Co. Limerick, Ireland. E-mail: conor.carroll@ul.ie. Telephone: + 353 61 202 984.

crisis management mantra of being open, honest and responsive. Despite this contrarian approach though, demand for Cadbury products soon normalized.

The information provided in this case comes from secondary published sources, including newspaper articles, recall notices, official Food Standards Agency briefings, press releases and television newscasts. It is necessary to rely on such sources because research into crises often creates challenges, in that stakeholders are reluctant to divulge information, particularly for legal and commercial reasons.

Cadbury's *Salmonella* Scare

The name Cadbury has been synonymous with chocolate for nearly two centuries. Founded in England in 1824, the company produces more than 50 different chocolate brands and employs 4500 people. To the UK public, Cadbury has the same iconic status as a Hershey® bar in the US. In addition to its widely known brand name, the company gained fame for its Bournville® project, which housed workers in designer communities close to the factory and was one of the world's most successful business innovations. Other leading business innovations initiated by Cadbury included provision for pensions, work committees, analytical laboratories and training and education for employees. The company has held its position as the leading chocolate brand in the UK for several decades. Its major tourist attraction, Cadburyworld, allows visitors to see chocolate being made, learn about Cadbury's history and, most importantly, buy chocolate.

Cadbury Schweppes plc merged in 1969; the combined company owns a variety of well-known brands such as Trident® gum, Snapple®, 7Up®, Halls® and Dr Pepper®. For decades, Cadbury had cultivated an excellent reputation and in this sense became an integral part of UK culture, beloved by customers, retailers and employees alike.

Cadbury currently holds 30.84 per cent of the UK confectionary market, and its popular Cadbury Dairy Milk® brand (begun in 1905) is the best-selling brand, with 8.75 per cent of the UK market. The brand also has developed an array of successful brand extensions and product variants. Cadbury produces more than 2500 different chocolate product variants, from a select stable of core brands. In 2004, Cadbury Schweppes was voted Britain's 'Most Admired' company by the business magazine *Management Today*. Figure 2.1 describes the UK chocolate market.

2006 *Salmonella* Outbreak

In the summer of 2006, Britain experienced a prolonged heat wave. On 22 June 2006, Cadbury issued a major recall of seven of its brands on sale in the UK and Ireland, in response to reports that traces of *Salmonella* were found in a variety of Cadbury chocolate bars, which represented a risk to public health. In total, more than 1 million chocolate bars were recalled and then destroyed. The company had to recall seven well-known Cadbury's branded product lines in the UK and two in Ireland. The products recalled included the iconic Dairy Milk® '8 Chunk' bars, Turkish Delight®, the 10p Freddo, Dairy Milk Mint®, Dairy Milk Caramel® and Dairy Milk Buttons® Easter Egg. But the debacle was not a straightforward recall, as first thought. During the spring of that year, the number of reported food poisoning incidents dramatically increased, which drew the concern

> ➤ Worth £4.7 billion.
> ➤ Experiencing growing demand for premium, organic and luxury chocolates.
> ➤ Heavy promotions of sharing packs.
> ➤ Dairy Milk sales account for £389 million.
> ➤ Top 10 confectionary brands are Cadbury's Dairy Milk 8.75%; Terry's 4.23%; Galaxy 4.10%; Malteasers 3.50%; Mars 3.33%; KitKat 2.93% Cadbury's Roses 2.65%; Quality Street 2.38%; Cadbury's Buttons 2.08%; Smarties 2.01%.
>
> **Key Competitors**
>
> | Nestlé | Masterfoods | Kraft |
> |---|---|---|
> | 19.72% of the UK market. Portfolio includes KitKat. Aero, Smarties, Milky Bar, Rolo, After Eight, Dairy Box and Quality Street. | 21.34% of the UK market. Suffering declining market share. Portfolio includes Galaxy, Mars, Snickers, Malteasers, Twix and Celebrations. | 5.39% of the UK market. Experiencing small but steady growth. Portfolio includes Terry's, Toblerone and Suchard |

Figure 2.1   UK chocolate market at a glance[2]

and attention of a number of public health watchdogs in the UK. *Salmonella* is a form of bacterium that, if present in food, can lead to severe food poisoning, diarrhoea and, in extreme cases, death for vulnerable members of the general public, such as children and the elderly.

The controversy grew when an independent laboratory asked the UK Health Protection Agency (HPA) to confirm a sample that it had tested was indeed a strain of *Salmonella*. When the HPA validated the findings and requested the source of the sample, the laboratory refused to identify its client. Alarmed by this lack of transparency, the HPA contacted the Food Standards Agency (FSA) to leverage its legislative powers and force the independent laboratory to reveal the client's identity. Only after this request did Cadbury come forward and admit that it had sent the sample for evaluation. The typical protocol for food safety authorities and crisis management experts suggested Cadbury should have contacted government agencies as soon as it recognized a potential problem, as mandated by legislation. However, Cadbury argued that the levels of *Salmonella* were so low that it did not warrant concern. A spokesperson further justified this stance, arguing, 'The level we found was so incredibly low that we decided there was no need to inform the FSA.'[3] Cadbury emphasized its opinion that the traces of the rare strain of *Salmonella* found were so minute, it would not be a risk to the general public. Was the company right, or was it just safeguarding its valuable Easter egg sales?

The UK FSA then revealed that traces of *Salmonella* had been found in Cadbury products as far back as 2002. Furthermore, the factory at issue suffered outbreaks in April and October 2002 but did not inform regulators. *Salmonella* poisoning causes severe stomach pains, cramps and diarrhoea, which can last for days, and in some cases, hospitalization is necessary. In total, more than 180 people were infected by the outbreak, according to the HPA. Although many people think of *Salmonella* exclusively as a poultry infection, the bacteria can thrive in chocolate, such that low levels can cause significant food poisoning incidents. Chocolate preserves the bacteria for a substantial period, due to its high sugar and fat levels. Food safety experts concur that there should be zero presence of *Salmonella* in chocolate products; in contrast, the food scientists at Cadbury deemed its presence a minimal risk and no real danger.

Cadbury agreed to a voluntary product recall only after notification from the HPA about high numbers of *Salmonella* food poisoning. A month after the product recall, the incidences of *Salmonella* poisoning dropped substantially. In the previous year, there had been 14 cases of *Salmonella* poisoning in the UK, most of them people returning from trips abroad. In 2006, 180 people had been struck down by food poisoning, many of them children, and many of those were under the age of four. The number of people infected could have been much higher, because many victims do not report stomach upset to their doctors. After the recall, Internet discussion boards were flooded with comments from hundreds of individuals who had been terribly sick, possibly as a result of eating Cadbury chocolate.

With this type of food poisoning, evidential links can be proven through laboratory testing, because each bug has a unique genetic fingerprint. A food poisoning outbreak can be traced to a geographic area, such as a restaurant or butcher shop, but when a contamination occurs in a product that is nationally distributed, it is harder to trace or form causal linkages. The HPA investigates all *Salmonella* cases occurring in the UK and attempts to establish causality patterns. First, it considers an affected person's dietary habits and detailed food consumption histories. Second, it rules out other possible causes, such as other common food brands, outlets, restaurants or food types. Cadbury was the one brand common to all the cases associated with the *Salmonella* outbreak. Following laboratory analysis of the samples taken from the Cadbury factory, the HPA was able to link the *Salmonella* outbreak to a chocolate crumb plant.

But the company had known about the production problem in January 2006, though it did not inform government sources until June – 23 weeks later. Cadbury also waited for 2 days after receiving the FSA's request for a full voluntary food product recall. Cadbury's response included a comprehensive cleaning of the factory, the introduction of a positive release system (product would be released for distribution only after testing negative for *Salmonella*), and increased sampling and testing to provide a higher degree of reassurance. Cadbury's official response noted though that, 'This is being done purely as a precautionary measure, as some of these products may contain minute traces of *Salmonella*.... Cadbury has identified the source of the problem and rectified it, and is taking steps to ensure these particular products are no longer available for sale.... *Salmonella* can come from any number of sources. Our view is that the chocolate was perfectly safe.'[4]

The *Salmonella* outbreak occurred at its plant in Marlbrook, Herefordshire. The plant blends milk, sugar and cocoa liquor to create a chocolate crumb, which gets transported to other production sites, where it is blended with cocoa butter to form milk chocolate. The company could yet face further legal challenges in light of the outbreak, from either affected consumers or public health agencies. In its defence, Cadbury claimed that the bug originated from the manufacturing process and that the contamination occurred in minuscule levels; its experts concurred that the occurrence was too small to be of any consequence. In contrast, the HPA found levels of *Salmonella* in the samples it took that represented a significant risk.

In the very early stages of the scare, the source of the problem was unknown. The source of the *Salmonella* outbreak might have been a composting company, close to the Cadbury plant, that accepted waste, such as carcasses of chickens, meat and unsold supermarket food produce, then churned it for 4 days and incinerated it at 70°C to kill any pathogens. The food waste is then sold as compost. Any rotting food that fails to break down sufficiently is spread out on fields. The compost company strongly denied any link or suggestion that it was associated with the problems. Although the media picked up this story, there was absolutely no link between the compost company and the outbreak. A potential third-party scapegoat thus failed to materialize for Cadbury; the responsibility for the incident was solely its own, due to a technical breakdown in its production systems.

Specifically, the outbreak occurred because of a leak in a pipe at the plant, whereby waste water, contaminated with vermin and bird waste, dripped into the chocolate mixture. Cadbury's independent testing procedure mandated sampling of ingredients from the production line and finished products three times a day at 8-hour intervals. These samples are tested by an independent analytical laboratory. The company obviously had concerns over food safety, yet still it did not tell food safety authorities immediately and continued to churn out chocolate in the run up to its busy Easter period. This delay and intransigence further infuriated food safety watchdogs, as did the lack of transparency. In response to these criticisms, a spokesperson for the company stated, 'The Food Standards Agency does not set a protocol for *Salmonella* testing in products. It is left to manufacturers to determine their own.... We based our protocol on sound science and at all times we have acted in good faith. We have worked with the authorities and now have a new protocol: If there's any contamination in a product, regardless of its level, the product will be destroyed.'[5] The actual text of the recall appears in Figure 2.2.

On July 4, 2006, the UK news network ITN carried a news report on the recall and highlighted some major inconsistencies between the company's viewpoint and the beliefs of independent health experts with regard to *Salmonella*. An investigate journalist

---

June 23, 2006

Cadbury is conducting a recall of seven of its products in the UK and two in Ireland. The products affected are:
250 gram – Cadbury Dairy Milk Turkish
250 gram – Cadbury Dairy Milk Caramel
250 gram – Cadbury Dairy Milk Mint (including 33 per cent extra free bars)
Cadbury Dairy Milk 8 chunk
1 kilogram – Cadbury Dairy Milk
Cadbury Dairy Milk Button Easter Egg – 105 gram
Cadbury Freddo 10p

This is being done purely as a precautionary measure, as some of these products may contain minute traces of salmonella. Cadbury has identified the source of the problem and rectified it, and is taking steps to ensure these particular products are no longer available for sale. Cadbury expects to have fresh stocks of these products back on the market in the near future. The decision was made in consultation with the Food Standards Agency with whom Cadbury has worked closely. 'We've been making chocolate for over 100 years and quality has always come first,' said Simon Baldry, the UK Managing Director of Cadbury. 'We have taken this precautionary step because our consumers are our highest priority. We apologise for any inconvenience caused.'

Figure 2.2   Cadbury's product recall

called Cadbury's telephone helpline and asked how many chocolate bars a person could eat before facing a risk of *Salmonella*. The spokesperson answered, 'As far as we have been advised you have to eat in the region of 60 standard bars.' This response directly contradicts the independent FSA, which argues there is no safe level when *Salmonella* is present in chocolate. Cadbury's UK Managing Director, Simon Baldry, also asserted, 'We, like the Food Standards Agency, have a common goal and that is to protect public health. Our products are perfectly safe and we say that because of rigorous testing based on solid science that we have done, therefore I can assure you these products are perfectly safe.'[6]

In August 2006, Todd Stitzer, CEO of Cadbury Schweppes, publicly apologized for Cadbury's quality assurance lapses in an interview with the BBC, 'We are truly sorry for that, we acted in good faith all the way through, but we have caused customer concern, and we apologize.'[7] The forlorn CEO used numerous public vehicles to publicly express a belated apology; Figure 2.3 contains a message posted on the company's website. The company also asserted that it was stuck in a 'regulatory catch 22' because the industry was self-regulatory, with insufficient safety protocols, which caused concern.

But Cadbury also claimed it had changed the rules (Figure 2.4). The disparity of opinion between the food safety regulators and the company about the appropriate remedial action during the scare raised many serious concerns for stakeholders. Such confusion could jeopardize consumers' confidence not only in Cadbury and its brands, but also in the statutory powers of government bodies responsible for food safety.

Ramifications of the *Salmonella* Outbreak

A tiny leak in a pipe had serious ramifications for the firm not only economically but in terms of reputational risk. It cost the firm £30 million, which was budgeted into its 2006 annual report. More than £7 million was collected from the company's insurance policies, so the real cost was in excess of £37 million. The company also set aside £5 million for a marketing communications campaign to rebuild consumer confidence. The

---

UK PRODUCT RECALL: A MESSAGE FROM TODD STITZER
August 2, 2006

Quality has always been at the heart of our business but the quality assurance process we followed in the UK has created concern about this in the eyes of our consumers. Although we have acted in good faith throughout, we have caused concern and for that I sincerely apologize.
I have instructed that changes to our manufacturing and quality assurance processes should be made so that this occurrence cannot happen again. We are now moving forward and are focused on delivering the very best confectionery and some great new products for our consumers this autumn.
The last few weeks have without doubt tested the strength of our relationship with our consumers and customers and the resilience of our colleagues. I'm grateful to our consumers and customers for their patience and loyalty and I'm grateful to our colleagues who worked nights and weekends to respond.

Figure 2.3   Cadbury's apology[8]

April and October 2002	Company discover traces of *Salmonella* in product range but does not recall or report the incident to the statutory governmental agencies.
January 20, 2006	Cadbury discovers *Salmonella* traces in its chocolate crumb factory. Contamination originated from a leak in pipe carrying waste water from cleaning machinery to the production line.
March 2006	HPA starts tracking the national increase of *Salmonella Montevideo* in the UK.
June 19, 2006	FSA learns of the *Salmonella* contamination in Cadbury products.
June 21, 2006	FSA requests Cadbury to issue a full recall, in the interests of public safety.
June 22, 2006	Cadbury issues a recall of seven brands.
June 23, 2006	Recall begins.
June 23, 2006	FSA documents say Cadbury posed unacceptable risk to public, 'All requests for information (to Cadbury) have to be reinforced.'
June 30, 2006	Documents supplied by Cadbury reveal that the same factory was infected with salmonella in April 2002.
July 3, 2006	FSA expert advisory committee criticizes Cadbury's product safety system as unreliable, out of date, and underestimating the likelihood of salmonella contamination. It also states that Cadbury did not conduct appropriate modern risk assessments of *Salmonella* contamination.
July 6, 2006	Representatives of the FSA, Cadbury and the Herefordshire council meet at the plant to discuss remedial actions. Cadbury enforces only a positive release system.
July 21, 2006	*Salmonella Montevideo* National Outbreak Control Team (OCT), comprising representatives from the HPA, FSA, Department of Food, Agriculture and Rural Affairs and local authorities concludes that Cadbury was the likely cause of the *Salmonella* outbreak that infected 180 people.
August 2, 2006	Chief executive Todd Stitzer announces he expects full-year cost of recall to be £20 million.
August 15, 2006	FSA announces intention to give money to a council investigating salmonella at Cadbury factory.
February 12, 2007	Cadbury issues further recall of several product lines due to a labelling error for a selection of Easter products that failed to carry warnings about potential nut allergies.
February 21, 2007	Cadbury Schweppes plc announces profits of £909 million, down 7 per cent from previous year. Market share remains steady. The scare cost the firm an estimated £37 million, £7 million of which was recovered by insurance policies.
July 17, 2007	Cadbury pleads guilty and is fined £1 million by a Birmingham Magistrates Court and Herefordshire Council.

Figure 2.4   Crisis timeline

recall cost the firm underlying profits of between £5 and £10 million. Despite these losses, UK revenues for Cadbury increased by 1 per cent. The timing of the recall minimized the impact; it was at the height of a torrid heat wave, when demand for chocolate was very weak, so customers barely noticed the recall, and retailers were not unduly concerned.

Possible civil claims and regulatory actions based on environmental health laws could have been taken against Cadbury, with costly ramifications. Cadbury thus faced a barrage of litigation claims from victims of the *Salmonella* outbreak. Solicitors' websites offered free legal advice, such as, 'Do you have a claim? If you or someone you know has been affected by *Salmonella* after consuming Cadbury's products, or any infected product, seek free legal advice immediately. Visit our product liability section.'[9] One possible claimant emerged in the national newspapers. A 62-year-old woman from Northern Ireland believed she fell victim after eating a Cadbury Caramel bar. She spent 5 days in a hospital's isolation ward with severe food poisoning. The company also faced litigation from government authorities because it had not informed them in time and pushed delays in issuing the recall. The FSA vigorously critiqued Cadbury's slowness in providing information about the incident. The reverberations of this event may lead to increased governmental scrutiny, fines and further amplified media exposure of this unsavoury affair, which could jeopardize Cadbury's reputation.

ACNielsen market researchers indicate that demand for chocolate declined by 5.5 per cent in the third quarter of 2006, attributed to both the record-setting warm temperatures and Cadbury's recall. The company's stock price dropped due to financial irregularities in its Nigerian subsidiary; the CEO and chief financial officer had fled the country, a debacle that cost the firm £53 million. The *Salmonella* scare also had global ramifications, affecting product sales in Ireland, Malaysia, the United Arab Emirates and Singapore. Malaysia issued a product recall; Singapore undertook precautionary checks and quarantined inventory to await inspection. In 2006, Cadbury Schweppes reported revenue growth of 4 per cent but profits down 7 per cent.[10] The company also dropped from third to ninth in the 'Most Admired' poll, evidence of the negative effects on Cadbury's reputation. The embattled CEO complained that, 'People want to focus on the negative when the bullets are flying.'[11] The company now insists that no product will be sold unless it passes rigorous *Salmonella* tests.

In February 2007 more troubles occurred for Cadbury, when it recalled a suite of Easter chocolate products because of a failure to warn allergen suffers of nut contents on the packaging. Brands such as the Cadbury Creme Egg were affected by the recall, and Cadbury was severely castigated in the press for another lapse in food safety protocols. The products themselves were safe for consumption; however, they posed a risk to consumers allergic to nuts, a potentially fatal allergy that means sufferers cannot eat any products that contain traces of nuts or that are manufactured on productions lines that might contain traces of nut. The failure could have been catastrophic for Cadbury, as such products aim almost exclusively at children. Once again, Cadbury hit the headlines for all the wrong reasons:

We're concerned about consumers with nut allergies. We've largely gotten the inventory back; we're stickering the packs. Some of you may have seen some of the stickered packs out on the shelves today. We think it will have an immaterial impact on our financial performance. Given the late Easter, we're turning around the whole thing so that we're back on shelf pretty quickly. So I don't think that's a big deal.

– Todd Stitzer, CEO Cadbury Schweppes[12]

The bad press continued, with more sound bites about the crisis released to the media. With the media spotlight on 'the nation's favourite' chocolate, the company suffered more intense scrutiny from its key stakeholders – consumers, the media, investors, and local and national government agencies. Politicians castigated the company in the UK Parliament for failing customers, and claims of a corporate cover-up were issued. The political pressure led to more investigations of the *Salmonella* outbreak, as well as a possibility of litigation. The scare was widely publicized in traditional media, online discussion boards and Internet blogs, which damned Cadbury's response and handling of the *Salmonella* scare (see Figure 2.5 for some sample editorial coverage).

Cadbury pleaded guilty to nine health and safety charges, including distributing unsafe chocolate to the general public and failing to inform authorities, in July 2007. In two separate rulings from the Herefordshire Council and a Birmingham Magistrates Court, the company was fined in a excess of £1 million. Specifically, it was fined £700 000, plus £52 000 in costs, in relation to the Birmingham charges, then fined £300 000, plus £100 000 in costs, for the Herefordshire charges.[13] The guilty plea meant the company

---

'Poisoned Choc Firm Bosses Could Face Massive Fines' – **The Sunday Mirror**

'A million "food bug" chocolate bars taken off shelves' – **The Times**

'Meltdown! 1m choc bars recalled over salmonella alert' – **The Sun**

'Chocalert; Cadbury Ordered To Recall 1million Bars...But Why Did It Take 5 Months?' – **The Mirror**

'Million chocolate bars withdrawn over salmonella' – **The Independent**

'Salmonella alert: Boss of Cadbury under fire' – **The Sunday Express**

'Chocolate bug cases spread' – **The Sunday Times**

'Cadbury needs its sweet image to shrug off the bad publicity'– **The Daily Telegraph**

'Salmonella scare: Chocolate may have poisoned more than 40: Watchdog says Cadbury's should have acted earlier: Contamination caused by leak of waste water' – **The Guardian**

'Chocs away: One million bars recalled after salmonella contamination' – **The Guardian**

'Salmonella chocolate still on shelves; Food Safety: Recall; Cadbury's reputation is on the line as it comes under fire for dragging its heels over a bug discovered months ago' – **The Sunday Herald**

'Unwrapped; How a leaking pipe poisoned Britain's favourite chocolate. Cadbury denies a cover-up, as millions of chocolate bars are removed from the shelves six months after contamination was detected' – **Independent on Sunday**

'Cadbury facing legal action: Consumer backlash catches chocolate giant by surprise as questions grow over health alert delay' – **The Observer**

'Cadbury's scare: Three in hospital' – **The Daily Mail**

'Cadbury salmonella alert has spread to 30 products' – **The Daily Mail**

'Cadbury's safety checks "unreliable"'– **The Guardian**

'Cadbury's "knew of bug" four years ago' – **The Times**

'Revealed: watchdog's damning verdict on Cadbury's over salmonella scare'– **The Independent**

'The secrecy that left a bad taste in the mouth and spoilt a reputation'– **The Sunday Express**

'Cadbury fined pounds 1m for poisoned chocs'– **The Mirror**

Figure 2.5   Typical news coverage of *Salmonella* scare

avoided a lengthy courtroom drama and minimized the potential for other damaging evidence being exposed to the media. Although it was a large punitive fine, the figure is of small consequence to the future heath of the company. This strategy of limiting the potential of further amplification of the crisis ensured that the *Salmonella* scare story was not prolonged in the media spotlight. In responding to the ruling, the company apologized for the incident and added that it had changed its production systems, such that 'the processes that led to this failure ceased last year and will never be reinstated'.[14]

In 2006, company profits fell, after accounting for one-off costs such as the Nigerian accounting scandal and the *Salmonella* recall. The UK market share for December actually grew by 2 per cent. Share remained at 34 per cent of the UK chocolate market, the same figure as in 2005, prior to the crisis. Sales for the Dairy Milk brand fell by 2.5 per cent in 2006, according to ACNielsen. However, in February 2007, Cadbury Schweppes reported that sales of Cadbury in the UK had returned to the level prior to the recall. It seems that the Cadbury brand weathered the storm successfully.

Cadbury put its long-term reputation at risk in a calculated gamble. If the food poisoning incident had created greater loss, anguish and dread among customers, it would have caused a mass furore that could have decimated the brand. The company would have been seen as putting profit before the public's health. The voluntary recall may also have amplified concerns unnecessarily amongst the general public. In summary, Cadbury was very lucky on a number of fronts – timing, severity of impact, lack of relentless media scrutiny and brand strength – which helped insulate the company from the impact of the crisis. The crisis management strategies deployed by Cadbury jeopardized a trusted and beloved brand, and the lack of transparency and timeliness in handling the crisis led to greater media and governmental scrutiny. But what can be learned from this controversy?

Lessons of the Cadbury case

This episode raises some very interesting questions and lessons in relation to crisis management within the food industry. By its very nature, the food/agricultural sector is exposed to various risks, some of which increase to the level of full-blown crises. Almost any company in the industry could suffer a breakdown in its food safety protocols. This case shows the need for companies to understand the nuances and dynamics of a food safety crisis and the possible crisis communication strategies that might be deployed to negate negative repercussions for the firm. Food companies can survive such an episode, yet they must consider the variety of stakeholder communication needs, as well as the short- and long-term ramifications of their crisis management response.

It is important to delineate the difference between an issue and crisis. Not every incident should be viewed as a full-blown crisis. Issues are commonly misconstrued as crises, but the definition is in the eyes of the stakeholder. If a food safety event raises dread about consuming a product and makes it a risky purchase, demand may disappear. Crises differ from standard issues, in that crises hyperextend the capabilities of the organization, whether in terms of resource availability, executive decision making or resource usage.[15] Fink argues that a crisis occurs when an event creates closer scrutiny of the organization by media or government, devalues the public image of a company, interferes with day-to-day operational activities and affect the firm's bottom line.[16] The *Salmonella* scare created all these results at Cadbury. Issues thus can emerge into full-blown crises. A simple leak in a production facility can snowball and jeopardize an entire brand. Issues between an organization and its stakeholders arise frequently in the modern business environment. Nonetheless, a minor issue often can be hyped into a frenzy as a result of sensational media coverage that raises public awareness of an issue and influences public perceptions. Media have a vicarious appetite for obtaining newsworthy material, so any material that is somewhat controversial, with conflicting points of interest, liable will be publicized therein. This likelihood increases the visibility of crises in the public domain. Companies must develop communications strategies that contain an issue, not amplify a crisis. However, if a company appears to act in a manner that jeopardizes stakeholder interests, the future repercussions could be severe.

Issues arise when one or more individuals attaches significance to a perceived problem or situation.[17] All a company's stakeholders could be affected by a crisis and become victims, suffering financially, mentally or physically as a result.[18] When a crisis breaks, affected stakeholders and the general public seek information on its cause and the responsible party. By their very nature, crises require organizations to communicate with their stakeholders, providing detailed and accurate information about causality, prognosis and rectification strategies. Stakeholders' expectations of a company's communications activities increase during a crisis event.[19] These stakeholders need more and better information about how the crisis event affects their interests. A crisis event also requires an extraordinary effort to identify and communicate with all potential stakeholders who may be affected.[20] The company should consider the informational needs of all affected stakeholders, including shareholders, employees, government authorities, suppliers, retailers and, above all, customers. Each of these stakeholders will use a different lens to evaluate a company's organizational response. Shareholders will be concerned about legal liability, retailers about out-of-stock situations and category impacts, government authorities about the public good and possible contravention of regulations, and customers will be concerned about whether the product is safe for consumption. Companies must ensure crisis communications strategies that meet the informational needs of their stakeholders to minimize the impact of a crisis.

This case demonstrates that crises have lifecycles that can be prolonged or shortened depending on the managerial reactions to the crisis. Theories of crisis management typically craft crisis lifecycles that begin with normality and end in normal operations. As Siomkos and Kurzbard note, crisis management involves the complex realms of perception, understanding and abatement.[21] Stakeholders' attitudes are shaped not only by crisis events but also by their pre-existing attitudes and the impact of the crisis after it terminates.[22] Crisis communications management should not been seen as dealing solely with a triggering event in isolation but rather as a key component in holistic reputation management. Crisis patterns are problematic, in that a crisis does not follow a clear predictable path due to its inherent unpredictability and contextual variables.

This crisis case study should not be viewed as a panacea for how to manage a crisis effectively, though notable lessons emerge. The crisis communication response strategies adopted by Cadbury were fraught with risk and could have been extremely detrimental to the brand. Standardized crisis management tactics dictate the need for timely, open and credible communication strategies with all affected stakeholders. This case demonstrates that the firm was less than forthcoming about the problem, and much of its reaction was measured, to curtail legal liability and further media amplification. The firm pleaded guilty to charges to ensure that the details of the debacle did not emerge for public consumption, which would have harmed the firm even further.

Conclusions

This type of controversy points at critical systems failures within the firm, which could have been severely detrimental to its future success and survival. Not only does a controversy of this nature damage consumer confidence in a brand, it intensifies both the media and legislative scrutiny of the firm and its management. In the event of future crisis incidents affecting the firm, it severely curtails positive goodwill that stakeholders might have exhibited toward the firm. A 'small leaky pipe' can have dire consequences. In the event of such incidents, firms need to have effective business continuity planning and crisis management plans in place.

The handling of this controversy also goes against the traditional rules for handling a crisis, that is, being open, transparent, honest and quick to respond to the threat. However, the newsworthiness of the controversy eventually diminished, and demand continued as normal. Several factors helped alleviate the impact of the crisis, including its timing, the relatively low numbers of people stricken by the food bug, and the company's reputation and once stalwart and pristine brand image. Some commentators would argue that Cadbury's handling of the scare was deplorable; others view Cadbury's crisis management strategies as appropriate, according to sales figures. Cadbury was extremely lucky to survive this incident, as just one or two factors could have tipped the balance to make it a complete catastrophe for the brand. This crisis case poses some very interesting points that warrant further debate and discussion regarding how to handle a crisis that affects a firm. Does a firm risk greater amplification of the crisis threat when it undertakes typical prescribed crisis communication response strategies, or is best to operate under a policy of intransigence? Every crisis should be put into context; other companies in the same position as Cadbury may not have survived such a damaging scare.

References

1.   Hickman, M. (2006), 'Revealed: Watchdog's damning verdict on Cadbury's over salmonella scare', September 23.
2.   Euromonitor (2006), *European Confectionary Market*. London: Euromonitor.
3.   Leake, J. & Walsh, G. (2006), 'Chocolate bug cases spread', *Sunday Times*, June 25, p. 8.
4.   Poulter, S. (2006), 'Salmonella in Cadbury's bars; chocolate giant orders a million bars to be cleared from shops', *Daily Mail*, June 24, p. 1
5.   Johnson, L. (2006), 'The secrecy that left a bad taste in the mouth and spoilt a reputation', *Sunday Express*, July 9, p. 10.
6.   ITN newscast (2006), 'Cadbury defends recall', June 24.
7.   BBC (2006), Radio interview with Todd Stitzer, CEO of Cadbury Schweppes PLC, August 2.
8.   http://www.cadburyschweppes.com/EN/MediaCentre/News/UK_product_recall_HTS.htm. (Accessed December 2006).

9.   Irwin Mitchell Solicitors (2006), press release, 'First legal action against Cadburys for suspected Salmonella poisoning considered', http://www.irwinmitchell.com/PressOffice/PressReleases/legal-action-cadburys-salmonella.htm.
10. Cadbury Schweppes PLC (2006), Annual Report.
11. Muspratt, C. (2007), 'Food Cadbury tastes 12pc profit cut as sales slump', *Daily Telegraph*, February 21, p. 4.
12. Cadbury Schweppes (2007), Transcript, 2006 Preliminary Results, Analyst Presentation, 20 February.
13. Tait, N. & Wiggins, J (2007), 'Cadbury in record £1m fine for unsafe chocolate', *Financial Times* (London), July 17, p. 2.
14. Ibid.
15. Pearson, C. M., Clair, J. A., Kovoor-Misra, S., & Mitroff, I. I. (1997), 'Managing the unthinkable,' *Organisational Dynamics*, Vol. 26, No. 2, pp. 51–64.
16. Fink, S. (1986), *Crisis Management: Planning for the Inevitable.* New York: AMACOM.
17. Crable, R. E. & Vibbert, S. L. (1985), 'Managing issues and influencing public policy', *Public Relations Review*, Vol. 11, No. 2, pp. 3–15.
18. Coombs, W. T. (1999), *Ongoing Crisis Communication: Planning, Managing, and Responding.* Thousand Oaks, CA: Sage Publications.
19. Ulmer, R. R. & Sellnow, T. L. (2000), 'Consistent questions of ambiguity in organizational crisis communication: Jack in the Box as a case study', *Journal of Business Ethics*, Vol. 25, No. 2, pp. 143–155.
20. Cowden, K. & Sellnow, T. L. (2002), 'Issues advertising as crisis communications: Northwest airlines' use of image restoration strategies during the 1998 pilot's strike', *Journal of Business Communication*, Vol. 39, No. 2, pp. 193–205.
21. Siomkos, G. J. & Kurzbard, G. (1994), 'The hidden crisis in product-harm crisis management', *European Journal of Marketing*, Vol. 28, No. 2, pp. 30–41.
22. Fink, op. cit.

3 Risk Communication and Food Recalls

BY SYLVAIN CHARLEBOIS* AND LISA WATSON†

Keywords

food recalls, risk management, risk perception, risk communication.

Abstract

In this chapter, we propose a framework for the analysis of risk communication and a list of recommendations to provide a new direction for industry pundits and policymakers. Specifically, we will present a conceptual model of risk communication; understand how risk communication efficiency can be enhanced by better accounting for the role of socio-political units; suggest how risk assessors should consider knowledge as a commodity; and, finally, present six recommendations for risk communication.

Risk Communication

It is widely accepted that consumers have a richer and more multifaceted definition of risk than do experts.[1] Risk perceptions differ between consumers and experts as well.[2] With regard to food safety concerns, regulators must cope with these perceptual asymmetries, particularly when dealing with food recalls. Industry and government have mismanaged public perceptions of potential environmental and technological health risks.[3] The risk of a communication vacuum may result in escalating consumer unease and plummeting credibility of authority figures. In addition, the lifestyle, expectations and tastes of many consumers have undergone substantial changes during the past decade, and these changes have had considerable impacts on food safety standards, practices and risk communication strategies.

* Dr Sylvain Charlebois, Faculty of Business Administration, University of Regina, 3737 Wascana Parkway, Regina, Saskatchewan, Canada. E-mail: sylvain.charlebois@uregina.ca. Telephone: (306) 337-2695.

† Dr Lisa Watson, Faculty of Business Administration, University of Regina, 3737 Wascana Parkway, Regina, Saskatchewan, Canada. E-mail: lisa.watson@uregina.ca. Telephone: (306) 337-2695.

The specific practice of interest in this study pertains to food recalls. Food recalls have increased in the past 10 years in North America,[4] largely due to microbial outbreaks. Manufacturers usually go beyond government standards, such as gaining ISO certification, to ensure their food products meet and exceed compliance with health and safety requirements.[5] However, the number of lawsuits is growing, and some argue that it is only a matter of time before these trends spread.[6]

Insurance policies also are increasingly becoming a concern for the food industry. Food packers and manufacturers have access to contamination coverage, but insurance is not an industry standard.[7] In many cases, insurance also does not make financial sense. For example, a $100 000 deductible may be required on a policy to cover one truckload of produce, with an average market price of less than $20 000. Thus, food recalls remain the preferred method for protecting the public and industry from microbial outbreaks.

AN INCIDENT

The September 14, 2006 recall of spinach drastically lowered the share price of spinach. The US reported an outbreak of *E. coli O157:H7* illnesses, later found to be associated with the consumption of bagged fresh spinach. The Canadian Food Inspection Agency (CFIA) followed with its own public advisory on September 15, 2006.[8] More than 200 consumers were hospitalized in over 26 states and some provinces in Canada. At least three direct deaths were reported in the United States. Even though food recalls usually have only a marginally negative impact on commodity prices,[9] the September 2006 recall wreaked havoc on the spinach farming industry. Sales fell significantly, down by more than 15 per cent 6 months after the outbreak.[10]

The CFIA is the main regulatory agency responsible for food safety in Canada, similar to the Food Safety and Inspection Service of the US Department of Agriculture. A recall requires removing the affected product from the market and encompasses all tiers of the affected product distribution system.[11] Food recalls in the US and Canada can be either mandatory or voluntary. A mandatory recall occurs declared when the Minister of Agriculture believes, on the basis of reasonable grounds, that a product regulated by the CFIA poses a risk to the public, as specified by Section 19 of the *Canadian Food Inspection Agency Act* (1997). A voluntary recall is initiated and carried out by the recalling firm, without ministerial order. Public warnings (that is, Health Hazard Alerts) are issued for recalls that require recall of a product to the consumer level. The US regulations for agricultural products, governed by the US Department of Agriculture (USDA), are somewhat similar to those in Canada. Most recalls are voluntary, with a great emphasis placed on the performance of plant managers in the industry.[12] In September 2006, most spinach bags available to consumers had been packed by California-based manufacturers, which voluntarily recalled all their spinach products.[13]

This study attempts to measure consumers' perceptions of the latest food recall of tainted spinach. We attempt to understand the relationship between food recalls and food safety perceptions. In addition, we provide some policy directions for food industry pundits and policymakers regarding how to establish more efficient risk communication strategies in the future. To reach these objectives, we first present a conceptual model of risk communication that depicts the roles of risk assessors and risk makers/takers in an aggregate environment. Next, we present empirical results from a survey. Supported by the conceptual framework and empirical findings, we offer some recommendations for practitioners and policymakers involved in food safety practices.

Risk Communication: A Model

Most linear risk communication models appear flawed, in light of their failure to consider audiences and additional holistic issues.[14] Furthermore, risk may be socially constructed, which implies that social systems can help bridge the gap between risk assessment and risk communication.[15]

The model we present in this chapter (Figure 3.1) suggests not only that risk communication relates to myriad levels of discourse, both within and outside the food industry, but also that risk is socio-politically constructed. In the model, social and political rules converge into one force that influences the distribution of power and dependency amongst nations and food safety regulators.[16] Insofar as it is possible, all socio-political issues should be explicitly considered and integrated in considerations of food safety crises.[17] Socio-political considerations also are key in food safety because the basis of many food safety recalls and alerts is an influence external to the industry.[18]

RISK ASSESSORS

Food safety regulators, labelled as risk assessors in Figure 3.1, are mainly concerned with the interface between the social and political units or agents of the market itself. Risk assessors are units and individuals linked to food safety structures and processes. Risk communication is socio-politically charged because it raises issues about access (or lack of access) to information, self-interested behaviour, legal interpretation, community and property rights and persuasion.[19] Even at its most basic, risk communication invokes a tangle of complex ethical issues.

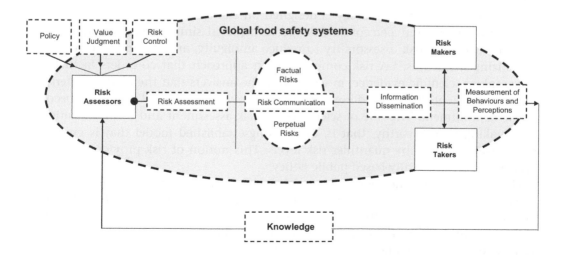

Figure 3.1   Risk communication: stakeholders and knowledge as a commodity

Social and political units in a global system attempt to control their environment and reduce uncertainty.[20] Global food safety systems extend the scope of markets beyond the borders of a single organization or a distribution channel to become regional, international or global in nature. With regard to food safety, external agents have great influence over the socio-political units of a given nation. Within global food safety systems, socio-political structures are defined by patterns of power and dependence relations amongst socio-political units. One given unit may exert more power over another with similar functions and responsibilities that is located in another region of the world. The environment confronts socio-political units with both dependency and uncertainty.[21] Dependency is created by the lack of alternatives or resources.

Nations tend to grant asymmetrical decision-making power to risk assessors, which insufficiently accounts for the complex socio-political production of risk. Risk assessors apply policies prescribed by political authorities, whether domestic or abroad. Domestic policies tend to have more weight for food safety regulations than do foreign policies, though recently increased global food trade has shifted the power balance. Specific incidences such as mad cow or foot-and-mouth disease are good examples. Regulatory agents or risk assessors that consider risk an exclusively empirical phenomenon may be regarded with suspicion or perceived as impersonal organizations that hide value judgments behind incomprehensible jargon. Values have a great impact on the technical choices of risk assessors. The inclusion of particular value assumptions depends in part on whether the question at hand is inherently or conditionally normative.[22]

Risk perceptions differ due to differences in values and experiences among consumers, and attitudes and social norms influence the perception of hazards and how consumers make preparedness decisions.[23] Value judgments get applied during the active assessment of risks. For consumers, the crucial element of risk communication is a dynamic process that encourages transparent dialogue and debate among stakeholders. Risk communication should be free of value assumptions, to the greatest possible extent. Risk communication processes also should allow risk assessors to adapt to the values expressed by consumers. Finally, risk controls provide current practices to monitor and manage risks domestically and internationally.

A risk communication strategy is designed on the basis of the outcomes of a risk assessment. Factual and perceptual risks get considered simultaneously. On the basis of factual evidence, risk assessors try to reduce ambiguity, anxiety and uncertainty by normalizing incidences.[24] A risk communication approach that considers factual risks only employs scientific evidence, in which case the answers that the method offers, the answers that scientists seek, and how scientists perceive the nature of resolution becomes a value commitment. The role of science-based risk assessment and communication in policymaking is noteworthy, that is, as the long-established model that is essentially probabilistic and thereby quantifies risk levels. This notion of risk provides an aura of objectivity to scientifically-based public policy.

Government announcements about risk that only use scientific language may not appear trustworthy though, as was the case with the *Bovine Spongiform Encephalopathy* (BSE) outbreak in the UK.[25] Attitudes towards uncertainty can be either implicit or explicit. Public authorities often do not address uncertainty about food safety because they fear overreactions by the public.[26] Many authorities' general tone in risk communications therefore is reassurance, employed in part because of their fear of general public dismay toward food supplies. Authorities thus may patronize the public regarding their assumed inability to cope with uncertainty. A scientific approach to risk communication is linear, with simple causes and effects, known as the technocratic approach.[27] A factual-based message thus is broadcast in a one-way relationship between experts and consumers. Within this paradigm, risk assessors intend to educate and influence consumers with regard to risks.

CONSUMERS AND RISK

However, public conception of risk is more nuanced. Consumers are not exclusively irrational or emotional in times of calamity; rather, people tend to view risk along a wide spectrum of emotional and intentional responses, such as anxiety, vulnerability and feelings of uncertainty, security or well-being.[28] Crisis events heighten risk perceptions and shape future behaviour in response to risk.[29] For risk assessors, the focus should be on perceptual risks in addition to factual risks. As is the case for factual risks, perceptual risks may be quantifiable and predictable.

By studying information dissemination, we address not only the extent of disclosure but also what gets disclosed and the means of that disclosure.[30] When information dissemination is excessive, it may cause the unintended effect of cognitive shutdown in the audience and thereby reduce protective responses.[31] A robust risk communication strategy must identify fields of intersection among government, industry, media and experts in the context of risk perception and science. Information dissemination should cater to both risk makers and takers.

RISK MAKERS

Risk makers are those who manufacture risks, usually organizations and individuals engaged in food production and processing. Consumers are rarely considered risk makers, but they should be. Manufactured risks are products of human activity in general rather than specific subgroups.[32] Risk takers usually assume the risk of purchasing and consuming food produced by the risk makers; they lie on a continuum between being risk taking and risk averse. Those who are risk taking do not fear pain or failure and are willing to take chances to gain benefits. Those who are risk averse are more cautious and avoid taking chances. Again, risk takers can include both consumers and organizations. The number of risk makers and takers varies by the scope of the risk communication strategy, which may be localized or internationalized.

Collaboration within a community can lead to a more efficient risk communication strategy, because the community agrees on an interactive process among risk assessors, makers and takers/avoiders according to their shared rules, norms and structures. These shared values may differ depending on the particular power-dependence forces that affect socio-political units within the community. The primary role of risk assessors is to maintain positive discourse about hazards with risks makers/takers.

By considering risk as a socio-political construction, knowledge production becomes implicit and included in the process. Two corollaries relevant to risk scholarship are evident. First, the gap between risk assessment and risk communication is eliminated, because the organic nature of this approach includes all socio-political units involved. Second, rather than considering consumers as passive audiences disconnected from decision-making processes, this approach involves the exchange and collaborative generation of knowledge, which improves the public policy, risk control procedures and value judgements that underlie risk communications. Because knowledge is a collection of perceptions agreed upon by a community, it becomes a commodity accessible to everyone, which addresses uncertainty and makes shared trust the dominant sentiment in the exchange.[33]

The model presented here highlights the multifaceted nature of consumer risk behaviour. Risk communication adapts technical communication to the realm of civic discourse. If regulators or risks assessors fail to build a communication bridge between themselves and risk makers/takers, undesired outcomes may occur. For example, consumers may not engage in the self-protective behaviour necessary to reduce personal or group vulnerability to a risk, because personal experience likely affects not only the recognition of risk itself but also the intention to engage in those self-protective behaviours.[34] An overoptimistic bias, or the unfounded belief that one's own risk is lower than that of others, also can become a significant barrier to self-protective behaviour.[35] For risk communication to be efficient, risk makers/takers must be receptive to risk assessors, but if people believe that a risk poses little or no threat to their well-being, they are less likely to seek information or attend to risk communication.[36] Grobe, Douthitt and Zepeda indicate that responsibility for the health of others is a greater motivator to take action than responsibility solely for oneself.[37] Risk communication strategies therefore should recognize the intertwining and active roles of consumers and their communities in risk perception.

In the next section, we present the methodology and findings of a survey conducted following the spinach food recall during fall 2006.

Methodology

This study attempts to test the effects of information dissemination on risk taker perceptions and behaviours, resulting from the spinach food recall during fall 2006. The study findings suggest some recommendations about relevant aspects of our risk communication model.

We employed a survey method and developed the instrument in two steps. First, we reviewed more than two dozen publications (for example, *British Food Journal, Food Policy and Journal of Public Affairs*) to learn about previous surveys of consumer attitudes about food recalls. Second, we undertook a questionnaire design to maximize response ease and encourage completion. To this end, the instrument consists of nine items and should take no more than 5 minutes to complete. We also made appropriate changes on the basis of comments from a pre-test group, prior to data collection.

Unless indicated otherwise, the survey questions employ bipolar (yes/no) scales. The first two questions ask whether the respondent had heard of the September 2006 spinach recall and were aware the recall had finished. Respondents who had not heard of the recall then simply completed the three demographic questions at the end of the survey. Next, respondents who had heard of the recall indicated if they had spinach in their homes at the time of the recall and, if so, if they had thrown it away. The fifth question asked these respondents to rate how their perception of the safety of spinach had changed since the recall, on a five-point scale of safer, just as safe, slightly less safe, somewhat less safe, and much less safe. After the respondents indicated whether they had eaten spinach since the recall, they noted their age group (younger than 15, 15–24, 25–34, 35–54, or 55 years and up), gender and whether they were the primary household grocery purchaser.

Respondents to the survey, conducted in classes at the University of Regina in Saskatchewan over an 8-day period in March 2007, 6 months after the recall occurred, included 839 part-time and full-time students. Although the demographics therefore skew toward a younger age group, we take this bias into account in our analysis. In our sample, 51 per cent of the respondents were women, and more than 90 per cent have a university education.

Some limitations affect the findings from this survey. First, the respondent pool reflects a convenience sample and may not represent the total Canadian population. Second, we did not screen for food taste biases, so consumers who do not eat spinach also appear in the survey. We chose to include these consumers in the sample because some may purchase spinach for other household members. Third, the survey was conducted 6 months after the recall was issued. Individual differences in long-term memory among respondents thus may have influenced the results.

Results

Descriptive statistics show that 75 per cent of risk takers sampled had heard of the September 2006 spinach recall, yet only 57 per cent of these respondents knew that the recall had ended. This percentage seems sizeable, considering that the recall ended in November 2006 and the survey was conducted approximately 4 months later. These findings therefore indicate that the CFIA, as risk assessor, was effective in disseminating information about the initial recall to the public, but it appears to have been unsuccessful in following through with relevant updates. The CFIA's methods of communicating with the public therefore require re-evaluation.

RECALL INERTIA

Another somewhat disturbing finding indicates that of the 25 per cent of respondents who knew about the recall and had spinach in their homes at that time, only 53 per cent threw it away, whereas 47 per cent did not, indicating a surprising lack of concern about safety. Of those who threw away their spinach, 77 per cent reported having eaten spinach since then, indicating that the recall did not diminish their perceptions about the safety of spinach. Furthermore, the respondents' perceptions of the safety of spinach were not significantly changed by the recall. Fifty-eight percent reported believing spinach to be just as safe as or safer than before the recall, and another 21 per cent believed it to be only slightly less safe. Only 14 per cent of respondents who knew about the recall perceived spinach as somewhat or much less safe than before the recall. These results were particularly unexpected, given the level of media attention focused on the more than 200 people who were hospitalized as a result of the tainted spinach. This result may suggest that different methods of communicating risk are required to ensure that risk takers' perceptions of risk match the factual risks.

Social marketing literature demonstrates that young consumers tend to consider themselves impervious to harm,[38] which warrants further examination of the sample demographics' influence on our findings. According to an ANOVA, gender and age have significant main effects on consumers' perceptions of spinach safety. Consistent with previous findings, women and older consumers are significantly more concerned with spinach safety than are men and younger consumers. Although serving as the primary grocery purchaser for a household does not significantly influence food safety perceptions, older women traditionally play such care-giving roles. As Grobe, Douthitt and Zepeda indicate,[39] our findings may suggest that responsibility for the well-being of others is a greater motivator to take action than when simply being accountable for oneself. However, results are not conclusive in this case.

We conducted a regression analysis to determine which factors had a significant impact on risk takers' perceptions about the safety of spinach. Knowing about the recall was not enough to have a lasting effect on safety perceptions. Four variables explained 26 per cent of the variance in respondents' spinach safety perceptions: whether they threw spinach away after the recall, whether they had eaten it since the recall, age categories and gender. Consistent with our intuitive logic, risk takers who had thrown away spinach and had not eaten it since were more likely to show risk aversion and be concerned about its safety. Consistent with previously reported ANOVA results, being older and female are indicators of risk aversion. Thus, young, male respondents are more likely to demonstrate risk-taking tendencies. This finding supports the previous suggestion that different communication methods may be necessary to ensure that risk-taking risk takers' perceptions better align with factual risks in food recall situations.

Inspired by our conceptual model and the findings of our empirical survey, we next discuss some practical implications for the future of risk communication.

Discussion

Even though our sample is not representative of the general population, our findings show that the CFIA successfully communicated a health hazard alert about American spinach to the public. Respondents also are not fearful about the safety of spinach since the recall. The data show that consumer confidence in the safety of food is fairly high, consistent with existing food safety literature.[40] Consumers seem to trust food supply chains, and though food has never been safer in Canada, the scope of each recall is becoming more and more difficult to manage. Enhanced surveillance techniques, demographics, increased potential for distributing contaminated food through mass production technology and many other contributing factors help explain the increasing number of food recalls.[41]

However, as a risk assessor, the CFIA failed to nurture an interactive and effective relationship with the risk taker subgroup. Overall, consumers seemed unresponsive to the food recall: Many who had spinach did not throw it out after the recall, and many did not know that the recall had been over since November 26, 2006, 4 months before the survey. This evidence leads us to suggest that the CFIA and industry officials should engage in fostering a more sustainable relationship with consumers through more effective communication.

THE MEDIA

In recent years, the CFIA has built rapport with the media, but the results of this cooperation have been mixed. Our findings suggest that the CFIA neglected consumers by failing to keep them informed. Although the media has focused on food recalls of late, its reporting focuses much more on potential dangers than resolutions. The spinach recall serves as a case study for disseminating an advisory through the media, and we have evidence that most Canadians knew about the recall. But the level of information flow between the CFIA and media is uncertain, as is the media's interest in disseminating follow-up information about food recalls after the initial danger has passed.

THE SCOPE

On the basis of our theoretical model and survey, we present some recommendations for encouraging more desirable outcomes from the practice of risk communication.

Recommendation 1: Global food safety systems should represent the scope for accurate risk assessment.

Global risks are viewed much differently today than in studies of preceding decades, because risk assessment has progressed from dealing primarily with risks inside one organization to an approach that analyzes the broader strategic objectives of an organization. Perhaps the most obvious changes involve the scope of risks addressed and the increased sophistication of food distribution systems employed by risk makers/takers. Increasing public distrust and reduced consumer confidence in food safety may have adverse economic effects on the food industry, as well as on national and international economies at the aggregate level.

SOCIO-POLITICALLY CONSTRUCTED RISK

The establishment of new regulatory bodies and national food safety agencies around the world makes trade and the establishment of food safety standards more sophisticated and inclusive of various social voices and interests. Hence:

Recommendation 2: Both risk assessors and risk makers/takers should recognize that risk is socio-politically constructed.

The power and dependence balance between social units should be considered in the design and implementation of a risk communication strategy. Power is a relational concept inherent to any exchange between social actors. Social systems should also be considered. Historically, food has been an ingredient of power and weaponry in international trade.[42] Countries that are self-sufficient in their food production are less likely to depend on other countries and tend to profit from the food reliance of their foreign trading partners. However, many governments, both historically and recently, have resorted to stringent food safety policies that arguably represent protectionist policies. Domestically, the power and control schemes of socio-political networks produce conflicting or cooperative relations amongst stakeholders, which means that risk assessors must consider where the power lies in an industry and who is dependent on whom. Because of the primacy of socio-political forces, which are constantly in flux, risk assessment is a dynamic process. What is true one day may not be the next.

FACTUAL VERSUS PERCEPTUAL RISKS

Evidence indicates that firms with multiple locations, which require more resources, perform worse than single-plant firms during food recalls. A greater awareness of socio-political networks may aid the CFIA in assisting industry to manage its food recalls without requiring supplemental resources. Risk assessors should become the centrepieces of an information-sharing network, though not necessarily the exclusive locus for information dissemination and food science.

Recommendation 3: Factual and perceptual risks should be considered concurrently in the process of establishing a risk communication strategy.

INFORMATION DISSEMINATION

Science provides a body of useful and verifiable knowledge, despite the uncertainty that pervades the interpretation of scientifically derived observations. By necessity, scientists work in conditions of uncertainty and scepticism. They tend to be at ease with uncertainty, which becomes an issue when they must communicate science to the public, industry, media and government. Not everyone copes equally well with uncertainty, which may lead to founded or unfounded fears. These perceptions should be addressed when communicating with these stakeholders. Risk assessors must deal with both fear and risk in a synchronized fashion.

Recommendation 4: Risk assessors should develop direct communication channels for information dissemination.

Because many respondents in our survey knew little about what happened after the initial recall, we note some concern that consumers make decisions without understanding the risks involved. Risk assessors should account for this gap in consumer awareness by relying on other communication channels, such as advertisements and public relations. Judicious strategies to communicate risks to consumers are vitally important as means to educate the public about the complexities of modern food distribution channels. It is not the quantity of recalls that will matter in the future but rather the quality of the relationships risk assessors nurture with their environment and, most important, with consumers. Risk assessors must pay careful attention to how they communicate risk instead of focusing entirely on the message itself. Better collaboration with food retailers, consumer-friendly websites, and newsletters may help risk assessors build new communication channels. These channels must remain flexible because the human mind, unfortunately, is not.

MEASUREMENTS

Recommendation 5: Apply strategies to measure risk makers'/takers' behaviours and perceptions.

What information do risk makers/takers expect from risk assessors? What do the risk makers/takers currently believe? Do risk assessors have the resources to communicate the message? Risk assessors should address these issues, because the demands of external stakeholders are likely to continue expanding in response to globalization and increasing consolidation in the food industry. Risk assessors also must convert the data gained from these types of investigations into practical knowledge. Active communication effectiveness, openness, transparency, action demonstration efficiency, fear treatment efficiency, effectiveness of sources used, context and the effectiveness of enabling self-responsibility are some of the variables that should be considered in measurement efforts. The relationship between consumer confidence and consumer behaviour also should be incorporated in the analysis. But to capture dynamic consumer perceptions and behaviours with regard to the safety of food products, a longitudinal approach is required. For ethical as well as practical reasons, these measurements and surveys should not be conducted by risk assessors themselves. An independent and impartial agency should be created to introduce greater neutrality into these studies.

For example, the BSE crisis in Britain led to the creation of the Food Standard Agency (FSA) in 1999. The main objective of this agency is to protect the public health from risks that may arise from the consumption of unsafe food and otherwise to protect the interests of consumers in relation to food. The agency is managed by consumers and reports directly to parliament. The FSA carries out its mandate of transparency by making all information readily available through mailed bulletins and through the use of innovative communication channels, such as user-friendly websites. Moreover, the agency appears to distinguish between theoretical and real risks, which reduces unfounded public uncertainty and improves communication and trust between risk assessors and risk makers/takers.

KNOWLEDGE AS A COMMODITY

Recommendation 6: Knowledge is a commodity that can be produced by risk assessors, makers and takers.

Although the spinach recall was imposed on the produce coming from only a few farms, it affected an entire industry for months. Following the recall, some consumers demonstrated self-protective behaviour by consuming less spinach, which explains why sales of spinach were down in North America. For those spinach processors and packers whose product was not at fault yet that still suffered a significant financial loss due to consumer fear or mistrust, insurance recovery is unlikely.[43]

Our call for increased stakeholder participation could take the form of consumer advisory committees or consensus conferences. Agricultural crises should generate a knowledge-enhancing process that includes the input of farmers, veterinarians, scientists, policymakers, economists, consumers and sociologists, to name only a few crucial perspectives.[44] A two-way dialogue between risk assessors and risk makers/takers is essential in any effort to confront and reduce market uncertainty.

By recognizing that we do not have all the answers, we can produce positive outcomes for risk assessors, given the connection between risk communication strategies and risk makers/takers. Discourse should not be driven exclusively by scientific judgment but also should absorb the personal insights that underlie normative assumptions. From this perspective, risk communication becomes a web, a network or an inclusive process of exchange. Because the range of assessment for food safety risk assessors in modern society transcends the borders of any given nation, it is vital to recognize the differences in social and political practices among the interconnected nations affected by risk.

Conclusions

We all eat, so food risks affect us all. Inevitably, future food crises will reintroduce high levels of uncertainty among the public and food industry. There is a need for greater public participation and increased cooperation between the public and risk assessors when addressing food safety events to mitigate the harmful effects of uncertainty. The public often does not participate in these decision-making processes but rather is represented by elected and industry officials who may be insufficiently aware of consumers' interests and concerns. Risk communication, or the process of honestly and effectively conveying the risk factors associated with a wide range of natural hazards and human activities, can be managed properly and thus foster mutual respect between stakeholders. In risk communication, both international and national organizations that encourage dialogue should be promoted to resist the adoption of a one-size-fits-all solution, in such a way that enables municipal and provincial levels to ensure their particular needs are met.

Our goal for this chapter has been to present a conceptual model of risk communication that suggests that risk assessors should interrelate more closely with their environment. The findings of our survey suggest that risk communication efficiency can be enhanced through better accounting for the role of socio-political units. It is also paramount that risk assessment procedures consider knowledge as a commodity, produced and shared among these units.

'Knowledge is of two kinds. We know a subject ourselves, or we know where we can find information on it.'

– Samuel Johnson (1709–1784)

This quote, from more than two centuries ago, remains applicable to food safety and risk communication. The choice of this may be perceived as controversial, because Dr. Johnson is usually associated with the Enlightenment, with its belief in scientific progress and rational authority – values questioned by this chapter. Yet the framework of international food trade has gone through innumerable changes in the past three decades, and more developments are imminent. Westernized society has reached a key juncture in its relationship to food supply and food policy, and both public and private risk assessors seem to be failing to grasp the extent of the challenge. The state, the corporate sector and society in general face difficult decisions. In food safety, public policy often lags behind the restructuring that takes place in the food system and thus is often reactive. Food manufacturers are driving the political agenda on food safety – arguably at the expense of producers. Most industries adopt a productionist paradigm, focusing mainly on output and trades and failing to synchronize production and consumption. Many agricultural public policies around the world currently concur with this paradigm.

Food agencies worldwide offer appealing prospects, at least in theory, for maintaining creative tensions in food governance between the establishers of food standards and the measures that they implement.[45] In the meantime, more crises similar to the spinach recall are bound to happen. The model and recommendations presented herein may apply to modern risk communication strategies, but their empirical value for such an implementation remains uncertain. Risk perceptions should be measured in the aftermath of future food recalls. In addition, measures of how much consumers value risks and safety may become more valuable than ever for both professionals and policymakers.

References

1. Marris, C., I. Langford, T. Saunderson, & T. O'Riordan (1997), 'Exploring the psychometric paradigm: comparisons between aggregate and individual analysis', *Risk Analysis*, Vol. 17, No. 3, pp. 303–312; Grobe, D., R. Douthitt, & L. Zepeda (1999), 'Consumer risk perception profiles regarding recombinant bovine growth hormone (rbGH)', *Journal of Consumer Affairs*, Vol. 33, No. 2, pp. 254–275.

2. Marris et al., op. cit.; Labrecque, J., S. Charlebois, & E. Spiers (2007), 'Is gene technology an emerging dominant design? An actor's network theory investigation', *British Food Journal*, Vol. 109, No. 1, pp. 81–98.

3. Powell, D. & W. Leiss (1997), *Mad cows and mother's milk: the perils of poor risk communication*. Montreal: Mc Gill-Queen's University Press.

4. Shang, W. & N. Hooker (2005), 'Improving recall crisis management: should retailer information be disclosed?', *Journal of Public Affairs*, Vol. 5, No. 4/4, pp. 329–342; Nganje, W., M. Slaplay, S. Kaitibie, & E. Acquad (2006), 'Predicting food safety losses in turkey processing and the economic incentives of hazard analysis and critical control point (HACCP)', *Agribusiness*, Vol. 22, No. 4, pp. 475–498.

5. Currie, L. (2002), 'Food safety: an industry perspective', *Canadian Journal of Dietetic Practice and Research*, Vol. 63, No. 1, p. F4.

6. Doering, R. (2006), 'Making hay out of spinach', *Food in Canada*, Vol. 66, No. 9, pp. 18–19.

7. Roberts, S. (2006), 'Huge spinach recall puts spotlight on contamination cover', *Business Insurance*, Vol. 40, No. 39, pp. 1–2.

8. Canadian Food Inspection Agency (2006), 'Various brands of imported fresh spinach may contain *E. coli O157:H7* bacteria', Health hazard alert, Food recall archives.

9. McKenzie, A. & M. Thomsen (2001), 'The effect of *E. Coli O157:h7* on beef prices', *Journal of Agricultural and Resources Economics*, Vol. 26, No. 2, pp. 431–445.

10. Food Marketing Institute (2007), *Report on food retail sales*, March.

11. Canadian Food Inspection Agency Act (1997), c. 6, C-16.5, assented to March 20, 1997.

12. Teratanavat, R., V. Salin, & N. Hooker (2005), 'Recall event timing: measures of managerial performance in U.S. meat and poultry plants', *Agribusiness*, Vol. 21, No. 3, pp. 351–373.

13. *The Wall Street Journal* (2007), 'Spinach recall weakens trust in vegetable safety', February 5, p. B2.

14. Grabill, J. & M. Simmons (1998), 'Toward a critical rhetoric of risk communication: producing citizens and the role of technical communicators', *Technical Communication Quarterly*, Vol. 7, No. 4, pp. 415–440.

15. Mirel, B. (1994), 'Debating nuclear energy: theories of risk and purposes of communication', *Technical Communication Quarterly*, Vol. 3, pp. 41–65.

16. Charlebois, S. & J. Labrecque (2007), 'Processual learning, environmental pluralism, and inherent challenges of managing a socio-economic crisis: The case of the Canadian mad cow crisis', *Journal of Macromarketing*, Vol. 26, No. 1, pp. 85–102.

17. Charlebois, S. (2005), *The impact of socio-political processes and structures to political economies: the case of the Canadian mad cow crisis*, doctoral dissertation, University of Sherbrooke, 281 pages.

18. Mitroff, I., P. Shrivastava, & F. Udwadia (1987), 'Effective crisis management', *Academy of Management Executive*, Vol. 1, No. 4, pp. 283–292.

19. Kostelnick, C. (2007), 'The rhetorical minefield of risk communication', *Journal of Business and Technical Communication*, Vol. 21, No. 1, pp. 21–24.

20. Benson, J. K. (1975), 'The interorganizational network as a political economy', *Administrative Science Quarterly*, Vol. 20, No. 2, pp. 229–259.

21. Dwyer, R. & S. Oh (1987), 'Output sector munificence effects on the internal political economy of marketing channels', *Journal of Marketing Research*, Vol. 24, No. 4, pp. 347–358.

22. Anthony, R. (2004), 'Risk communication, value judgement and the public-policy maker relationship in a climate of public sensitivity toward animals: revisiting Britain's foot and mouth crisis', *Journal of Agricultural and Environmental Ethics*, Vol. 17, pp. 363–383.

23. Grobe et al., op. cit.

24. Ward, R., D. Bailey, & R. Jensen (2005), 'An American BSE crisis: Has it affected the value of traceability and country-of-origin certifications for US and Canada beef?', *International Food and Agribusiness Management Review*, Vol. 8, No. 2, pp. 92–114.

25. Tacke, V. (2001), 'BSE as an organizational construction: a case study on the globalization of risk', *British Journal of Sociology*, Vol. 52, No. 2, pp. 292–306.

26. Klint-Jensen, K. (2004), 'BSE in the UK: Why the risk communication strategy failed', *Journal of Agricultural and Environmental Ethics*, Vol. 17, pp. 205–423.

27. Plough, A. & S. Krimsky (1987), 'The emergence of risk communication studies: social and political context', *Science, Technology and Human Values*, Vol. 12, No. 2&3, pp. 4-10.

28. Slovic, P. (1987), 'Perception of risk', *Science*, Vol. 236, pp. 280–285.

29. Kasperson, R. (1992), 'The social amplification of risk: progress in developing an integrative framework'. In *Social Theories of Risk*, Sheldon Krimsky & Dominic Golding, (eds) Westport, CT: Praeger.

30. Beretta, S. & S. Bozzolan (2004), 'A framework for the analysis of firm risk communication', *International Journal of Accounting*, Vol. 39, No. 3, pp. 265–276.

31. Aldoory, L. & M. Van Dyke (2006), 'The roles of perceived shared involvement and information overload in understanding how audiences make meaning of news about bioterrorism', *Journalism and Mass Communication Quarterly*, Vol. 83, No. 2, pp. 346–362.

32. Beck, U. (1992), *Risk society: towards a new modernity*. New Delhi: Sage.

33. Grabill & Simmons, op. cit.; Perez-Floriano, L., J. Flores-Mora, & J. Maclean (2007), 'Trust in risk communication in organization in five countries of North and South America', *International Journal of Risk Assessment and Management*, Vol. 7, No. 2, pp. 205–221.

34. Ehrlich, I. & G. Becker (1972), 'Market insurance, self-insurance, and self-protection', *Journal of Political Economy*, Vol. 80, pp. 623–648; Weinstein, N. (1989), Effects of personal experience on self-protective behavior', *Psychological Bulletin*, Vol. 105, No. 1, pp. 31–50.

35. Sparks, P. & R. Shepherd (1994), 'Public perceptions of the potential hazards associated with food production and food consumption: an empirical study', *Risk Analysis*, Vol. 14, No. 5, pp. 799–806.

36. Schafer, R., E. Schafer, G. Bultena, & E. Hoiberg (1993), 'Food safety: an application of the health belief model', *Journal of Nutrition Education*, Vol. 25, No. 1, pp. 17–24.

37. Grobe et al., op. cit.

38. Whalen, C., B. Henker, R. O'Neil, A.H. Hollingshead, & B. Moore (1994), 'Optimism in children's judgments of health and environmental risks', *Health Psychology*, Vol. 13, pp. 319–325.

39. Grobe et al., op. cit.

40. De Jong, J., L. Frewer, H. Trijp, & R. Jan Renes (2004), 'Monitoring consumer confidence in food safety: an exploratory study', *British Food Journal*, Vol. 106, No. 10/11, pp. 837–846.

41. O'Neil, J. (1999), 'Food safety recalls: There's a right way', *Defense Counsel Journal*, Vol. 66, No. 3, pp. 424–445.

42. McDonald, K. A. (1999), 'Debate over how to gauge global warning heats up meeting of climatologists', *Chronicle of Higher Education*, February 5, p. A17.

43. Roberts, S. (2006), 'Huge spinach recall puts spotlight on contamination cover', *Business Insurance*, Vol. 40, No. 39, pp. 1–2.

44. Bennett, P. & K. Calman (2002), *Risk communication and public health*. Oxford: Oxford University Press.

45. Lang, T., E. Millstone, & M. Rayner (1997), *Food standards and the state*. London: Centre for Food Policy.

4

Food Safety, Quality and Ethics in Supply Chains: A Case Study of Informing in International Fish Distribution

BY PER ENGELSETH,* TAKEO TAKENO† AND
KRISTIAN ALM‡

Keywords

food supply chains, fish (mackerel) product transformation, ethics and food supply, information transparency, societal and environmental contexts.

Abstract

Supply chains take joint responsibility for providing safe and high-quality products. In food distribution, ethics involves product flows and information transparency. The analysis in this chapter points to the demand for adequate information, the value of freedom and human well-being as ethically pertinent issues internal to mackerel supply chains, which feature interaction between a focal supply chain and its business, societal and natural environmental context. The proposed models depict individual food supply chains as mediating resources in their context. Ethics therefore creates a new understanding of how information can help secure safe, quality food supplies.

*   Dr Per Engelseth, Ålesund University College, Serviceboks 17, 6025 Ålesund, Norway, E-mail: pen@hials.no. Telephone: + 47 7016 1261.

†   Dr Takeo Takeno, Iwate Prefectural University, 152-52 Sugo, Takizawa, Iwate, 020-0193, Japan, E-mail: take@iwate-pu.ac.jp. Telephone: + 81 19 694 2646.

‡   Dr Kristian Alm, BI Norwegian School of Management, Oslo 0442, Norway, E-mail: kristian.alm@bi.no. Telephone: + 47 4641 0481.

Introduction

The issue of food safety has become prominent in supply chains due to recent events, including animal disease and human poisoning through food consumption. To counter these problems, EU regulations have widened from food safety control measures in flows of goods to include demands associated with information about foods, as well as the traceability of both food products and consumer packaging. The focus of this chapter is information about foods as a means to achieve safe, quality food supplies.

A single case study describes mackerel caught wild, processed in Norway, and then distributed to Japan as a packed frozen product. This study was originally conducted as an explorative quest with the general aim of providing an overview of the focal supply chain. Empirical material also reveals some ethically pertinent issues and thereby suggests a better understanding of the interplay between ethics and food supply, with a specific focus on information about foods. Information represents a technical resource that links actors to the physical supply of products and thus reflects actors' perceptions of the interaction between product supply and the technical flows of goods.

This study adopts a complete food supply chain perspective. The case starts with the catch of fish at sea in Norway, follows the goods through production and logistical and marketing activities in relation to distribution, and extends to their final sales in restaurants or retailers in Japan. This study also considers the societal and environmental contexts of the supply chain. The aim of this ethics-based research approach is to provide a greater understanding of the ethics of food supply, especially as it pertains to information about foods in a fluctuating and therefore often incomprehensible context.

Literature Review

This section provides a frame of reference and the foundation for collecting and analyzing the data.

FOOD PRODUCTS, SUPPLY AS SYSTEMS AND ETHICS

'Ethics' provides guidance regarding right or wrong, good or evil, and responsibilities in food supply chains. The supply chain is a business-driven entity with the aim of economical product supply. It is therefore vital that foods both be safe and have quality features to ensure customer satisfaction. These features also must be attained in a cost-efficient manner. However, friction may occur between the two different aims of achieving food quality and ensuring food safety. In this chapter, quality food supply also denotes food products that are delivered economically, which helps retain customers and suppliers. Safe food refers to the predominately societal aim of enhancing human welfare through the physical attributes of the product.

Food chains that attempt to supply quality and safe foods consist of various cooperating actors; they are conglomerate business entities. Supply chains also are influenced by business competition, which may entail struggles between actors within the chain or between complementary or competing chains. In addition, natural environmental and societal concerns create challenges for economical food supply chains. Economic, societal and environmental factors have ethical weight; that is, actions within the food supply chain can lead to more or less good or bad and positive or negative consequences in and between supply chains, for society, and for the natural environment.

According to logistics and supply chain management (SCM) literature, food supply involves a system that represents a particular view of reality. Ethics in the context of food supply systems reflects Lilienfeld's idea of an interlinked component (parts, links, goals and feedback mechanisms) in a chain structure.[1] A systems approach to food supply involves a focus on both the technical and the organizational aspects of distributing foods through links, feedback mechanisms and boundaries.[2] Supply chain boundaries are usually determined according to the core logistical flows of goods, supported by an information flow.[3] Individual food supply chains connect multiple actors with differentiated needs within a single flow of goods. Supply chain actors in most cases also manage and operate multiple interlinked flows of different products.[4] This feature increases in importance further downstream in food supply chains, such as in retail settings that display an abundance of different products for sale. Moreover, food supply chains must adapt to environmental constraints. Raw material supply, consumer preferences, the competitive structure and government legislation regarding food safety and quality often change. To counter this challenge, incremental development in supply chains helps companies survive in the short term, even in an ever-threatening context. From a systems perspective, managing supply chains requires the adequate coordination of resources within the chain, as well as coordinating the complete chain with its externalities. This chapter employs this fundamental understanding of food supply.

FLOW OF FOOD PRODUCTS

The flow of foods is the core value-creating entity in a supply chain context; therefore, information in a supply chain should predominately reflect this flow. However, from a logistical and marketing standpoint, food products are no simple resource. The peculiarities of food product supply distinguish it from supply chains for other types of goods.[5] As Thompson notes, food chains are 'long-linked technologies'.[6] Sequential dependencies are characteristic of value creation in the physical distribution. For example, activities that transform downstream goods, such as during their physical distribution, depend on preceding activities (for example, transport, storage, production, materials handling). Damages and timeliness, from raw material supply to transport, continually affect subsequent food production activities. Production failures also influence the quality of retail products. These examples demonstrate the sequential logic prominent in food supply.

From a marketing channels perspective, Alderson offers a little used but potentially fruitful framework that imagines sequential technical activities as directed by intermittent decision-making 'sorts.'[7] Sorts refer to events that use information to both control and direct the flows of foods. The supply chain's purpose relates to securing safe and quality product placement in the hands of an end-user. The scope of the supply chain in this view is end-to-end or complete. The supply chain encompasses a starting point in an 'original' upstream state in which goods are 'conglomerate resources' (for example, raw materials, rudimentary components). Goods are then transformed to satisfy various intermediary and end-supply chain actors (including consumers), through the sequential provision of the form, time, place and possession utility of products.[8] Alderson also argues that at sorts, actors assign goods to operations, and laying the ground for transforming products.[9]

Information thus is a vital tool that binds actors to the flow of products. Value creation through food product transformation is a step-by-step process by which products interact with sets of heterogeneous resources (including actor competence, information, supply facilities and products), managed by sequentially interrelated supply chain actors.[10] This functional way of thinking may be viewed as a rediscovery of the framework for studying product supply – intellectually 'going back to the future'[11] – as illustrated in Figure 4.1 as a sequential and piecemeal view of product transformation through 'flow.'

Alderson's and Thompson's views represent a complete chain perspective regarding the dependencies among the activities involved in product supply, from raw material to consumption.[12] Food safety and quality depend on the cumulative effect of activities that precede the final supply of a product to an end user. Supply chains also interact with other business supply chains in the increasingly competitive global marketplace.[13] Moreover, individual supply chains are affected by natural environmental and societal constraints.[14] This network describes the complexity involved in achieving a safe, high-quality food supply.

FLOW OF INFORMATION ABOUT FOODS

Information serves as a resource that directs the value-creating process and helps communicate about products. Thus, information needs to be organized. From a structural perspective, information flows in supply chains consist of interlinked information systems used by different actors. Each information system is an individually implemented medium for recording, storing and disseminating linguistic expressions, as well as drawing conclusions from such expressions.[15] Information systems may also be regarded as social systems in which the understanding of product supply affects the information system design and functions.[16] Food supply practices also exhibit multiple, weakly interlinked information systems that manage a common flow of foods.[17]

Information about foods comprises product transactions that indicate the price of the product. Price negotiations represent the immediate context for communications as a means to reach a price that satisfies both the buyer and the seller. Price provides a type of product information communicated to purchasers. In addition to product utility, price communicates product quality and safety, which affects customers' expectations of their potential satisfaction through purchase. Transactions then lay the groundwork for information that directs the flows of goods.

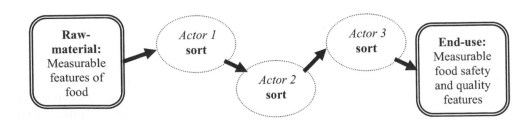

Figure 4.1   Food product transformation as sequential flow

Information in a food supply chain offers a management and operational tool to direct the flow of goods and control product supply, in accordance with marketing and logistics plans. In complex, multiple interacting systems, the divergent purposes of information about foods may become obscured, even if not intentionally. Product information transparency depends on the efforts of supply chain actors to coordinate the flow of information though mutual investments that secure information quality. Information systems play important roles in facilitating the use of food safety and quality control mechanisms, such as hazard analysis and critical control point (HACCP; http://vm.cfsan.fda.gov).

Information about foods functions to (1) assign goods (for example, future time, place and form of a product), (2) track goods (current location of goods), and (3) trace products (detailed history of the time, place and form features of a product during supply).[18] These informational activities all involve communication within the context of business relationships in supply chains. Information connectivity accordingly is a vital feature of the information flow, especially for achieving efficient control of food safety and quality and securing product traceability.[19] Food product traceability encompasses the entire scope of an individual supply chain. Therefore, a complete chain approach may account for the complete set of activities that lead to a finished product. Because information about products encompasses a complete food chain, transparency in supply chains, which depends on information quality, becomes vital. Managing the flow of goods enables not only the provision of safe, quality goods, which is the core value-creating element of food supply, but also strong information quality to support this task, as a distinct and measurable entity.

INFORMATION ABOUT FOOD SUPPLY AND ETHICS

Delivering safe food products of high quality to end-users is in itself an important ethical issue, because quality is a multidimensional sign that distributors take responsibility for the health of numerous consumers. Through this action, supply chain actors avoid recurring product failures; to do so, they must strive to learn how to develop marketing and logistics activities that secure the delivery of safe and quality foods to end-users. In addition, this challenge is complemented by qualities associated with the flow of information that support the core function of physical food supply.

Supply chains comprise flows of goods, flows of information and the connections of these flows through goods identification. Such supply chain qualities affect information transparency and product quality, measured by the end-user upon delivery. Therefore, these three aspects of supply quality have ethical relevance. Ethics encompasses a set of theories about responsibility and directs attention to the features of supply chain actors, as well as how they manage and operate in relation to other actors within the same supply chain, which itself is embedded within a wider environmental context.[20] The joint responsibility belongs to supply chain actors. In line with Boatright's definition, joint responsibility is as an expression of three basic values:[21] fairness (or justice), freedom and well-being. These values represent the joint accomplishment of supply chain actors who seek to provide consumers with safe and quality goods, dependent on information quality.

Information guides products, and the control of goods can update information about products. The features of this information provision and use depend on the connections of information systems managed by different supply chain actors. Such connections relate to how supply chain actors, together and over time, develop information connectivity to secure quality information exchanges.

However, there is no guarantee that information quality will be secured in a supply chain; rather, an embedded risk exists that information may be insufficient, distorted or even manipulated. Therefore, supply chain actors are responsible for detecting discrepancies in information content for the parties potentially affected by these manipulations. Information discrepancy accordingly may affect a supply chain actor's ability to act freely.

The content of responsible action and its consequences and the presuppositions for such actions in people's attitudes and personalities represent the core of ethical behavior.[22] The vast complexity of modern societies and individuals embedded in them have moved ethics away from the notion of the free individual to the post-modern challenges of sharing responsibility with others. Because managing supply chains involves the coordinated technical efforts of multiple supply chain actors, this development increases the complementary relationship between supply chain research and ethics. In this setting, ethics provides guidance for handling potential information obscurity, linked with supply network complexity.

The Mackerel Case

METHODOLOGICAL CONSIDERATIONS

At its starting point, this study attempted to discover potential research topics associated with the export of Norwegian fish to Japan. This quest demanded greater research focus, which gradually became a single case study with a product focus, specifically, mackerel products that moved from Norway to final Japanese consumers, with a consideration of societal and environmentally pertinent issues.

The actual case study was conducted in autumn 2006 with the assistance of 16 key informants, including a wide range of actors directly or indirectly linked to the actual mackerel supply chain. The initial interview with a Norwegian exporter of pelagic fish suggested a synopsis and means to locate other key informants in the focal supply chain. The subsequent interviews in Japan involved fish importers that had developed business relationships with the initial Norwegian exporter, marketing consultants who aided the Norwegian exporter and academic researchers involved in ongoing or previous research projects pertaining to fish distribution. The study also involved observations from two wholesale fish markets in Japan. Finally, a previous study of the pelagic fish supply from Norway to Holland served as the basis for the case description regarding raw material supply, production, goods handling, storage and transport of mackerel.[23] This previous study used the same firm for which the initial Norwegian exporter interviewed worked. The interviews were semi-structured and open ended, allowing a conversation-style interview that created an atmosphere in which informants were motivated to provide the researcher with new understandings that would lead to further inquiries.

This case narrative describes how mackerel is transformed from its raw material source into products sold for consumption in Japan, with a focus on the technicalities of providing goods and how information provides support. The resulting analysis combines the logistics and marketing channels aspects of mackerel supply. Ethics are not explicitly considered within the narrative text. The product-oriented narrative functions mainly as a basis for analyzing ethical aspects of mackerel exporting from Norway to Japan.

MACKEREL PRODUCTS

The mackerel market has high export volume and low prices per kilo. The average price of mackerel for Norwegian fishermen in 2006 was 8.26 NOK (Norwegian Kroner), approximately equivalent to one Euro. Japan is by far the largest market for Norwegian mackerel exports, amounting to almost 600 million NOK in 2006, though down from 1200 million NOK in 2005. This decline was mainly due to decreasing demand in the Japanese market, and the lower level of mackerel exports continued through 2007. Mackerel are pelagic fish, which travel together in schools and migrate over large distances. Pelagic fish are also relatively fatty and therefore considered a highly nutritious food product. An average Norwegian-caught mackerel may weigh about 400–500 grams.

Mackerel rapidly deteriorates; it is the only cured form of sushi in Japan. Mackerel is caught wild at sea during various seasons. In Japan, mackerel provides the raw material for several processed food products, including surimi (fish paste) and kamaboko (fish cakes), and is displayed as a 'fresh' product in retail settings. In this case, the 'fresh' designation indicates that distribution packaging has been opened so that the retailer can defrost the fish and then repackage it (usually at the store) into consumer packages, labelled with best-before dates that count from the time the distribution package was unfrozen. In Japan, mackerel is consumed in various ways as different types of finished products, both in home-prepared meals and in restaurants.

PRODUCT TRANSFORMATION

Distributing mackerel from Norway to Japan requires a supply chain with multiple cooperating actors who play different roles in achieving logistical and marketing aims. These include fishermen, industrial producers in Norway (who pack the fish into distribution-level packages), industrial producers in China or Japan (who produce consumer-adapted products, often packed in consumer-level packages), traders in Norway and Japan who never actually handle the fish, Japanese wholesale distributors, Japanese fish markets, Japanese retailers or restaurants and a range of logistics service providers.

Due to its limited durability, mackerel can be caught only 1–2 days before processing on land in Norway. The mackerel catch is therefore a coastal form of fishery. A portion of the products distributed to Japan are gutted, cleaned and cut in accordance with Japanese customer specifications, prior to being packed and frozen. Most of these goods are frozen 'round' (rinsed but not gutted). Most mackerel products then are packed into 20kg., plastic-lined distribution packages that are palletted and frozen. The fish is then stored in cold storage until transported by reefer containers on ships to Japan. About 50 per cent of the mackerel destined for intermediary production is sent through China for processing, due to the lower production costs there.

The import of most fish species to Japan, including mackerel, demands import licences. At present, a surplus of Norwegian mackerel import quotas exists because the profitability of importing this fish product is low. Japanese fish import quotas are government regulated and seldom shift hands when given to specific trading houses. Upon arrival in Japan, Norwegian fish get distributed through a unique Japanese distribution system, represented by the extensive use of different types of fish markets, organized as 'layers' in a distribution channel. The system includes (1) seaside markets that handle the catch from Japanese vessels, (2) wholesale markets for trading both domestic and imported fish and (3) consumer markets that also sell fish to private consumers. Figure 4.2 depicts various forms of direct and traditional distribution forms for Norwegian fish in the Japanese market:

Wholesale markets are the focal actors in distributing Norwegian-caught mackerel in Japan. These regional institutions are usually administered by local or regional governments. Japanese wholesale fish markets consist basically of a facility that provides the physical room for the storage and display of fish, as well as information and banking resources that support fish-based transactions. The size of the facilities varies, from the enormous Tsukiji market in Tokyo that covers several blocks of the city centre, to smaller regional facilities. This study considers the relatively small Morioka facility, located in the Iwate prefecture north of Tokyo; this fish market shares the same building with fruit and vegetable markets. According to one of the interviewed Japanese mackerel importers, 40 per cent of its Norwegian mackerel imports are distributed through Japanese producers, 30 per cent through wholesale markets, 10 per cent direct to retailers and 20 per cent through wholesalers to retailers. This importer noted a trend towards the increased use of direct distribution, which omitted the use of intermediaries such as wholesale markets. Most of the other interviewed professionals shared this opinion.

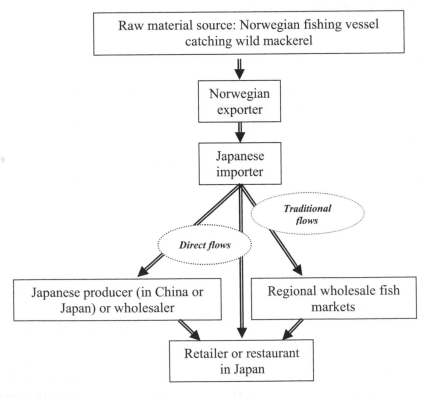

Figure 4.2   Norwegian fish distribution to Japan

Japanese importers also conveyed some positive features of 'traditional' distribution structures. Specifically, they indicated that wholesale fish markets allocated a vast number of fish species from many different suppliers (domestic and imported) to a large number of customers. This benefit was clearly evident when observing these Japanese fish markets. The number of species and classifications seems amazing to an observer accustomed to fish markets with far more limited species available. In addition, the traditional distribution system, which relies on tacit capabilities for quality control, has proven efficient in securing quality fresh fish supply for consumers. Norwegian fresh airfreight salmon takes 3–4 days after harvest to reach the end-user, such as a sushi shop anywhere in Japan. There have been no reports of quality scandals pertaining to fish distributed through this system.

INFORMATION PROVISION AND USE

In the case, Norwegian producers and exporters and the downstream Japanese retailer chains are also the primary users of advanced information and communication technology (ICT)-based enterprise resource planning (ERP) systems that supports the supply chain for distributing Norwegian mackerel to Japan. These different advanced systems are weakly interlinked and do not provide an electronic, 'seamless' information flow throughout the mackerel supply chain.

Regarding the features of goods control and data capture, Norwegian suppliers label distribution packaging (for example, 20 kilogram cartons) and transport-level packaging (for example, pallets of distribution-level fish cartons) with relatively advanced GS1 (www.gs1.org), standard transport labels that include barcodes to facilitate package scanning to identify goods. In Japan, distribution-level packages are commonly hand-marked only with the date of catch, product quantity and the name of the fish species. The name of a producer or distributor is usually printed on the distribution-level carton.

Business relationships between Norwegian exporters and Japanese importers in this exporter's supply chain are well developed. Formalized agreements, including the use of written contracts, are limited. Contracts may be written and used for guidance, but they are usually not signed. In relation to the trading of both mackerel and salmon, Japanese importers in the chain often send representatives for prolonged stays in Norway. During the main mackerel catching season, representatives of seven large and five smaller fish importers remain present. Approximately 25 Japanese representatives in all are hosted by the focal Norwegian pelagic fish producer, located at three different production facilities in the northwestern region of Norway, close to Ålesund municipality. The seasonal presence of Japanese representatives substantiates the importance of quality control of the raw material from the Japanese perspective. These Japanese representatives also negotiate the allocations of portions of the production to the different importers.

A topic of dispute in this case relates to setting the price of goods through negotiations. Norwegian producers express their wish for greater predictability, both in purchasing raw material through from the Norwegian monopoly (Norges Sildesalgslag: www.sildelaget.no) and in their relations with their customers. Legislation regulates the prices of fish raw material. Excellent catches of mackerel and herring in the past few years have allowed Norwegian fishermen and fishing vessel owners to profit from this situation. Pelagic fish quotas are set annually through negotiations between the government and fishing vessel owners, based on scientific estimates of the resources of this species. Japanese importers seek though sales negotiations to gain the lowest possible price to remain competitive. Both the Norwegian producers (exporters) and Japanese importers get squeezed between the Norwegian fishermen and Japanese retailers. The retailers in turn are pressured by Japanese consumers, who increasingly purchase cheaper fish for their meals rather than more expensive, quality mackerel products. Norwegian exporters and Japanese importers describe an option to bargain with other suppliers or customers on a global market. Norwegian mackerel has become established in the past 10 years as a differentiated product on the global market, due to fish quality (species/handling), which limits the impact of price on product demand.

According to two Japanese importers, importing Norwegian mackerel to Japan currently provides low profit margins. In addition, Japanese mackerel importers feel obliged to secure continuous deliveries of Norwegian goods to customers at 'reasonable' prices. This situation reflects a vital aspect of the Japanese distribution system, namely, a willingness to secure long-term business relationships, even if they may be unprofitable in the short term. Recently, five different Norwegian pelagic fish producers, including the focal firm in this study, merged into one firm to counter low profitability in their industrial sector.

Traceability requirements for fish products in Japan are limited to documenting the country of origin. One informant recalled an incident involving repacked mackerel labelled as 'Norwegian' when it actually was a Korean catch. Counterfeit labelling has occasionally been detected among fish products destined for Japan, when fish products considered of low value get labelled as with a country of origin associated with higher value.

Although EU regulations demand full internal and chain traceability of food products, such formal measures only exist for oysters and beef in Japan, two products that have been subject to food scandals and affected Japanese consumer health and welfare. These two products are regulated by 'experimental' measures (www.maff.go.jp/trace/guide_en.pdf). Full chain traceability of Norwegian fish gets lost moving through the traditional fish market distribution system, because this market form does not identify goods in a formal manner when packages and documents arrive at the facility. Goods may be traced accurately, because they remain packed and labelled. Fish production is appropriately registered in both Chinese and Japanese fish-producing plants, securing internal product traceability; these records are accessible to other supply chain actors on demand.

A key feature of fish product traceability in the supply chain of Norwegian fish exported to Japan is that this capability is not highly demanded by consumers or professional actors in the supply chain. Rather than representing efforts to sustain food safety and quality, informants conveyed the view that implementing full chain traceability of foods is a slow process in Japan that mostly plays the role of a formal or courtesy gesture. Japanese fish importers are aware that this perception of fish traceability will presumably change in coming years, due to changes in the societal and natural environment. One informant said that if a fish scandal were to occur, the speed of implementation for fish traceability legislation in Japan would increase, leading to Japanese product traceability regulations comparable to those used in most other developed nations.

Another feature of product traceability is that each actor maintains 'private' records of the transformation of goods in the leg of the supply chain for which it is responsible. These records are poorly interconnected, creating a supply chain that may be described as islands of information. Each island uses relatively more advanced ICT, but communications among firms are basic. In practice, supply chain actors are interconnected by relatively primitive forms of mainly manual communication, such as fax, telephone and pdf file e-mail attachments.

As this study revealed, when Norwegian mackerel arrives in Japan, the data capture procedures are rudimentary. The Norwegian catch is labelled with GS1 (www.gs1.org) transport labels that use bar codes but cannot be used at the fish markets, which lack ICT functionality to support electronic data capture. In direct distribution, the goods (both cartons and pallets labelled with barcodes) are manually counted. At the Morioka fish market, a wide variety of fish products are displayed by early morning. Then, through a market ritual, buyers and the seller who occupies a specific area of the fish market move from product to product to negotiate sales. Only rapid sensory product control occurs. The Japanese catch is marked on the package with the name of the species, the weight of the contents and the name of the supplier. In Japanese supermarkets, packed mackerel is displayed with a country-of-origin label and 'best before date' information. This date is measured from when the goods were defrosted. No documentation follows fish from wholesale markets to restaurants and retailers; only the package label may be used. In restaurants, customers must ask the waiter for this information.

The wholesale fish markets predominately use price mechanisms, combined with product quality control, to coordinate fluctuations in supply with fluctuations in demand. Payment in this system is settled immediately, which in practice means within 1–3 days of purchase. Sales to retailers often involve credit terms that delay the payment up to 3 months. The wholesale markets are traditional structures that keep the use of modern technology seemingly to a minimum, as evident in the Morioka wholesale fish market, which opened in May 2001. The use of ICT is limited to accounting and financial transactions within the facility. The main criticism of the fish market relates to the government-regulated price margins; according to Japanese fish importers, the fish markets involve unnecessary, cost-increasing middlemen. Direct distribution of mackerel is most common when goods get assigned to industrial production involving large consignments.

SUPPLY CHAIN AND NETWORK CONTEXT

Production of mackerel and other pelagic fish in Norway currently suffers from overcapacity, because the catch is strictly regulated to sustain these species. When different Norwegian producers compete for the quota-limited catch, the price of raw material increases. Norwegian mackerel must also compete with alternative mackerel-type fish products, usually offered at lower prices, on the global market, such as domestic (Japanese-caught) mackerel and Tasmanian mackerel, fish species with lower fat content than the Norwegian product.

Due to previously limited catches of domestic mackerel, Japanese consumers have become accustomed to the larger and fattier Norwegian mackerel product. Scottish fishermen use trawlers to catch the same type of raw material mackerel as the Norwegian fishermen, a more cost-efficient method than nets used by the Norwegian fishermen. Trawls, however, damage fish to a greater degree than nets through bruising and cuts, influencing the durability of the fish. The catch of mackerel in the last few years has been exceptionally good but faced problems associated with illegal catches by EU fishing fleets. This problem supposedly has been solved.

Finally, the societal element of the focal mackerel chain pertains predominately to changes in Japanese consumer preferences. In Japan, fish consumption trends are in slow decline, and older people mostly remain the loyal mackerel fish consumers. There is also a potential for switching between Norwegian and competing mackerel products on the Japanese market. For this business-to-business supply chain, this study is limited because it can address only superficially the more detailed aspects of Japanese mackerel consumption.

Discussion

The case narrative describes how Norwegian mackerel gets transformed in a supply chain context by interrelated actors managing and operating the physical transformation of mackerel products by providing information and supplying safe and quality products. The interrelationship with the external environment of the mackerel supply chain is also significant; it includes issues pertaining to overfishing and product traceability legislation. The focal ethical aspects, according to the preceding case narrative, are based on the literature review. These ethical aspects are the demand for adequate information, the value of freedom and human well-being. The following discussion therefore describes two models that may provide an understanding of food supply chains from an ethical perspective. These models are meant to complement a dominant, business-managerial perspective of food distribution prominent in marketing and logistics research.

DEMAND FOR ADEQUATE INFORMATION

The demand for adequate information restricts the use of false and misleading information. This issue is more problematic regarding the transfer of title to products. A seller in this setting is not regarded as responsible for providing information that may be perceived as valuable to the customer. The purchasing actor (customer) accordingly must actively seek trustworthy information to secure viable transactions. Responsibility for information in product transactions thus is shared between buyer and seller. The exact nature of this shared responsibility depends on the situation and should not be generalized without caution. Consumer legislation demands for adequate information are formalized to protect consumer well-being. However, in the upstream business-to-business environment, professional business actors supposedly do not need this protection, because they are 'professionals'.

Several information-related issues emerge from the case narrative. Different supply chain actors have individual economic interests that may conflict, centred on issues of opportunism and power struggles. The centrally located actors, Japanese fish importers and Norwegian fish exporters, express unhappiness with the price of goods purchased or sold. Price in this respect is a symbolic measure of the value of the product. However, these central actors mainly provide price information coupled with quantity, within a slim range of quality variation, when negotiating transactions. Uncertainty about raw material supply and final consumer demands in Japan are relatively low for Norwegian mackerel, because of the standardized features of this processed and packaged product in the important central coordinating part of the chain. These standardized product features represent a buffering mechanism that reduces organizational complexity. Given the routine character of these business transactions in the central part of the chain, information about products should be relatively simple. Yet it is not. Transactions in business relationships always have a distinct time and situation-specific character, which means the revealed patterns must be finite.

The case narrative also suggests a picture of how the focal food supply chain is a network of actors that are embedded in a network of interrelated multiple supply chains. A perception of component entanglement emerges; flows of title, information and goods from different supply chains interweave, criss-cross and interface in different ways. In addition, multiple supply chains have complex, changing and differentiated relationships with one another. These flows do not exhibit an orderly, parallel distribution through flows of goods, as is often anticipated in supply chain management literature.[24] For instance, a common electronic market provides a hub to distribute multiple types of pelagic fish products in Norway. Goods may also be transported to an export market prior to sales, regardless of the risk this represents.

The entanglement of different and interlinked supply chains renders them difficult to observe. In a competitive business environment between supply chains and within the chains, product authenticity must be controlled because supply chain actors may choose to misinform others about product features to attain a higher product value. Controls involve documenting product features in detail through a combination of sequentially transformed time, place and form features of the product. Unethical practices may result, at least in part, from a lack of informational visibility within a complex supply network setting that consists of multiple, interrelated supply chains. Sorts may be regarded as ethically pertinent decision-making events; in this study, sorts appear as parts of a supply system, so the ethical features of one sort are interdependent with comparable features of other sorts in the same supply chain. The decisions to record information at one sort have an impact on the ability to trace and track products at a subsequent sort.

At the final stage of the supply chain, a professional seller meets the consumer in retail or restaurant settings. From the perspective of the end-user, this last professional actor is responsible for the delivery of safe and quality products at the site of the final sort prior to product use. In restaurants, this sort coincides with the end-use phase itself. Menus are the main source of information for consumers regarding the food offering, though most do not provide product traceability. In a retail setting, a product label is the common source of food product information for consumers. This site is the place where food safety and quality ultimately is measured in a supply chain context. However, supply chain actor responsibility stretches into the use phase of a product, within Japanese households. In the vital final interplay between retailer and customer in a store setting (or consumption in a restaurant setting), a product label minimally should depict, in a complete and trustworthy manner, where the mackerel has been caught, initially produced, its 'best before' date, and the contents of the consumer-packed product. Moreover, the consumer-packed product should provide a linkage upstream to suppliers responsible for product safety and quality, as well as information about how to make contact upstream in the food chain in case of queries.

VALUE OF FREEDOM

The value of freedom in the studied mackerel supply chain pertains to manipulating and misleading information, both in transactions and for other purposes. Freedom relates to decision-making capabilities at sorts and consumer purchase of mackerel products based on the adequate provision of information by other supply chain actors. This question also involves features of ICT use that affect the information content for the mackerel product. Information may become lost or distorted through the supply chain, reducing the resource value of information content. In the context of the mackerel supply chain, dissatisfaction results from low profit margins for the central actors, and the case narrative depicts how these actors recognize that they are squeezed between the more profitable raw material supply and retailing parts of the same supply chain.

Less directly related to transactions is the issue of product traceability. The ability to provide detailed product information may influence a purchasing decision by a consumer in the Japanese marketplace, which primarily represents a threat to future business rather than a current reality. Weak information connectivity between supply chain actors leads to insufficient product traceability. Developing the technicality of information connectivity is clearly a joint responsibility. The sequential and dependent nature of transforming goods means that the features of final product for end-use purposes depends on the accumulated actions of all supply chain actors responsible for transforming the interrelated time, place and form features of these goods. This effort involves joint development of information systems, which allows supply chain actors to identify and process information about goods, including how information systems in the food supply chain are interlinked. This interlinking then affects the quality features of the information exchange that encompass product traceability.

In traditional wholesale fish markets in Japan, product safety and quality do not depend solely on information connectivity that secures efficient goods control mechanisms. That is, business culture, in addition to instrumental traceability routines and control mechanisms, contribute to secure, safe and quality fish products for Japanese consumers. This point highlights an important issue for further study: Modern rational-analytical management paradigms, often expressed through statements like 'planning, operations and control,' may be interpreted as an exhibition of poor trust in the business relationships. This common approach to management, from an ethical perspective, may also be interpreted as the result of weak joint and individual responsibility by actors within supply chains.

HUMAN WELL-BEING

Well-being entails the impact of products in a societal and natural environmental context. Regarding the social context, well-being involves mackerel product safety, such as from poisoning or contamination, and product quality related to an adaptation to user needs in mackerel consumption. Regarding the natural environment, examples of pertinent issues include overfishing and the impact of supply chain activities on pollution. Intermediary production in China may be analyzed in relation to environmental production controls and the impact of increased transportation distance when using a third location for intermediary production.

The mackerel supply chain from Norway to Japan provides products that are traditionally in demand by Japanese consumers. The Norwegian product substituted Japanese mackerel products when domestic resources previously were depleted. The development of the studied mackerel supply chain into a substantial economic entity accordingly results from previous Japanese overfishing of its own mackerel resources, providing Norwegian fisheries and exporters with a new group of customers in a global market. This situation poses a pertinent ethical question regarding how Norwegian fish exporters have profited from the depletion of the natural resources of other fish species. Different supply chains intertwine; two competing supply chains may in times of depleted resources complement each other by providing substitute products suitable to meet specific consumer preferences. Consumer preferences also change in response to both their knowledge of the environmental impact on raw material supply and the actual depletion of resources, which affects price and demand. The question actually may relate to how Japanese consumer preferences for mackerel have changed after several years of depletion of the Japanese mackerel species. This question is especially pertinent because this resource is again widely available for consumption. A possible dependency may exist between raw material supply in the natural environment and consumer preferences in a societal context, with supply chains serving an intermediating factor. In addition, the impact of structural changes in supply chains, as they have adapted to alternative sourcing of mackerel, requires consideration. This shift includes probes into the wider supply network of alternative and competing suppliers, often from different countries.

Conclusions

This case study exhibits how information plays a fundamental role in securing supply chain transparency. It also reveals that transparency depends on information connectivity among different actors and various physical resources to secure food product supply. Information, according to logistics and supply chain literature, has long been a core component in supply chains. This study indicates that information provided and used in food supply has its own ethical characteristics that influence the actors and is influenced by and influences the flow of goods. This understanding of information as a complex resource suggests using information as one of several core components in analytical frameworks to develop food supply. An understanding of the role of the food supply chain as a mediator suggests a pathway to recognize good and bad, right or wrong, and responsibility. This approach should be developed further to create a basis for increased understanding and develop safe, quality food supply practices in an environmentally sustainable manner.

The view of a single, complete food supply chain in this case indicates a conglomerate of components with the collective purpose to achieve safe and quality product supply. This predominately business entity, the complete food supply chain, should be viewed as a mediating resource among the pressures from (1) raw-material supply, (2) features of end-use and (3) competition with other supply chains. The key logic of a supply chain is economical – a collective, business-oriented rationality aimed at the efficient use of resources to reduce costs and secure incomes through recurrent sales, in competition with other supply chains. However, the logics of other supply chains, including the natural and social environments, differ. This paradox requires attention from relatively divergent actors in supply chains, who must balance the externalities with the internal workings of supply chains. This balance may be pictured as the realm of challenges facing food supply management today, that is, balancing the inner supply chain coordination to face external pressures. This mediating aspect of supply chains is depicted in Figure 4.3.

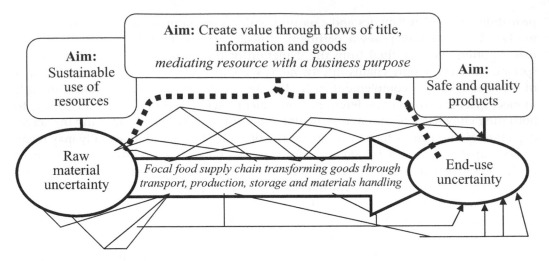

Figure 4.3   Food supply chains as a mediating resource

Figure 4.3 depicts how product safety and quality can achieved through the cumulative effect of raw material supply and sequential goods transformations through transport, production, storage and materials handling. These processes in practice combine logistical and marketing activities to transform a food raw material into a finished product and thus represent core aspects of value creation in food supply chains. The mackerel supply chain interacts with three different types of contexts to achieve value creation through collective actor effort. The model in Figure 4.4 depicts this interactive logic.

The supply chain in this model is depicted as the mediating entity among the different types of environments, each functioning in accordance with different logics. The mackerel supply chain, because it mediates different external pressures, may be regarded as an organizational resource; it is a knowledge-based entity that, by interacting with contexts, creates value for a multiplicity of business actors, according to how these actors interpret the purpose of the business. In this picture, supply chains emerge as mediating resources that interact with business, societal and natural environments; each environment reflects paradoxical expressions of purpose that different interacting managers in the food chain supply must take into account. The importance of supply chain management thus becomes paramount; a business logic that should facilitate ideas for coping with paradoxes to achieve economical, high-quality supply, including providing safe foods through integrated supply chains. In addition, supply chain management should be widened to encompass coping with supply chain environments.

In this arena of complex food supply, the idea of ethics directs attention to food safety as the fundamental objective of the complete food supply chain. This view clarifies that attaining this aim depends on the internal business-related workings of supply chains, as well as the workings of supply chains as business entities in their wider context. This approach also opens alternative views of managing and operating supply chains as complete entities. According to this ethics-based systems perspective, marketing and logistics managers need to contemplate how achieving safe (and the more business-related objective of quality) foods is not limited to applying limited economic rationality to their business. Individual business profitability is only one of many components involved in food distribution. Achieving safe and quality foods is now, through legislation, explicitly coupled with product traceability.

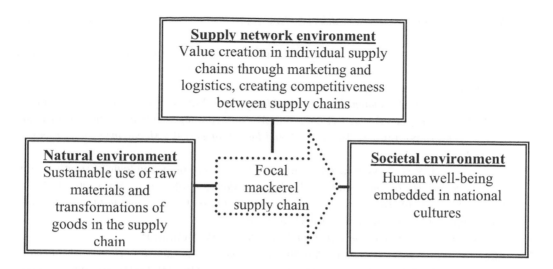

Figure 4.4 Food supply chains interacting with different contexts

Therefore, transparency in the chain is an important feature, achieved by creating quality information about products. The dual development of safe and quality product supply, matched with quality information about products, is complex. From a single-actor perspective, it involves coupling intra-organizational with inter-organizational capabilities to interact efficiently in a societal, natural environment and competitive business context.

In increasingly volatile environments, food supply chains face increasingly difficult challenges to attain the multiple aims of safe, quality and profitable food supply. This multiplicity calls for further research that explicitly approaches the complexity of food supply. Research on food supply in fluctuating and therefore uncertain business, societal and natural environments should analyze component features and component interrelationships not bound by given system functions and system borders. Achieving resource flexibility emerges as a focal issue to cope with the challenges associated with food supply. Resources in food supply involve actor competences, products, information and a range of product and information transforming facilities, which represents an almost unaccountable number of resource components. However, each of these components also has a local environment that may be mapped. As directions for further research and process development in food supply, a combined network and radical process thinking approaches may be worthwhile.[25] These extensions would require directing analytical attention to developing a more local and time-specific approach based on the understanding of food supply contexts and how these contexts function. The dynamics of food supply chains as arenas for mediating both within the supply chain borders and between these supply chains and their environmental contingencies could be further revealed and developed through such an approach.

References

1   Lilienfeld, R. (1978), *The Rise of Systems Theory. An Ideological Analysis*, New York: John Wiley & Sons.

2   Arbnor, I. & B. Bjerke (1997), *Methodology for Creating Business Knowledge*. Thousand Oaks, CA: Sage.

3   Heskett J.L., N.A. Glaskowsky Jr., and R.M. Ivie (1973), *Business Logistics, Physical Distribution and Materials Management*. New York: The Ronald Press Company; Arlbjørn, J.S. & A. Halldorsson (2002), 'Logistics knowledge creation: reflections on content, context and process', *International Journal of Physical Distribution and Logistics Management*, Vol. 32, No. 1, pp. 22–40; Lambert, D.M., M. Cooper, & J.D. Pagh (1998), 'Supply chain management: implementation issues and research opportunities', *The International Journal of Logistics Management*, Vol. 9, No. 2, pp. 1–20.

4   Håkansson, H. & G. Persson (2004), 'Supply chain management: the logic of supply chains and networks', *The International Journal of Logistics Management*, Vol. 15, No. 1, pp. 11–26.

5   Engelseth, P. (2006), *The Impact of Freezing Foods on Information Exchange in the Supply Chain*, Proceedings of NOFOMA Conference, Oslo, Norway.

6   Thompson, J.D. (1967), *Organizations in Action*. New York: McGraw Hill.

7   Alderson, W. (1965), Dynamic *Marketing Behaviour. A Functionalist Theory of Marketing*. Homewood IL: Richard D. Irwin; Bowersox, D.J. (1969), 'Physical distribution development, current status, and potential', *Journal of Marketing*, Vol. 33, pp. 63–70; Angell, I.O. & S. Smithson (1991), Information Systems Management: Opportunities and Risks. Basingstoke: Macmillan Press; Lambert et al., op. cit.

8   Alderson, 1965, op. cit.; Bowersox, op. cit.

9   Alderson, W. (1957), *Marketing Behaviour and Executive Action*. Homewood, IL: Richard D. Irwin.

10   Jahre, M., L-E. Gadde, H. Håkansson, D. Harrison, & G. Persson (2006), *Resourcing in Business Logistics–The Art of Systematic Combining*. Malmö, Sweden: Lieber and Copenhagen Business School Press; Thompson, op. cit.

11   Gripsrud, G., M. Jahre, & G. Persson (2006), 'Supply chain management? Back to the future?' *International Journal of Physical Distribution and Logistics Management*, Vol. 38, No. 8, pp. 643–659.

12   Alderson, 1965, op. cit.; Thompson, op. cit.

13   Persson, G. & S.E. Grønland (2002), *Supply Chain Management, En flerdisiplinær studie av intergrerte forsyningskjeder* [A multidisciplinary study of integrated supply chains], Research Report 09/2002. Oslo: BI Norwegian School of Management; Christopher, M. and H. Peck (2004), 'Building the resilient supply chain', *The International Journal of Logistics Management*, Vol. 15, No. 2, pp. 1–13.

14   Engelseth, P. & M. Abrahamsen (2007), *The demise of traditional fish distribution structures in Japan? A case study of fish supply chains from Norway to Japan*, Proceedings of the 23rd IMP Conference, Manchester, UK.

15   Langefors, B. (1973), *Theoretical Analysis of Information Systems*. Lund, Sweden: Studentlitteratur.

16   Angell & Smithson, op. cit.

17   Engelseth, P. (2007), *The Role of the Package as an Information Resource in the Supply Chain, A Case Study of Distributing Fresh Foods to Retailers in Norway*, Series of dissertations, Vol. 1. Oslo: Norwegian School of Management BI.

18   Angell & Smithson, op. cit.

19   Ibid.; Dreyer, H.C., R. Wahl, J. Storøy, E. Forås, & P. Olsen (2004), *Traceability Standards and Supply Chain Relationships*, Proceedings of the NOFOMA Conference, Linköping, Sweden.

20   Donaldson, T. & Werhane, P. (1996), *Ethical Issues in Business, A Philosophical approach*. New Jersey: Pearson.

21   Boatright, J. (2006), *Ethics and Conduct of Business*. Englewood Cliffs, NJ: Prentice Hall.

22   De George, R. (1995), *Business Ethics*. Englewood Cliffs, NJ: Prentice Hall.

23   Engelseth, P. & A. Nordli (2006), *Using the TraceFish Code in a Frozen Fish Supply Chain*, Proceedings of the 7th International Conference on Management in AgriFood Chains and Networks, Wageningen, The Netherlands.

24   Lambert et al., op. cit.

25   Håkansson, H. & A. Waluzewski (2002), *Managing Technological Development – IKEA, the Environment and Technology*. London: Routledge; Stacey, R.D. (2001), *Complex Responsive Processes in Organizations. Learning and Knowledge Creation*. London: Routledge; Griffin, D. (2002), *The Emergence of Leadership: Linking Self-Organization and Ethics*. London: Routledge.

5

Is Fresh Milk Powdered Milk? The Controversy Over Packaged Milk in Vietnam

BY VIRGINIE DIAZ PEDREGAL* AND NGUYEN NGOC LUAN†

Keywords

milk marketing, Vietnam, food controversy, consumer behaviour.

Abstract

This chapter deals with a controversy that flared up in the milk marketing sector in Vietnam. In late 2006, leading Vietnamese dairy companies were accused of providing incorrect content information: Bottles of 'pure fresh milk' actually contained 40–100 per cent powdered milk. This chapter considers how the crisis was managed by different stakeholders.

Introduction

This chapter deals with a controversy relating to food and agriculture marketing that flared up in Vietnam's milk marketing sector. In late 2006, leading dairy companies in Vietnam were accused of providing incorrect content information on their fresh milk product labels.

Many consumers, dairy distributors and journalists complained to the Ministry of Health and other related departments about the quality of certain fresh milk products. An enquiry by the State revealed that bottles of 'pure fresh milk' actually contained from 40 per cent to 100 per cent powdered milk, far from the 10 per cent authorized in

---

*   Dr Virginie Diaz Pedregal, Researcher at CIRAD-MALICA (Centre for International Cooperation in Agriculture Research for Development–Markets and Agriculture Linkages for Cities in Asia) in Vietnam. E-mail: virginiediaz@yahoo.fr.

†   Mr Nguyen Ngoc Luan, Researcher at IPSARD-RUDEC (Institute of Policy and Strategy for Agriculture and Rural Development—Rural Development Centre) in Vietnam. E-mail: nguyenngocluan@gmail.com.

most developed countries and the 1 per cent stipulated by Vietnamese law (according to Norm TCVN 7028/2002). A large-scale press campaign censured the practices of big milk companies and spread the scandal throughout Vietnam, with consequent repercussions for the communications and marketing strategies of dairy firms in the country.

This controversy may not seem serious, because it was not really of crucial importance for the Vietnamese people. Nevertheless, the issue of the consequences on the public health was raised. Comments suggested – though not scientifically proved – that the most vulnerable population groups, such as young children who consume milk daily, may have been affected by the lack of nutrients in the powdered milk compared with fresh milk. This debate is not inconsequential, because it indicated that the overall supply chain of fresh milk needed to be rethought, because there are not enough dairy herds in Vietnam to meet the growing demand for fresh milk.

The controversy also is particularly interesting because it occurred in a country in transition, where legal rules governing the duties and rights of consumers and firms are still in a state of flux. The crisis was 'treated' and 'managed' by different stakeholders.[1] In other words, this scandal illustrates the value of introducing business plans to a country in economic transition, attempting to be economically reliable on the international scene since its entry into the World Trade Organization in 2006.

After describing the background of the controversy, the literature review, and the methodology used in this research, we detail how each stakeholder reacted upon being confronted with the issue. In this chapter, 'stakeholder' refers to any actor directly or indirectly involved in the sterilized milk crisis – dairy firms, producers, distributors, consumers, consumer protection associations and government regulators. We analyze how each stakeholder managed the conflict and then look at how the scandal subsided. The chapter concludes by outlining a few managerial recommendations that will enable marketers to recognize emerging trends in milk consumption in Vietnam and anticipate future developments. After completing this chapter, readers should be able to understand the mechanisms of a food controversy in a developing country and apply these findings to other fields of research or areas of marketing.

Background to the Controversy

Vietnam is a country in economic transition. Although it is classified as having one of the lowest levels of dairy product consumption in the world (8 litres/person in 2005, compared with 50–60 litres in Malaysia and China and 100 litres in South Korea), the situation is changing rapidly. The Vietnamese are quickly taking to Western tastes in food. The consumption of fresh milk, yogurt, condensed milk, ice cream and other dairy products is growing very fast, especially among the younger, well-educated, well-to-do class living in urban environments (a segment that grew by 45 per cent between 2000 and 2005). Although the national production of fresh cow's milk increased from 50 million litres in 2000 to 190 million litres in 2005, growth is still too slow to satisfy demand that has skyrocketed to 650 million litres. Vietnam therefore has to import 70–80 per cent of its milk volume to meet its needs in the context of globalization and opening markets.[2]

Since 2000, the Vietnamese government has been trying to encourage milk production for three main reasons. First, the government wants to get away from importing powdered milk, which inflates costs and contributes to an overdrawn balance of trade. Second, the Vietnamese government wants to improve the income of rural communities through the dairy industry, as well as promote rural employment. Third, increased fresh milk consumption is in keeping with public health programs launched by authorities. In 2007, the Ministry of Industry approved an investment plan worth US$137 billion to meet the demand for milk products of 10 litres/person/year by 2010 and 20 litres/person/year by 2020, meaning that the dairy industry must grow by 5 per cent to 6 per cent per year during the next few years.[3]

Various obstacles stand in the way of this policy. First, it is difficult to raise dairy cattle in Vietnam due to its hot, tropical climate. Production would have to be concentrated in the more temperate central and northern regions, which immediately limits its scope. Second, dairy animals need a lot of feed, which is hard to come by in Vietnam. Third, the country's medical sector is still weak. Veterinary care is often deficient because of the lack of practitioners able to handle dairy animals. Fourth, the genetic dairy program chosen in the 1990s did not fulfil government expectations. Vietnam imported dairy cows with a high milk yield potential – Holstein Friesian and Jersey – but these breeds were unfit for the conditions in Vietnam. Animals died, and small farmers were ruined. Today, genetic programs that cross local and foreign races have been undertaken to improve milk production in Vietnam.

In this context, big firms attracted by the flourishing milk market are unable to source all the fresh milk they want. And milk sourced from numerous smallholders is not of uniform quality. Companies therefore are inclined to add powdered milk to fresh milk as a quality measure. Because there is no immediate, known health danger associated with it, the practice of adulterating fresh milk with milk made from powder is quite widespread.

In turn, we need to identify the reasons behind this debate over milk in Vietnam. It has been unfolding in the context of heightened concern over the quality of food products. In recent years, Vietnam has experienced a number of crises linked to food scandals and health risks, including the much publicized avian influenza outbreak that affected chickens and ducks, foot-and-mouth disease in cattle and swine, toxic chemicals in water, and so forth. Urban consumers no longer grow their own food and must rely on unknown producers and processors for their daily fare. In this context of apprehension over food, a mere rumour of poor product quality can cause much harm throughout the commodity chain.[4] In Vietnam, as in other Asian countries, people perceive a clear linkage between food and health; the Vietnamese people are commonly preoccupied with their health and therefore with the quality of their food. They remain distrustful of their government's capacity to ensure food quality, whereas imported food products, especially from Western countries, are considered safe.

Literature Review

No scientific literature pertains to the controversy about processed milk in Vietnam, because it is a relatively recent crisis. However, numerous newspaper articles document the media's discovery and analysis of the story. Articles from Vietnam Net, Vietnam News and Tuoi Tre highlight the reactions of consumers and the response of the dairy firms.[5] Furthermore, they demonstrate that the price of a litre of reconstituted liquid milk is close to the price of locally produced raw (or fresh) milk. Prices and the market are the key factors for understanding this controversy.

Academic papers and detailed reports from non-governmental organizations also provide information about the history of and trends in dairy production in Vietnam.[6] As these sources show, dairy production was not a historical tradition in Vietnam; dairy cattle breeding began only in the 1920s, during the colonial era, and the production was intended for French residents. From the 1950s to the mid-1980s, dairy cattle husbandry in Vietnam was restricted to a few large state-owned farms. Today, dairy farms or cooperatives are becoming increasingly privatized and more competitive.[7]

Moreover, scientific research delineates the food risks and consumer behaviour in the face of such risks in Vietnam. For instance, researchers have studied the perceptions of food-related risks by consumers in depth, mainly with regard to vegetable and meat consumption.[8] Research on quality chains also reveals that food risks remain very high in Vietnam, despite consumer health concerns.[9] Nevertheless, in the controversy over quality products, no real link can be drawn to informal businesses, despite the prominent position occupied by the informal dairy production sector in Vietnam (19 per cent, according Hemme).[10] The controversy over fresh milk instead involved something else.

The controversy centres primarily on the usage of powdered milk, added to fresh milk, with no indication of the addition on the packages. This practice amounts to cheating the consumer and brings into play the social responsibility of firms[11] because consumer protection is still weak in Vietnam. Firms have been accused of committing trade fraud – a far cry from the ideal of 'ethical marketing'.[12] Moreover, this scandal gave fresh breath to a widespread uneasiness with regard to food products.[13] It thus begs the issue of marketing strategies in developing countries.[14]

Methodology

To find answers to the issues we have identified, we used qualitative methods to collect data. Just as other researchers have suggested in their own surveys,[15] we believe that in-depth interviews are the most effective way to understand sensitive concepts, because they allow for the exploration of implicit meanings.[16] We established no pre-set number of interviews but instead used a theoretical sampling approach. That is, our goal was to hear the point of view of all the actors involved in the controversy: the state, the firms implicated, milk producers, distributors, and, of course, consumers.

We conducted semi-structured personal interviews with two representatives of the Department of Food Safety, Ministry of Health (MoH), two marketing agents of the Moc Chau and Hanoi Milk firms, three milk producers from Bac Ninh province, two representatives of the distributor Big C, two independent milk distributors in small shops in Hanoi, one representative of Vietnam's consumer protection association (VINASTAS), and 18 end-user consumers in urban (Hanoi municipality) and rural areas (Thai Nguyen and Hai Duong provinces). End-users included persons who purchased milk for home consumption (all of whom were women, as might be expected in Vietnam). A member of the French Embassy's Economic Mission working on dairy production was also interviewed to obtain an expert's point of view. The interviews were conducted at the place of work or residence of each respondent, recorded, and transcribed in their entirety. Table 5.1 lists the interviewees' characteristics.

Table 5.1   Interviewees' characteristics

Order	Interviewees	Characteristics
I	**Consumers**	
	Mrs T.	
	Age	30 years old
	Address	Ai Quoc district – Hai Duong province
1	Profession	Cooks for a company
	Household size	3
	Number of children under 10 years old	1 (5 years old)
	Household income	2 000 000 VND/month
	Mrs H.	
	Age	25 years old
	Address	Cau Giay district – Hanoi
2	Profession	Small trader
	Household size	3
	Number of children under 10 years old	1
	Household income	3 000 000 VND/month
	Mrs N.	
	Age	32 years old
	Address	Gia Lam district – Hanoi
3	Profession	Housekeeper
	Household size	3
	Number of children under 10 years old	1
	Household income	4 000 000 VND/month
	Mrs M.	
	Age	35 years old
	Address	Thai Nguyen province
4	Profession	Agricultural engineer
	Household size	4
	Number of children under 10 years old	2
	Household income	4 000 000 VND/month
	Mrs V.	
	Age	34 years old
	Address	Thai Nguyen province
5	Profession	Worker in a factory
	Household size	4
	Number of children under 10 years old	2
	Household income	3 000 000 VND/month
	Mrs I.	
	Age	29 years old
	Address	Cau Giay district – Hanoi
6	Profession	Small trader
	Household size	3
	Number of children under 10 years old	1 (2 years old)
	Household income	7 000 000 VND/month

Table 5.1   *Continued*

Order	Interviewees		Characteristics
	Mrs K.		
		Age	29 years old
		Address	Cau Giay district – Hanoi
7		Profession	Instrumentalist of Vietnam traditional operetta theatre
		Household size	3
		Number of children under 10 years old	1 (3 years old)
		Household income	4 000 000 VND/month
	Mrs R.		
		Age	26 years old
		Address	Hanoi Agricultural University N°1 – Hanoi
8		Profession	Works in Hanoi Agricultural University
		Household size	3
		Number of children under 10 years old	1 (8 months old)
		Household income	5 500 000 VND/month
	Mrs P.		
		Age	33 years old
		Address	Dong Da district – Hanoi
9		Profession	Trader
		Household size	4
		Number of children under 10 years old	2 (5 and 9 years old)
		Household income	5 000 000 VND/month
	Mrs W.		
		Age	43 years old
		Address	Dong Da district – Hanoi
10		Profession	Housekeeper
		Household size	4
		Number of children under 10 years old	1 (7 years old)
		Household income	8 000 000 VND/month
	Mrs Y.		
		Age	29 years old
		Address	Gia Lam district – Hanoi
11		Profession	Staff of a governmental organization
		Household size	5
		Number of children under 10 years old	1 (3 years old)
		Household income	10 000 000 VND/month
	Mrs L.		
		Age	30 years old
		Address	Gia Lam district – Hanoi
12		Profession	Secretary
		Household size	3
		Number of children under 10 years old	1 (9 years old)
		Household income	5 000 000 VND/month

Table 5.1 *Continued*

Order	Interviewees		Characteristics
13	**Mrs J.**		
		Age	28 years old
		Address	Hoan Kiem – Hanoi
		Profession	Researcher
		Household size	4
		Number of children under 10 years old	1 (5 months)
		Household income	6 000 000 VND/month
14	**Mr X.**		
		Age	29 years old
		Address	Cau Giay district – Hanoi
		Profession	Accountant
		Household size	4
		Number of children under 10 years old	2 (2 and 4 years old)
		Household income	5 000 000 VND/month
15	**Mrs RR.**		
		Age	37 years old
		Address	Gia Lam district – Hanoi
		Profession	Staff of a governmental organization
		Household size	4
		Number of children under 10 years old	2 (9 and 2 years old)
		Household income	5 500 000 VND/month
16	**Mrs TT.**		
		Age	33 years old
		Address	Hoan Kiem district – Hanoi
		Profession	Small trader
		Household size	4
		Number of children under 10 years old	2 (1 and 5 years old)
		Household income	5 000 000 VND/month
17	**Mrs HH.**		
		Age	34 years old
		Address	Gia Lam district – Hanoi
		Profession	Accountant
		Household size	4
		Number of children under 10 years old	2 (2 and 9 years old)
		Household income	4 000 000 VND/month
18	**Mrs NN.**		
		Age	26 years old
		Address	Tu Liem district – Hanoi
		Profession	Accountant
		Household size	3
		Number of children under 10 years old	1 (1 year old)
		Household income	3 000 000 VND/month

Table 5.1   *Continued*

Order	Interviewees	Characteristics
II	**Distributors**	
1	**Mrs A.**	
	Age	36 years old
	Professional Address	Hanoi Agricultural University – Hanoi
	Profession	Small trader
2	**Mrs D.**	
	Age	42 years old
	Professional Address	Gia Lam district – Hanoi
	Profession	Small trader
3	**Mrs C.**	
	Age	28 years old
	Professional Address	Big C supermarket
	Profession	Responsible of fresh products at Big C
4	**Mr CC.**	
	Age	25 years old
	Professional Address	Big C supermarket
	Profession	Employee for fresh products at Big C
III	**Producers**	
1	**Mr Q.**	
	Age	58 years old
	Professional Address	Tien Du district – Bac Ninh province
	Profession	Dairy raising farmer
	Household size	4
	Household income	5 000 000 VND/month
2	**Mr S.**	
	Age	47 years old
	Professional Address	Tien Du district – Bac Ninh province
	Profession	Dairy raising farmer
	Household size	5
	Household income	4 000 000 VND/month
3	**Mrs U.**	
	Age	44 years old
	Professional Address	Tien Du district – Bac Ninh province
	Profession	Dairy raising farmer
	Household size	5
	Household income	5 000 000 VND/month
IV	**Dairy Milk Companies**	
1	**Mr O.**	
	Age	51 years old
	Professional Address	33 Cat Linh, Hanoi
	Profession	Vice-director of Moc Chau

Table 5.1 *Concluded*

Order	Interviewees		Characteristics
	Mr F.		
2		Age	32 years old
		Professional Address	Cau Gia, Hanoi
		Profession	Marketing agent at Hanoi Milk
V	**Consumer Advocacy Association**		
	Mr G.		
1		Age	87 years old
		Professional Address	214/22 Ton That Tung, Hanoi
		Profession	Retired
VI	**Political Representatives**		
	Mr B.		
1		Age	42 years old
		Professional Address	138 Giang Vo, Hanoi
		Profession	Representative of the Department of Food Safety – Chief of the Norms and Technology Department
	Mrs E.		
2		Age	38 years old
		Professional Address	138 Giang Vo, Hanoi
		Profession	Representative of the Ministry of Health (MoH)
VII	**International Expert**		
	Mrs Z.		
1		Age	45 years old
		Professional Address	57 Tran Hung Dao, Hoan Kiem, Hanoi
		Profession	Economic expert on milk production and consumption at the French Embassy's Economic Mission

A thorough search of Vietnamese print media returned various articles written in Vietnamese, English and French. We collected all available information about dairy activities and dairy controversies in Vietnam in recent years. We also consulted government decisions and changes in legislation made in the wake of this controversy.

For the qualitative analysis of the transcripts, both authors participated in the interpretive process. We first analyzed all collected data (for example, interviews, documents) individually, then compared and contrasted the analyses in an iterative process to reach a consensus about their meaning. The identification of the interviewees used just a letter to maintain confidentiality.

Throughout this paper, 'packaged milk' refers to milk pasteurized or sterilized (that is, heated to destroy germs), and then bottled or put in cartons. We use the term 'fresh milk' to refer to liquid milk that comes directly from the dairy and has not been dehydrated in any way to make powder. 'Fresh milk' differs from 'powdered milk', although fresh milk may have been 'pasteurized' or 'sterilized'.

The Controversy and its Management: Findings

HOW THE CONTROVERSY STARTED

All respondents interviewed had already heard about the controversy. Although stakeholders had differing levels of knowledge and points of view, nobody had ignored it completely.

The controversy about fresh milk adulterated with powdered milk started in September 2006, though it is unclear who first mentioned the problem. Some respondents believed that consumers were the first to complain about the flavour of so-called fresh milk, whereas others, including the Hanoi Milk marketing agent, asserted that journalists with the Tuoi Tre newspaper were the first to make an enquiry into and report on the matter.

In all cases, the respondents expressed doubts about the origin of fresh milk: 'In reality, in Vietnam, there are not many herds of milk cows. If powdered milk is not imported from foreign countries to produce sterilized milk, the demand cannot be met. The misunderstanding between journalists, consumers and milk producers was that some producers put "fresh milk" on the package, whereas it wasn't all fresh milk,' explains Mr F. Mr O, from Moc Chau, sums up the situation of dairy firms, '20 to 30 per cent of milk in Vietnam comes from local production, 70 to 80 per cent is imported, mainly in powder form. So if companies are short of fresh milk, they reconstitute powered milk to produce milk.'

The leading dairy companies were suspected of being parties to the fraud. Currently, approximately 20 companies collect and process milk and dairy products in Vietnam. The first one to come under fire was Vinamilk – Vietnam's biggest dairy company – which has captured more than 70 per cent of the dairy market. It is 50.01 per cent state-owned, and foreign investors hold a 30 per cent share in it. The other key Vietnamese (Hanoi Milk, Elovi Food) and foreign (Dutch Lady Vietnam, Nestlé) dairy groups also came under scrutiny, though not all of them were guilty of labelling improprieties.

We begin by assessing consumer reaction to the scandal, which manifested itself very quickly. We will then analyze the attitudes of distributors, the positions of the producers, company strategies, the responses of consumer advocacy associations and the actual response by the state government.

CONSUMER REACTION

Before describing consumer reaction to the scandal, a few words on consumer milk habits are in order.

Consumers' habits

Milk consumption in Vietnam, though very low, is increasing rapidly, especially in urban areas. Consumers possess a very positive image of cow's milk. According to Mrs L, 'In my opinion, cow's milk is very nutritious. I am not a nutritionist, but I think that there are more calories in cow's milk. For children, I think it's good. Soya milk is good for the health, but it is not as nutritious.' Many consumers do not have a clear idea of the health advantages of milk but assume that 'it helps children to grow and to be more intelligent' and 'it provides energy and calcium to strengthen bones and increase height' as Mrs T and Mrs K affirm. It is well known that the state wants to promote the idea of a 'taller and healthier' nation. However, the population segments that most urgently need to improve their nutritional status do not have easy access to dairy products because of the price and the remoteness of the areas where they live.[17]

Vietnamese consumers buy both fresh and powdered milk. Powdered milk is mainly used for baby food, because it may be enriched with elements such as calcium, which is viewed as important by Vietnamese mothers. Powdered milk comes mainly from foreign countries and has been very popular in recent years, despite the government's promotion of breastfeeding. This state policy is motivated by health reasons, as well as economic reasons, in so far as nearly all powdered milk for infants is imported. Fresh milk is consumed by children and adults alike.

Quality, convenience, habit, brand name and price appear to be the key factors influencing the purchase of fresh milk. One consumer declares, '[I chose this milk] because I heard people say that the quality is quite good. I am familiar with this milk and it is quite cheap.' To make sure that the milk they buy is safe, many consumers pay great attention to the product expiry date. Moreover, consumers express trust in the person they buy milk from on a daily basis. Freshness is a major concern for Vietnamese consumers.[18] Thus, (non-enriched) powdered milk is not viewed as equal to fresh milk, 'Powdered milk is not as good as fresh milk. I think that when it is preserved for a long time, it is not as good as when it was fresh,' says Mrs L. People prefer to buy milk in small, 180 ml packs, which keep it 'fresh' more easily. Many consumers do not have a refrigerator, so maintaining freshness can be a problem in this tropical country.

Children are very important in Vietnam. Even in relatively poor families, they can dictate the household's purchasing behaviour, 'I buy any kind of milk that my child likes. Sometimes, if the fresh milk is in a package with exciting colours that he likes, or there is an appealing advertisement on TV or a special offer (like a Superman or a 3D house), I buy it. I buy milk for my son regardless of the company that produces it,' explains a mother whose family income is only 2 000 000 VND per month (US$125).

Many suffer from a lack of information about dairy products. Consumers buy all kinds of fresh milk but do not always know the difference between the brand name and the dairy firm that processed the milk. For instance, Mrs T says she mainly bought Izzy fresh milk but did not know that it was produced by Hanoi Milk. This lack of knowledge, as well as inadequate regulations governing proper labelling, has been advantageous for the firms involved in the controversy.

Because of this information shortfall and the general context of corruption in Vietnam, consumers openly state their distrust of the quality of food products sold in their country, as explained by Mrs L, an urban consumer, 'There was recently a problem involving imported milk, XO of South Korea, which is very popular here. It contains some powdered milk. Well-off people bought it. But there were big problems because the lead content was too high. It was really dangerous. Mothers were afraid. It is difficult to have total confidence in products here … because brand names, labels, can be sold. You make milk, you put a brand on it and you sell it. Nobody's going to check.' Consumers trust foreign and imported products but are very suspicious of Vietnamese practices, 'I have never trusted the quality of food sold in Vietnam,' says Mr X.

Consumer behaviour during the controversy

The controversy became quite widely known among consumers, though details were scarce. The respondents 'had heard about it but didn't know exactly' what it involved. They were usually aware of the problem and knew that it involved misleading labelling, 'Vinamilk's fresh milk label didn't have the right information on it. It said pure fresh milk, but in fact the product was made from powdered milk,' claimed Mrs J. Most of the respondents remember that the story was big news 'in late 2006', 'I remember this because at that time it was a controversial topic; Vietnam had just become a member of the WTO.'

Many of the consumers interviewed also said they first heard about the scandal on television. The newspaper and television media in Vietnam play a major role in people's perceptions of and behaviour toward food risks.[19] 'The neighbourhood grapevine' was also mentioned by rural and urban consumers as a way to get information and find out the degree of risk this issue may have had for their health. The Internet is not yet a widespread information channel for the great majority of Vietnamese households.

Most consumers believed the controversy over the milk did not harbour any danger for them, 'The problem is only the label, the low quality of the product, but it is not harmful to our health,' explains Mrs H. Some consumers declared they did not change their buying practices, whereas others felt trapped by the dairy companies. As Mrs L states, 'All companies were involved in this problem. It affected almost all dairy companies. So there was no choice.'

A person's circle of acquaintances can dictate attitudes. In school, mothers look at what other mothers do, 'I saw a number of other mothers bringing their children to the nursery school who usually buy Izzy fresh milk. I don't know what company produces this milk [Hanoi Milk], but it has very catchy ads on TV. Children are impressed by that, so I think that is the reason most children have a box of Izzy in hand before going to class,' says Mrs T.

A lot of consumers, especially those in urban areas, significantly reduced or even completely stopped their milk consumption for a few weeks. For instance, Mrs M explains that she had not bought Vinamilk milk since she heard about the problem. The same was true for Mrs J, 'With this story, I changed my brand of milk right away.' Then both consumers mentioned their consumption of new brands that appear safer, such as Moc Chau, 'Before last year, I had never bought Moc Chau milk. I didn't drink Moc Chau products and I didn't see them in stores. Now, Moc Chau does more advertising. They launched new products. They are trying to make their products known and widen their market,' adds Mrs J.

The reaction of rural consumers to the crisis was more limited. Many rural dwellers kept buying fresh milk or decreased their milk consumption only a little. Nevertheless, milk consumption in rural areas remains low in Vietnam.

ATTITUDE OF THE DISTRIBUTORS

Distributors were affected by the scandal for a couple of months when their milk sales decreased. The affected distributors include informal sellers who sell milk of different origins on the street, small distributors who sell dairy products in little shops, and big distributors such as Big C and Metro. They all sell fresh milk from big companies such as Vinamilk or Hanoi Milk.

Informal distributors are independent buyers who buy milk from different sources (for example, big firms, independent farmers) and sell it under various names at low prices, not subject to any quality control. This informal sector (milk traders and informal shops) accounts for about 19 per cent of the dairy business.[20] It can lead to quality deterioration in dairy products, because the transportation and storage of highly perishable products is not well controlled. However, the controversy over fresh milk did not start because of failings on their part, though it affected their activity. Informal distributors were directly affected by the decline in milk consumption and therefore decided to sell other products during the crisis.

Small distributors adopted a similar strategy. In Hanoi, small distributors sell all kinds of goods, from food products to detergent, nails and light bulbs. They do not specialize in dairy products and sometimes do not know the difference between the brand name of the milk and the dairy firm that produces it. Similar to the end-user consumers, they first heard of the controversy on television news, but they felt particularly concerned about it because it threatened their business. The strategy adopted by small distributors to avoid a loss of income was quite simple, as Mrs D explains, 'At that time, I waited to hear the final verdict from the authorities and cut down on the number of different kinds of milk I was selling in my shop.' The small distributors interviewed said that they discussed the matter with consumers, 'It was the talk of the town. Consumers often mentioned it.' In the end, business losses were not as bad as expected, 'In the wake of the controversy, the volume of milk sales did go down. However, this situation didn't last very long because of consumer habits and the demand for milk,' asserted Mrs D.

Finally, the big distributors are quite new in Vietnam. Supermarkets have come onto the scene only in the past few years in the major cities of Hanoi, Hai Phong and Ho Chi Minh City. Consumers consider supermarkets safe places to purchase products. Although goods sold in supermarkets are more expensive than those sold in little shops, they are perceived as offering good quality, reliable and guaranteed by the reputation of the distributor.[21] However, sales of fresh milk in supermarkets such as Big C decreased greatly during the period. Mrs C, the sales manager for fresh products in Big C, explains, 'When we look at the number of sales, we see that consumers did not want to buy Vinamilk products anymore. Vinamilk sales started dropping in November. The decrease was 26 per cent in November for fresh milk. It didn't change for yogurt. In December, sales decreased another 20 per cent. Other suppliers have seen their sales drop too, but Vinamilk was particularly affected. Sales of Moc Chau products didn't change in Big C. It was said on TV that these products contained 100 per cent fresh milk, so sales didn't change, but actually increased a little. For the past 3 months [since early April 2007], Vinamilk has got back to normal.'

POSITION OF THE PRODUCERS

The milk producers interviewed in this survey have been raising dairy cows for less than 10 years. As Mr S, a producer in Bac Ninh province, affirms, in the late 1990s, 'the movement toward raising dairy cattle was strong and enabled us to increase our income, so I changed from agricultural production to dairy farming.' Nowadays, income from dairy cattle accounts for up to 90 per cent of the total breeder household income, and dependency on this production is virtually complete. Milk producers in Vietnam are smallholders. Livestock is raised on 95 per cent of small-scale private farms in Vietnam, and most of the producers own from one to 15 cows for milking and breeding. Their average income is quite high for rural inhabitants; they may earn up to 5 million VND a month (US$315). A few are big farmers with more than 50 cows and earnings of up to 16 million VND a month (US$1000). Dairy development so far mainly involves well-off farmers, due to the high initial investment and technical capacities required. Most farmers are not organized into professional associations and lack empowerment and advocacy rights. There is no umbrella dairy farmers association in the country.[22]

Producers seemed well aware of the fresh milk scandal. Unlike the majority of stakeholders, producers took advantage of the controversy though. Producers usually sold their milk to dairy companies at prices ranging from 3500 to 5000 VND/litre (US$0.22–0.31). The milk would then be sterilized, packaged, transported and sold in shops. At the end of the chain, the consumer paid 14 000 to 20 000 VND (US$0.88–1.25) for a litre of milk. While the controversy raged, some consumers were apprehensive about the milk processing methods used by dairy firms, so they chose to buy milk directly from the producer, at a lower cost for them but higher prices for the producer. Consumers then pasteurized the milk themselves at home, which was more profitable for the dairy farmers, as Mr S explains, 'As a result of this problem, I was able to sell much more unsterilized fresh milk, because many consumers came to my house to buy it rather than from agents.'

STRATEGIES ADOPTED BY MILK COMPANIES

The milk firms adopted several different strategies, often depending on the extent of their guilt. Vietnam's largest company, Vinamilk, was stung the most in this scandal. Many consumers linked the scandal directly to Vinamilk, 'I remember that on TV they said Vinamilk brand fresh milk was made from powdered milk' and 'Vinamilk's factories make fresh milk from powdered milk. However, on its labels, Vinamilk states that it is pure sterilized fresh milk.' Vinamilk indeed resided at the centre of the controversy. The company's representatives chose to limit contact with the media during the controversy and did not respond to independent enquiries, such as ours, either during or after the affair. To restore consumer confidence, the firm simply acknowledged its mistakes in late 2006.

Labels on Hanoi Milk products were not misleading, but the company lost sales because of general consumer distrust of dairy products and the positions taken by those companies. According to the Hanoi Milk sales manager,

'The media made too much ado of the problem of fresh milk and powdered milk. Of course, our sales have been affected. We didn't do anything because our firm never had the wording 'fresh milk' on its milk packages. The only firm that benefited from the problem was Moc Chau. Our turnover decreased between 25 and 50 per cent from September 2006 to February-March 2007. In fact, though, it depends on the season. There is a big difference in milk consumption between winter and spring. In spring, we sell more milk. In winter, it's cold, so consumers prefer hot beverages but don't like hot milk very much. But this year, our turnover increased only 10 per cent compared to the same period last year, whereas other years, it increased by 20 per cent. That's the general situation for all dairy companies in Vietnam. It creates good opportunities for the sale of foreign products in Vietnam.'

Only Moc Chau benefited significantly from the scandal; it took clear advantage of the situation. The policy of Moc Chau has been well-established for 50 years, as the sales manager explained,

'Moc Chau went into the dairy business in 1957. Our production and marketing are carried on in a closed circuit. That means that all breeders of Moc Chau are members of our company, are stockholders of Moc Chau. They work for the firm but also for themselves. That is why the quality is so high. Moreover, our milk cow herds are well adapted to the Moc Chau climate. At the outset, we imported cows from China. But they weren't suited to the climate here, as Moc Chau is 1200 m above sea level. So we decided to import milk cows from Cuba, the United States, Australia and the Netherlands, and it's working well.'

The crisis enabled Moc Chau to put its product out front and pursue its market development strategy. Before the scandal, Moc Chau declared in its advertisements that it produced '100 per cent pure milk' that is, 'pure fresh milk.' After September 2006, Moc Chau fresh milk sales jumped 200 per cent, which is remarkable in this recessive context.

COMMENTS FROM THE CONSUMER ADVOCACY ASSOCIATION

In Vietnam, the only consumer advocacy association designed to protect consumer interests is the Vietnam Standard and Consumers Association (VINASTAS). Unlike Western consumer advocacy associations, VINASTAS is closely linked to the communist government. It works with a limited budget. Nevertheless, its position during the controversy was important, because VINASTAS was very prominent in the media. Rather than providing information to help consumers choose among different brands of milk, VINASTAS worked to spread news of the scandal. Mr G explains the role of this association, 'We interviewed consumers. The media also interviewed people. We studied consumer reactions. It was after disclosure that there was a problem. VINASTAS did not uncover the scandal. In these kinds of controversies, VINASTAS always works with the press to inform consumers.'

The matter of the ingredients present in powdered milk was raised by Mr G, 'Here, we're not talking about the quality of the powdered milk used for making the fresh milk. In fact, their characteristics are different. For example, powdered milk does not contain fats, unlike fresh milk. And when we use powdered milk to make liquid milk, we have to add ingredients.' VINASTAS lacked enough qualified members and funds to make a comparison of the quality of different milk brands, as might be done in other countries. Mr G only noted that, 'An expert conference on milk took place. They spoke about the process of making liquid milk from powdered milk. In this process, milk quality is always preserved. Maybe it is a subjective point of view. In VINASTAS, we don't have enough tools to judge quality. If we want to make tests or estimations, we have to ask laboratories. VINASTAS therefore has a challenging job to protect consumers. We mainly function as a media sounding board to inform consumers. Obviously, in the milk crisis, consumers had the right to be informed, and this is in VINASTAS's charter, but it wasn't respected. In this controversy, dairy firms were not compliant with the first rule of trade – 'the client is king'.[23]

WHAT THE STATE DID

While the media was creating a furore over the scandal and dairy prices were dropping, the Department of Food Security, Ministry of Health (MoH), intervened to set up expert committees to evaluate the situation.

Before the crisis, national regulations required that processors show the ingredients on their product labels. Because the labelling law applies to all kinds of manufactured products, it did not stipulate specifically that milk containers had to indicate the percentage of pure fresh milk compared with powdered milk. Norm TCVN 7028/2002 stipulates that fresh milk cannot have more than 1 per cent powdered milk in it, but it does not mandate that dairy companies must indicate the exact percentage of powdered milk mixed with so-called fresh milk on their packaging.

The brunt of this food scandal fell upon the MoH. After the problem was disclosed by the media and input was provided by VINASTAS, the ministry dispatched a group of experts to investigate major dairy company practices. Generally speaking, the MoH conducts two types of inspections to control food product safety. Regular inspections occur periodically, perhaps twice a month. Spontaneous inspections are carried out when a scandal breaks out or when a major risk appears. For both types of inspections, companies are warned before the arrival of the government inspectors. Regarding the milk controversy, Mrs E, an expert from the MoH Food Safety Department, related, 'We made inspections in November and December 2006. There were consumer complaints because the fresh milk label of some companies did not comply with labelling standards. It claimed to be fresh milk but it really wasn't fresh milk.'

Inspection groups consisted of representatives of the Food Safety Department, the National Institute of Nutrition and the Department of Goods Management, under the Ministry of Technology and Science, which is in charge of standards and labels in Vietnam. The inspection group leader was from the MoH. The two or three inspectors who make up an inspection group inspected six dairy companies in six days, going directly to the dairy companies. They looked at the formula used to produce the milk, compliance with current standards and legislation, and so forth. They also analyzed samples at random to check the composition of the milk. According to the MoH, the inspections were performed to enforce sanctions on dairy companies whose labels were not compliant with the legislation, not to preclude a repetition of the problem by clarifying the law on dairy product labelling. There still is no specific decree about milk labelling in Vietnam.

The Controversy Winds Up: A Discussion

Stakeholders largely feel that the controversy is now over. How did that come about? A simple settlement about product labelling resolved the crisis. First, firms guilty of not complying with the rules, such as Vinamilk, explained their position in the media. They acknowledged their mistake in labelling their dairy products,[24] and they agreed to label their products more clearly. Vinamilk decided to change the wording 'pure sterilized fresh milk' to 'sterilized milk' if the content were not entirely fresh cow's milk. But the company refuses to reveal in detail what its products contain, claiming it is a 'private formula' and internal 'production know-how'.[25]

Dairy firms that suffered economically launched a series of new products. After the Tết period (Vietnamese New Year in January–February), the consumption of goods in general decreases because people spent lavishly in the weeks preceding the holiday. Vinamilk chose this period to boost its sales, launching a new 100 per cent sugar-free 'pure sterilized milk' and a new 95 per cent sweetened 'pure sterilized milk,' in packages of 1 litre, 200 ml, and smaller. Advertisements on television grew aggressive, with much colour, singing and repetition: 'One hundred per cent, one hundred per cent, Vinamilk fresh milk is one hundred per cent pure fresh milk! Day in, day out.'

This kind of advertising works quite well, as attested by Mrs J, an urban consumer, 'Vinamilk has more attractive packaging now. It has more colours. I think it is nicer than before. They changed the design too. On TV, their advertisements are catchier. They have songs on TV; it's more colourful.... They also launched many new products.' Nevertheless, after the scandal, dairy companies tended to increase the price of their dairy products, including fresh milk, to cover the financial losses subsequent to the scandal. Although the producers are not earning more, prices have increased for consumers. Thus, 'The scandal is over now. We asked for better labelling. This is solved. But now, the most important issue is the price of the milk,' confirms Mr G from VINASTAS. Some consumers therefore remain unsatisfied with the current policy of dairy companies, 'Although the controversy has stopped, I am still not satisfied with them. During 2007, the price of milk increased a lot. The main reason is that they had to spend so much on advertising in order to compete with other milk enterprises, not because of the high cost of imported ingredients.'

The government also made some recommendations for labelling. In the early stages of the controversy, the government issued Decree 89/2006/ND-CP, dated September 30, 2006, which represents as the highest level legal document on this matter. It includes all types of goods, but regarding foodstuffs, it differentiates 'ingredients' from 'quantities of ingredients' instead of using the general term 'composition' as before. Moreover, the ingredients must be shown in weight order from highest to lowest.[26] However, there is no mention in the new decree of better product quality or a commitment to health, social welfare or protection of the environment, despite such calls by individual consumers and consumer associations.

Each company tends to follow its own rules, as Mr G from VINASTAS observes, 'It's true that the exact ingredients are not still written on packaging. Rather, the content of vitamins, iron, proteins, and so on, is shown. Vinamilk's products show 95 per cent fresh milk, sugar, and so on, but the other ingredients are not detailed. Labelling requirements are not very precise in Vietnam. The government does not determine what information must appear on labels. Thus every company presents the information in its own way.' The Ministry of Technology and Science is in charge of this issue, whereas the Ministry of Health only deals with offending firms.

Consumers are therefore still asking for a tighter legal framework and fairer dealings in the dairy market. The respondents for this study believe the state should exert more regulations and greater control in this area. 'We need better monitoring of milk from the cattle herds down to consumption. We must have greater traceability. Controls at all levels are needed. If things are done properly at the processing stage, but producer practices are substandard, you can't have a safe product at the end,' says Mrs L. According to Mrs H, 'To ensure consumer confidence, firstly, the authority must have clear regulations and a method of investigation in order to properly control the enterprises. Secondly, the enterprises need to guarantee the product quality and ingredients shown on the label.' Small distributors note the doubt that remains in the minds of some customers, 'The problem has blown over now. But some customers still think that there is no fresh milk,' explains Mrs A. 'The legal system in Vietnam is not strong enough to protect the consumer! In terms of laws in Vietnam, I don't feel that the legal system is strong enough in all fields! I don't think the laws can be relied upon to protect consumers. I would feel more confident in milk from abroad, from Europe, for instance. Milk from France, yes, I would trust it,' adds Mrs J. Corruption is also an issue, 'Corruption is very common here. There is no real control on the quality of the milk sold. Farmers pay the inspectors to be quiet. I don't trust State controls,' said Mrs J, expressing her lack of trust in government's ability to protect consumers.

More broadly, consumers, small distributors and even producers remain largely unaware of the legislation governing milk products. They do not know what may have changed in the legislation regarding food labelling in recent years, nor what obligations are incumbent on milk companies in terms of the information on the packaging. The absence of communication may be the crux of the matter, 'I believe that if people aren't talking about this problem anymore, it means that fresh milk is now guaranteed to be 100 per cent pure fresh milk,' declares Mrs H.

Neither consumers nor producers appear vindictive toward the dairy companies. Powdered milk may not be as good as fresh milk but, 'the scandal didn't have serious implications, because mixing powdered milk with fresh milk is not toxic,' noted Mrs L. She went on to say, 'If I am told that there is X per cent of powdered milk in my fresh milk, I surely won't like it, but powdered milk is not harmful to my health. If they do what needs to be done to protect my health, I accept it. But I prefer that they be honest rather than making false claims about their products.' Mr S, a farmer from Bac Ninh province, declares that, '[companies] have admitted their mistake and pledged not to do it again. So it is acceptable.' Most of the consumers interviewed offer a similar assessment, 'I think it is not very important if it is fresh milk or powdered milk. But for business ethics, it is not correct. It was necessary to fight against this misrepresentation, but it has no dramatic consequence on consumers. It is unacceptable in terms of ethics,' says Mr G.

Big distributors are still distrustful of dairy firms. Employees from Big C present their company as an active stakeholder in management of the conflict, 'The Big C buying group asked to suppliers to change their packaging. The government also did so. If not, Big C returns the product to the suppliers. The government legislated concerning product shelf-life, date of manufacture, and so on. Big C faxed the law to all suppliers. Government inspectors used to come to Big C to verify compliance with the law. If something is wrong, they fine the wrongdoer. In this case, the department manager pays the fine because he is responsible for his shelf. When a new product is launched, it is necessary to be careful. Producers also pay attention because they don't want to pay to have goods transported for nothing. Today, labelling rules have improved; things are more precise and clearer,' says Mrs C. The government appears to fulfil its role when controlling products sold in the supermarket, 'Government representatives verify the labelling, the weight of the products sold, and so on. It is not possible to falsify the weight of milk sold. You must have 1 litre of milk, not 950 ml. Some suppliers make containers of 1100 ml or 950 ml. They have to be cautious. Government inspectors often come from the Ministry of Health. They come several times a month. During the period of bird flu, they also came a lot. With eggs, chicken, pigs, they have many opportunities to come,' declares Mrs C.

'Anyway, the controversy is not over yet because firms don't have enough fresh milk,' warns Mr O from Moc Chau. 'If we went on to thoroughly investigate 100 per cent fresh milk products, other scandals may come to light. Because if we count the number of milk cows in Vietnam, we'll see that some fresh milk is actually made from powdered milk,' argues Mr F from Hanoi Milk. Vietnamese dairy farmers are quite optimistic about their future. All producers interviewed plan to increase their herds, which makes good sense as demand continues to increase.

Conclusions and Managerial Recommendations

Big firms lost face as a result of this controversy. To restore their image, they recognized their mistakes and launched new products with timely, powerful advertising. It worked, as the demand for milk products quickly recovered after the crisis. But dairy firms have two challenges to face in the milk market: milk supply and consumer confidence.

The nagging issue for dairy firms is the lack of fresh milk available from local dairy farmers. To solve this problem, dairy companies might work in close cooperation with the producers with regard to, for example, the choice of the races of dairy cow, determining a 'fair price' for production, and so forth. Such cooperation would enable firms to ensure producer trustworthiness at the time they sell their products. To that effect, signing exclusivity contracts with producers might be a welcome development, as already adopted by Moc Chau.

To ensure the growth and sustainability of the consumer market, Vietnamese dairy firms should realize that competition is going to grow quickly with Vietnam's entry into the WTO. Foreign companies are victims of corruption in the country, but they enjoy several advantages with regard to the Vietnamese market. Mrs Z from the French Embassy's Economic Mission explains, 'Foreign companies must be watchful. Agreements have been made between importers and customs officials – by means of corruption – to allow the entry of dairy products into Vietnam. A foreign company told me it had to pay four times between the Saigon airport and its warehouses. Nevertheless, as the market liberalizes, foreign companies are in a position to launch very active advertising campaigns, promoting products with a guaranteed content of 90 per cent fresh milk. This is possible with Vietnam's entry into the WTO. These campaigns will work because the Vietnamese consumer is extremely concerned about having quality products to ensure good health.' As we have seen, Vietnamese consumers also have a strong preference for foreign foodstuffs.

To that end, packaging is not the only way to give consumers a feeling of security. Communication regarding sustainable development and the creation of organic milk production directed to the young, well-to-do population of urban Vietnam could be interesting approaches for diversifying products and ensuring quality. For now, very little information about corporate environmental policies is available, making it a possible means to ensure consumer confidence and fidelity. Further research on the willingness of consumers to pay for safe, organic dairy products is needed. This recommendation could easily extend to other food products as well, because organic production is minimal in Vietnam – only 0.07% of the total agriculture area[27] – but has strong potential for marketing to well-to-do consumers concerned about their personal health.

Programmes in schools to educate children about the importance of dairy product quality also might be launched by the dairy firms. Children exert considerable influence on their households' shopping habits. With the help of the state – which is possible because of the government policy toward milk consumption – dairy companies could gain a reliable, stable market. Nevertheless, professional ethics and respect for consumers, especially if they are young, represent absolute conditions to keep markets moral.[28]

As in other countries of Southern Asia, such as Laos, Cambodia, Malaysia and Thailand, as well as in China, the fresh milk market in Vietnam is far from being mature and saturated. There is plenty of room for modern, socially conscious enterprises in search of business that will offer quality products and invest in not only transaction-based but also person-based relationships with their consumers.[29]

References

1. Latour, B. (1987), *La science en action* [Science in Action] Paris: La Découverte.
2. Tauziède, B. (2007), *Le marché vietnamien des produits laitiers* [The dairy products market in Vietnam], reference report, the Hanoi Economic Mission.
3. Vietnam News (2007), *Plan gouvernemental visant à augmenter sensiblement la consommation de produits laitiers au Vietnam* [Government plan to strongly promote dairy product consumption in Vietnam], May 7.
4. Apfelbaum, M. (1998), *Risques et peurs alimentaires* [Food risks and fears] Paris: Odile Jacob.
5. Tuoi Tre Online (2006), 'Su that cua "sua tuoi nguyen chat"' [The truth about 'pure fresh milk'], October 15; Tuoi Tre Online (2006), 'Sua tuoi nhung khong phai la sua tuoi' [Fresh milk but not fresh milk], October 15; Vietnam Net (2006), 'Who processes fresh milk from powdered milk?', October 13.
6. Chu Thi Kim Loan, Yokogawa, H., & Kawaguchi, T. (2004), 'Dairy production in Vietnam: opportunities and challenges', *Journal of the Faculty of Agriculture, Kyushu University* Vol. 49, No. 1, pp. 179–93.
7. Bourgeois Luthi, N., et al. (2006), *Review, analysis and dissemination of experiences in dairy production in Vietnam*, VSF/CICDA, AI, CGCA, FAO, Hanoi (July), pp. 16–17.
8. Moustier, P., Bridier E., & Loc N.T.T. (2002), 'Food safety in Hanoi's vegetable supply: some insights from a consumer survey', in: Hanak, E. et al. (eds), *Food safety management in developing countries*. Proceedings of CIRAD/FAO International Workshop, CD-ROM; Figuié, M. et al. (2004), 'Hanoi consumers' point of view regarding food safety risks: an approach in terms of social representation', *Vietnam Social Sciences*, Vol. 3, p. 101.
9. Cadhilon, J.J., et al. (2006), 'Traditional vs. modern food systems? Insights from vegetable supply chains to Ho Chi Minh City (Vietnam)', *Development Policy Review*, Vol. 24, No. 1, pp. 31–49.
10. Hemme, T. et al. (2005), *The economics of milk production in Hanoi, Vietnam, with particular emphasis on small-scale producers*. International Farm Comparison Network, FAO. Draft PPLPI Working Paper.
11. Lindgreen, A. (2005), *Relationship marketing: design, implementation and monitoring*, Hyderabad, India: ICFAI Hyderabad University Press.

12. Pastore-Reiss, E. & Naillon, H. (2002), *Le Marketing éthique: Les sens du commerce* [Ethical Marketing: Trade Directions]. Paris: Village Mondial; Duffy, R., Fearne, A., & Hornibrook, S. (2003), 'Measuring distributive justice and procedural justice: an exploratory investigation of the fairness of retailer-supplier relationships in the UK food industry', *British Food Journal*, Vol. 105, No. 10, pp. 682–94.

13. Beardsworth, A. & Keil, T. (1996), *Sociology on the menu: invitation to the study of food and Society*, London: Routledge.

14. Nguyen Phuong Lien. (2003), *Thuong hieu voi tien trinh phat trien va hoi nhap* [Brand names in the development and integration process]. Hanoi University of Foreign Trade.

15. Hingley, M., & Lindgreen, A. (2004), 'Supplier-retailer relationships in the UK fresh produce supply chain', 20th Annual Conference of the Industrial Marketing and Purchasing Group, Copenhagen, Denmark.

16. Sayre, S. (2001), *Qualitative methods for marketplace research*. Thousand Oaks, CA: Sage Publications.

17. Bourgeois Luthi et al., op. cit., p. 111.

18. Figuié, M. (2004), *Perception of food-related risks by consumers in Hanoi, Vietnam*, Working paper. CIRAD/IOS, MALICA, Hanoi.

19. Vietnam Net (2005), "Ai' huong loi 'gi' tu truyen thong' ['Who's getting "what" out of communication?'], June 21.

20. Hemme et al., op. cit.

21. Moustier, P. et al. (2006), *Participation of the poor in supermarkets and other distribution: value chains – synthesis*. Project Report, 'Making markets work better for the poor'. Malica, ADB, Hanoi.

22. Bourgeois Luthi et al., op. cit., p. 12.

23. Tap chi Cong nghe (Technology Magazine) (2007), 'Around the issue of milk quality', (January), p. 58.

24. Vietnam Net (2006), 'Vinamilk admits mistakes in dairy product labelling', October 20.

25. Department of Science and Technology, Dong Nai (2006), 'Vinamilk shares set price record despite labelling violations', July 12.

26. Ho Tuong Vi (2007), 'Labelling: a key ingredient to protect consumers'. Available at http://www.vision-associates.com> (accessed 15 August 2007).

27. Willer, H., & Yussefi, M. (2006), *The world of organic agriculture. Statistics and emerging trends 2006*, IFOAM, FIBL, Germany.

28. Blank, R. & McGurn, W. (2004), *Is the market moral? A dialogue on religion, economics and justice*. Washington: Brookings Institution Press.

29. Lindgreen, op. cit.

Agri-food Systems, Product Innovation and Assurance

Quality Assurance Schemes and Food Marketing in the European Union

BY STEPHAN HUBERTUS GAY,* FATMA HANDAN GIRAY,* PÉNÉLOPE VLANDAS* AND MONIQUE LIBEAU-DULOS*†

Keywords

food quality, quality assurance schemes, agri-food chain, food marketing.

Abstract

Quality assurance schemes are increasing in importance all over the European Union. They have considerable influence on the marketing of food products. Two main types can be identified: in-chain quality management schemes and market differentiation schemes. The first group has a considerable influence on the food chain, whereas the second group is larger in numbers. This chapter discusses perceptions of stakeholders and the functioning of selected schemes using two sets of the data gathered from stakeholder hearings (and panels) and economic analyses of eight case studies. Quality assurance schemes appear to be an integral part of current food marketing in the agri-food chain; a fundamental change in the future is not likely.

Introduction

Quality assurance schemes (QAS) play an ever increasing role in food policy and, in particular, in the food supply chain. In principle, QAS are defined as codes of practice,

*   Dr Stephan Hubertus Gay, Dr Fatma Handan Giray and Ms Monique Libeau-Dulos; European Commission, Joint Research Center, IPTS – Institute for Prospective Technological Studies, Edificio Expo, C/Inca Garcilaso 3, E-41092 Sevilla, Spain. E-mail: hubertus.gay@ec.europa.eu. Telephone: + 34 95 448 8314. Pénélope Vlandas; Institute Manager, UCL Energy Institute, Room 228, 2nd Floor, North Cloister, Wilkins Building, Gower Street, London, WC1E 6BT.

†   The views expressed in this paper are those of the authors and do not necessarily correspond to those of the EC.

standards or sets of requisites that enable stakeholders to guarantee compliance by adhering to that which is declared. This adherence is signalled to the end or next user, underlined by some independent verification process that adds authority to the stakeholders' statement.[1] Participation in QAS is entirely voluntary, though some schemes in some countries have gained a quasi-mandatory status. Although QAS are homogenous in their aims and orientation, their structure varies wildly across the EU. Some QAS are confined regionally and hence affect a very small volume of agricultural produce; others operate on a national or even global level. Other differentiations include that some are private and others public, and some are regulated by national law and others by European law.

The perception of the function and importance of QAS by stakeholders varies widely. This confusion needs some clarification. On the one hand, the proliferation of QAS is seen as a problem, but on the other hand the aim of most QAS is to differentiate with regard to quality. Generally, consensus suggests that no competition can function based on food safety, which is secured by basic Food Law. Some QAS stress that they encourage compliance with the Food Law by enforced controls, which creates an implicit competition related to the food safety concerns of consumers. These examples show the wide range of issues related to QAS in the field of food marketing. This chapter attempts to present these QAS-related controversies in food marketing in the framework of the experience in the European Union. It focuses on stakeholder perceptions and adds information obtained from case studies of different QAS.

This chapter is based on a pilot study carried out by the European Commissions' Joint Research Centre, Institute for Prospective Technological Studies (JRC-IPTS), initialized by the European Parliament. The Directorate General for Agriculture and Rural Development requested the JRC-IPTS to carry out a project on 'Quality assurance and certification schemes managed within an integrated supply-chain.'[2]

In the first section, this chapter provides a description of quality assurance schemes, followed by a detailed discussion of stakeholder perceptions of the current situation and future developments. The third section highlights findings regarding the economics of QAS based on case studies. Finally, conclusions are drawn with regard to the impact of QAS on food marketing.

Quality Assurance Schemes

A central element of all QAS is quality. Quality is the biggest 'marketing' tool these schemes have, though with regard to user-oriented quality, product quality is highly subjective and difficult to measure.[3] Therefore, quality has different meanings in different contexts and for different stakeholders, ranging from intrinsic quality aspects and quality attributes via food safety and processing guarantees to authenticity as a means to signal quality. Therefore, quality and quality assurance are multidimensional constructs, and the variety in quality notions implies that many different aspects can be part of a QAS.

However, QAS can be broadly classified into two main types:

1. those aiming to standardize and guarantee certain aspects or requirements of the company or production unit; and
2. those aiming to differentiate and guarantee the product according to some peculiar characteristics of the product, production process or production factors.

The QAS belonging to the first group are often multinational in scope (disseminated over several countries). They are relatively few but appear in all EU member states to different degrees. These schemes always have a reference regulation (regulated quality) and almost always refer to requirements dealing with the organization of the company, production unit or production process (for example, quality management system, environmental management system, occupational health and safety management system), not the product's intrinsic characteristics. These QAS apply to business-to-business (B2B) markets and often are not communicated to final consumers.

The demand for better food quality and greater awareness of consumers about how products are produced and processed has led to the emergence of private (and public) standards as an increasingly dominant instrument of governance in the agri-food chain, both nationally and internationally. At the same time, it raises challenges for policymakers in defining appropriate responses to emerging food safety and quality issues.[4] Food legislation and QAS developed rapidly in the 1980s and 1990s in Europe, triggered by food-borne illnesses and food contaminants, combined with increasing awareness of animal welfare and the environment.[5] These trends led to more QAS, focussing especially on traceability and the production processes and practices. Private agri-food standards appear increasingly of concern in the agri-food sector; as Fulponi put it: 'Those who do not meet the standard may be excluded from markets in the short run and may eventually be forced to exit the sector.'[6] In addition, the balance within the agri-food chain regarding influence on the development and application of private agri-food standards has to be considered. Codron et al.[7] show that in France, the involvement of public authorities depends on the type of product, generally higher in the meat sector than for fresh produce.

Concerns exist not only within the EU agri-food sectors but also beyond. The World Trade Organization (WTO) has hosted a discussion on private standards and developed the Sanitary and Phytosanitary (SPS) Agreement.[8] At the moment, no defined measures exist, but the issue is under discussion, and the outcome might influence the further development of QAS in the EU. As the overall focus of this book is controversies in food and agricultural marketing, the focus in this chapter remains on concerns, which should by no means undermine the importance of QAS in the integration of the agri-food chain and the success reflected in their wide usage throughout the EU and beyond. Jaffee and Henson reflect on the issue of standards and the effect on agro-food exports.[9] The findings indicate that standards, as set by QAS, are less problematic than expected a priori.

The second group of QAS aims to highlight the differences between a product and its competitors. These schemes mostly tend to guarantee claimed product characteristics (for example, GMO-free, certain chemical composition, humane production techniques). They are usually local or national in scope and use labels to signal product and process qualities to consumers. Thus, they also appear in business-to-consumer (B2C) markets.

In the second group, almost 800 PDO (Protected Designation of Origin), PGI (Protected Geographical Indication), and TSG (Traditional Speciality Guaranteed) are covered by Council Regulation (EC) No 510/2006 and Council Regulation (EC) No 509/2006. Some research attempts to determine the value of PDO/PGI labels and indicates that the combination of the region of origin, labelling and quality perceptions is most important.[10] A related question considers whether the same effects might be achieved for the majority of PDO and PGI. Figure 6.1 highlights the much greater uptake in the southern EU member states, partly due to the history of geographical identifications,

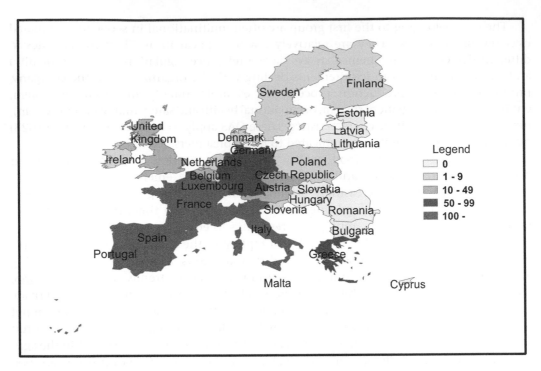

Figure 6.1   PDO/PGI and TSG in the European Union[11]

which for wines started by the nineteenth century. Other products emerged in the 1930s in Italy and France.

In addition, approximately 350 private and public QAS in the European Union belong to the second group. The exact number is difficult to detect because there is no official registry, and distinguishing between QAS and brands can be subjective. With regard to the first group, the amount of QAS is much smaller, but these QAS have generally a much higher market penetration. Some QAS comprise elements of both objectives, which makes it difficult to draw a clear line between the groups.

Stakeholder Perception

To capture the position of the different stakeholder categories, workshops were held during 2005, concluding with a structured hearing in 2006. This section includes the stakeholder groups' answers to prepared questions. The hearing featured a set of panels, each representing a given stakeholder category: farmers/producers (5), traders (6), food processors (5), certification bodies (5) and retailers (7). A panel of academic experts (10) on food quality schemes ensured that the discussion addressed pending issues and substantiated the statements.[12] Each panel received a background paper and specific questions,[13] listed in Table 6.1.

The following sections discuss stakeholder perceptions based on the information collected from both the panel presentations and the contributions received after the event. Furthermore, the discussions during the hearings, among each stakeholder panel, the expert panel and the audience (90 participants), also are considered.[14]

Table 6.1 Specific questions addressed at the stakeholder hearing, 2006[15]

SPECIFIC QUESTIONS / PANELS	Producers	Traders	Processors	Certification bodies	Retailers
How would you describe the roles of supply-chain stakeholders in developing and managing QAS?	√	√	√	√	√
Can you provide a brief assessment of the main costs and benefits for stakeholders of the present state of QAS in Europe? Are QAS achieving their intended aims in the supply chain?	√	√	√	√	√
What is the main impact of QAS on competitiveness, employment and sustainable development, in particular in rural areas?	√	√	√	√	√
In your opinion, what are the most important quality attributes that should be covered by QAS?					√
To what extent do you consider that QAS effectively contribute to providing consumers with reliable information?					
What is the present impact of QAS on vertical and horizontal relations between supply-chain stakeholders?	√	√	√	√	√
To what extent do you believe that QAS are turning into compulsory private standards?	√	√	√		√
To what extent do QAS, in which your sector participates, also involve farmers/producers? What are, in your view, the main reasons for not involving them?		√	√		
To what extent are small-scale producers, whether or not organized in groups, able to attain adequate bargaining power by means of the mechanisms considered to ensure fair and reasonable conditions of contract with larger entities in the chain?	√		√		√
What are the most important drivers of change for the development of QAS?	√	√	√	√	√
How do you envisage the future development of QAS in Europe?	√	√	√	√	√
In particular, do you believe that endogenous trends will push towards rationalization of QAS (for example, mutual recognition and benchmarking)?	√	√	√	√	√
In the light of the main challenges that you have identified, to what extent do you believe that an EU-wide framework for the development of QAS would help in addressing those challenges? In particular, what is your opinion with respect to the following options concerning QAS: No intervention at EU level Regulation of mutual recognition and benchmarking Standardization of existing quality assurance schemes/general implementing rules European registry of quality assurance logos European logo confirming compliance with EU regulations Further development of existing EU schemes Other options	√	√	√	√	√

GENERAL PERCEPTION AND EXPECTATION OF FOOD QUALITY ASSURANCE AND CERTIFICATION SCHEMES

According to several members of the farmers' panel, the most important issue is that QAS are feasible, take into account the structure of agriculture and do not create unnecessary or excessive additional burdens. Regardless of participation in a QAS, every link in the food chain must fulfil its legal requirements. Compliance with legal standards is therefore a basic requirement for every QAS. According to the same panel members, consumers' growing need for food safety and transparent and traceable production and handling processes has resulted in comprehensive, strict legislation (pertaining to food safety, traceability, residues, labelling, the environment and animal welfare). With the introduction of QAS, the agri-business sector and the food industry have taken on voluntary obligations to achieve better product quality and added value.

Agri-food traders stated that they support independent standards for good production practices and supply chain activities and the certification thereof. Traders expect QAS certification to:

- contribute to moving toward sustainable production and high quality;
- prevent consumer health scares by providing transparency and information;
- help industry comply with legislative requirements and avoid legal problems and litigation;
- optimize management and traceability from field to fork;
- keep the supply base diverse with fair competition;
- protect the reputation of brand or private labels;
- become more flexible and help implement innovations;
- link tracking and tracing to competitive supply chain activities.

Food processors also pointed out that the wide variety of QAS have diverse characteristics in the supply chain. However, they believe that the distinction between food safety and food quality requirements is essential to address the industry's role in QAS.

Food safety, along the entire food chain, is a fundamental and non-competitive prerequisite before a product may enter the market. The industry does not compete on food safety. Food safety is a joint responsibility shared by all the stakeholders in the food chain and requires combined efforts.

Quality requirements are a private affair and can provide a competitive advantage. The market dictates food quality, and it is the responsibility of the industry.

According to certification bodies' representatives, standards of the agri-food system may be classified into two categories: horizontal and vertical. The first can be applied by each player in the food chain; if applied by all stakeholders, certification gives the consumer guarantees from farm to fork. The second category includes standards for individual activities (for example, agricultural activities, processing). The combined implementation of the first and second categories of standards could lead to overlaps. The intended goal therefore is 'certified once, accepted everywhere'.

Retailers also present two broad categories of schemes: B2B and B2C. These main points are summarized in Table 6.2.

Table 6.2   Business-to-business (B2B) and business-to-consumer (B2C), presented by retailers[16]

	B2B schemes	B2C schemes
Examples	Post-farmgate (BRC/IFS), Pre-farmgate (EurepGAP)	PDO, PGI, organic
Focus on	Verification of practices Mainstream product offering	Product differentiation Usually forms part of a retailer's niche market
Application	Applied globally Common basis for safe, lawful product (BRC/IFS) and sustainable product (EurepGAP) Pre-competitive between retailers Not visible to consumers	National or regional application Focus on differentiating quality attributes, for example, provenance, organoleptic Offers potential competitive advantages for retailers Visible to consumers
Visibility	For retail label products (own brands) Part of supplier partnership agreements and/or commercial contracts	Offered by retailers to increase choice (market segmentation) Branded offering: Usually no retail input to product development

In the past 10 years, B2B schemes have fallen in number as a result of the trend toward rationalization in the retail sector, while B2C schemes have increased as a result of increased market segmentation and product differentiation in Europe.

ROLES OF STAKEHOLDERS IN THE SUPPLY CHAIN AND PUBLIC AUTHORITIES IN MANAGING DIFFERENT SCHEMES

Members of the farmers' panel considered QAS developed and implemented in accordance with the needs of the market. They also stated that it is very important that all stakeholders are able to participate in the development of a QAS. But this point raises the question of who the stakeholders are, particularly in schemes for wider geographical areas. Farmers' representatives must be fully involved in QAS that set the criteria for farm production. The European Council of Young Farmers (CEJA) sees a positive role of supply chain stakeholders in developing and managing QAS: 'There can be no doubt that we need to become better at producing for the market, and this is exactly why we need to get in better touch with the market!'

In traders' opinion, QAS are effective and useful only when stakeholders have an opportunity to provide input (for example, expertise). The structure of various QAS differs markedly, as does the extent of stakeholder involvement. In addition, in a few QAS, stakeholders play some role in technical committees and other such bodies. Consequently, stakeholders have some influence over the standard and its requirements. Traders believe that they must motivate suppliers to comply with QAS and offer effective management solutions to ensure that involvement in QAS provides competitive advantage to suppliers.

According to food processors, one stakeholder in the supply chain takes the lead in developing each QAS, based on its own interest. However, it is essential to involve all stakeholders who will apply the QAS when developing the scheme. This participation helps create confidence and commitment from different stakeholders in the supply chain. Audits are the responsibility of accredited external and independent bodies.

Certification bodies pointed out that for some standards, not all stakeholders in the food chain segment are included in the working groups. According to them, the various schemes in horizontal food chain segments should be harmonized, and those for each food chain segment should be reduced. The first step would require mutual recognition between similar schemes.

Retailers work with suppliers to establish, maintain and improve standards, which may be perceived as an imposition on other stakeholders or look as if the retailers provide the catalyst. Although they have kick-started many QAS, there has been increasing involvement of the supply side in those schemes. Moreover, according to retailers, the standard owners are working with them to reduce duplicated audits. In their view, it is very important to ensure consistency for the consumer throughout the entire chain.

IMPACT OF QAS ON VERTICAL AND HORIZONTAL RELATIONS BETWEEN SUPPLY CHAIN STAKEHOLDERS

Young producers (CEJA) believe horizontal relations could be improved through greater collaboration between producers within the same QAS. Regarding vertical relations, QAS have positive influences on relations between farmers and the industry, especially at local and regional levels. Relations with retailers and consumers still require improvement. Other members on the farmers' panel (COPA/COGECA) believe QAS support the network between stakeholders in the food chain by making information about the producer available to the next customer, which brings transparency to the supply chain. Horizontal relations in particular could improve and grow stronger if producers within the same QAS were to act together. In any event, horizontal relations provide peer pressure and self-regulation to protect the reputation of the industry and help with recovery after a loss of confidence in some sectors following crises. In their opinion, QAS significantly strengthen vertical relations in the supply chain. The main advantage is the more direct relationship with other stakeholders in the supply chain and a more direct link to both retail and the consumer, which represents an essential advantage of QAS.

Traders believe that vertical integration allows for a better focus on customer demands. They worry third parties providing services to retailers may end up competing with suppliers. Traders also believe that horizontal integration improves the chances that suppliers can work together and share relevant best practices. Although QAS have increased the chain approach in recent years, the traders believe that in the future, the distance between traders and farmers/producers will narrow or even disappear, as growers incur traders' costs and traders enjoy growers' benefits. Finally, the traders are involved in implementing QAS in cooperatives.

According to food processors, QAS make relations between stakeholders easier. On the food safety side, QAS improve vertical relations as well as horizontal relations. The food processors also stress that quality requirements relate to the individual company and depend on supplier-client relations. Thus, collaboration aims more toward vertical than horizontal relations.

FROM QAS TO COMPULSORY PRIVATE STANDARDS

According to members of the farmers' panel (COPA/COGECA), across many countries, food scandals, increased global trade and a larger share for retailers' own brands are driving retailers to develop QAS to ensure food safety and product quality. Compliance with these standards is often a prerequisite for supplying products to the market. Producers that export to a country in which buyers have their own QAS may be forced to gain certification under several QAS. Young farmers (CEJA) believe that as a result of the concentration of retailers and the processing industry, private standards get imposed to increase traceability along the chain. Although CEJA favours increasing traceability on the market, it argues that this traceability must be transmitted to the consumers, including the identification of the origin of all food products.

Traders also believe that QAS are turning into compulsory private standards, raising expectations amongst both traders and suppliers. Heavy reliance on QAS results from the desire to reduce risk and liability. In addition, participation in such schemes has become obligatory to satisfy the demands of the retail sector. Many retailers demand varying requirements, such that traders and suppliers need multiple certifications.

According to food processors, a large number of retailers require certification on the basis of their own standards, which suggest specific guidance about good practices and compliance levels. Thus, producers that want to trade products may need to comply with three or four different 'compulsory' private standards. Comparisons of these individual standards show that the bulk of the requirements are at least similar and often identical. The industry's distinction between regulatory requirements and commercial (or quality) requirements thus could help strengthen QAS and limit the specific requirements imposed by individual standards.

Retailers stress that only regulatory standards are truly compulsory. Because B2B schemes define requirements to meet regulatory standards and translate legal requirements into more precise measures, they represent an important step toward a 'level playing field' for all sources of supply.

FUTURE DEVELOPMENT OF QAS IN EUROPE

For young farmers, the future development of QAS depends largely on the different stakeholders' willingness to collaborate and give each partner the possibility to develop the QAS. Some members of the farmers' panel also believe that QAS will be widespread and generally used in the future, leading to more market segmentation and higher-quality performance by the overall supply chain. They also believe that in the future, QAS should stay autonomous, dynamic and market-oriented, based on a legal framework, and, if needed, interconnected through mutual recognition.

Traders stress that the number of requisites is increasing, but sometimes the increase fails to focus enough on improving the quality management system, which implies that some QAS are merely compliance systems. These developments may exhaust smaller enterprises. The effort put into multiple certifications can be great, which would limit the resources allocated to product development, research and development, or other technical activities that improve competitiveness. Traders do not believe that endogenous trends will push toward a rationalization of QAS; so far, the opposite is taking place. According to these traders, the retail sector continues to demand certification according to individual requirements, as well as schemes with wider participation.

Considering the costs entailed in QAS, food processors see the need for rationalization and focus on the implementation of legal requirements. The number of such schemes, in the context of B2B, should decrease (or has already decreased), and greater transparency in the requirements will enhance this trend. However, food processors also believe that schemes that focus on differentiating quality attributes represent responses to consumer demand and reflect important product differentiation in the EU.

Certification bodies imagine greater acceptance in the short term through standardization or mutual recognition. In the medium term, they anticipate a smaller number of schemes in each segment of the food chain.

Retailers envision an ongoing process of harmonization of QAS on the free market, with some initiatives (for example, checklists) already in progress and gaining pace. According to them, the number of B2B schemes has declined in the past 10 years. For example, 31 retail members now use the pre-farmgate EurepGAP[17] scheme in their supply chains. In contrast, B2C schemes have increased in number, which reflects market segmentation and product differentiation within Europe.

Table 6.3 summarizes the positions of the different stakeholder categories regarding the benefits and costs of QAS. In general terms, market access appears to provide the most important benefit, whereas the administrative burden is considered a major obstacle. The main driver behind the development appears to be consumer demand. In this regard, the unavailability of consumer organizations for the hearing make the results incomplete. The translation of consumer demand into QAS normally is conducted by other stakeholders.

Table 6.3   Summary of benefits, costs and drivers of QAS by stakeholder[18]

	Farmers	**Traders**	**Processors**	**Certification bodies**	**Retailers**
Benefits	Market access Greater market segmentation Possibility to profile their products, which may command a premium price Ensures compliance with EU legislation Improving production quality and efficiency	Market access, satisfying retail demands Improving technical capabilities of the suppliers Customer-oriented approach Improving awareness and control in raising quality Crisis management options Greater brand protection Increasing liability for suppliers	Maintain and increase consumer's confidence Add value to the product Improve the production process Provide documented proof, in case of product liability Reduce the cost of controls Provide transparency in the market	Food safety Sustainability and consumer confidence Access to clients and the European market	Fewer standards and fewer duplicated audits due to harmonization of B2B QAS Provide suppliers with clear criteria for market access Obey transparent. non-discriminatory rules Fundamental for opening markets in developing countries

Table 6.3   Concluded

	Farmers	Traders	Processors	Certification bodies	Retailers
Costs and shortcomings	Inspection costs Costs for complying with standards Costs for increased administration Overlap between private and public control	Financial implications of gaining certification in primary agriculture Needs to be certified by many schemes Long process in the acceptance of a scheme Overlap between private and public control	Unnecessary costs and duplication of efforts, if suppliers have to comply with different standards Implementation of QAS and adaptation of management structures generate fixed costs Companies with different facilities have to certify each of their plants	Accreditation costs of approval for each scheme and qualification of auditors Maintenance costs of such approval. Implementation, certification, and certificate maintenance costs for suppliers Duplication of audits is a problem	Food legislation is not fully harmonized, which creates implementation problems and extra costs
Drivers of QAS development	Market conditions Consumers Legislation/ policy Technology	Preferred suppliers, expectations, lower transaction costs Inter-changeability of suppliers Socio-demographic factors and consumers Technology (more virtual trading) Legislation	Consumer confidence Competitive advantages Management tool for improving process within global production systems Documented product liability Reduction of costs regarding audits	To reduce duplication in the same segment of the food chain To lower costs by combining or reducing the number of audits and harmonizing schemes	Loss of consumer confidence Retailers are legally the 'producers' of retail-label products Globalization of production and retailing Chain approach to establish, maintain and improve standards Costs Challenge by civil society and the media

Economic Case Studies

The goal of the case studies was to obtain a thorough understanding of the key stakeholders, processes and performance of a broad variety of QAS in different value chains.[19] The case studies include:

- Parmigiano-Reggiano cheese, a PDO in Italy, a well-established, successful QAS with large market share.
- Comté cheese, a PDO in France that generates benefits for farmers by establishing a product with a strong reputation in the market.

- Dehesa de Extremadura, a PDO in Spain, exists in a niche market for cured Iberian ham.
- Baena olive oil, a PDO in Spain, competing against industrially produced olive oil and other PDOs in a very competitive market. Label Rouge is a well-established scheme in France backed by the Ministry of Agriculture and organized by key actors in the chain. It is complex and has significant market share (for example, chickens).
- The Neuland scheme, in Germany, is a very small scheme that explicitly addresses the issue of animal welfare.
- Boerenkaas, in the Netherlands, is a Gouda-type cheese produced from raw milk on-farm. This case involves different schemes and therefore a number of specific issues. It is a niche product with several strengths and weaknesses.
- Red Tractor/Assured Produce Standard (UK), in which the section for potatoes sets the standard for potato production in the UK.
- EurepGAP is a worldwide applied scheme, establishing a production standard.

To add to the discussion of the stakeholders, the case studies help analyze the functioning of a selected group of existing QAS within the European Union. This exercise attempts to understand the implications, in terms of costs and benefits, for different stakeholders and the dynamics and structures within the QAS. Only the final outcomes are summarized here, including the costs and benefits, followed by a combined discussion of the findings of the economic case studies.

COSTS AND BENEFITS

The cost side first contains direct costs related to QAS in terms of certification, membership fees and control costs. In some instances, governments absorb some of these costs, but they have to be paid increasingly by the participants of the QAS. Notwithstanding many complaints by farmers, these direct costs are usually not more than 1–3 per cent of the price of the product. The indirect costs of QAS are usually substantially higher due to restrictions on agricultural practices (for example, herd density, animal feed, plant variety) and processing practices (for example, minimum maturing time, technique used). Another indirect cost is the additional administrative paperwork that QAS create, which is difficult to differentiate from regular administrative paperwork.

Benefits accrue to producers and consumers, as well as the general welfare. Because many QAS protect traditional agricultural products, they are important but offer unquantifiable benefits in terms of protecting the cultural heritage and rural landscape. Consumers benefit from being assured of the authenticity of the product – it really is what it claims to be. This point has particular importance for product characteristics that cannot be detected by inspecting the product itself. The benefits for producers are mainly in the form of attracting a price premium in the marketplace, which should at least cover the higher production costs per unit produced. Another important benefit is the easier market access granted to participants.

FINDINGS OF THE ECONOMIC ANALYSIS

The QAS analyzed in this study fall in two groups: in-chain quality management schemes (first group, mostly B2B), represented by EurepGAP and the Red Tractor/Assured Produce

Standard, and market differentiation schemes (second group, B2C), represented by Parmigiano-Reggiano, Comté, Dehesa de Extremadura, Baena, Neuland and Boerenkaas. The first group targets the whole market, whereas the second aims to differentiate products and thus focuses on segments of the whole market. In other words, the objective of the first group of QAS is to become the standard product, whereas the objective of the second group is to be differentiated. Thus, a comparison of their respective market shares does not offer a good instrument for comparison of the QAS groups.

The growth of buyer-driven, in-chain quality management schemes reflects a general shift in power toward the end of the value chain. As a general conclusion, it appears that in most cases, rents accrue more strongly at the end of the value chain (wholesale and retail) than at the beginning (farmers).

FINDINGS RELATED TO THE FIRST QAS GROUP (IN-CHAIN QUALITY MANAGEMENT)

The first group of QAS provides the minimum standards on which additional quality assurance measures can be built. In some cases, a convergence between in-chain quality management and legal requirements can be observed.

The successful implementation and enforcement of the first group of QAS result from effective collective action and/or market power wielded by QAS leaders. EurepGAP is becoming a de facto international standard for in-chain quality and safety management, likely because of the effective collective action of the retailers behind the scheme and because their market power enables them to enforce the standards, as well as to pass on a significant portion of the costs to producers.

An ongoing process (in many countries at least) appears to shift some erstwhile public responsibilities to the private sector. Agricultural extension and animal health once were considered public tasks but increasingly are seen as private sector responsibilities. A similar shift seems to be taking place with regard to in-chain quality and safety management, by which at least some tasks devolve to the stakeholders in the chain.

The strict requirements demanded by many of in-chain quality management schemes (first group of QAS) promote upgrading to (inter)national standards and therefore encourage process innovation. In contrast, the objectives of the market differentiation schemes explicitly or implicitly aim to maintain traditional practices. A key challenge is therefore to integrate the different objectives and approaches of the two types of schemes.

FINDINGS RELATED TO THE SECOND QAS GROUP (MARKET DIFFERENTIATION SCHEMES)

Successful market differentiation QAS appear to enjoy well-functioning professional intermediate organizational structures, like the French Intra-Chain Gruyere and Comté Committee (CIGC) or the Consorzio del Formaggio Parmigiano Reggiano (CFPR). Moreover, these organizations have mandates and undertake activities that go far beyond managing the PDO, which means it is not always easy to separate the PDO activities from other activities of these organizations.

Promotion is a key condition for establishing the QAS product as a brand in the market. Successful schemes have significant budgets for promotion, which effectively create the brand. The power of brands clearly has been illustrated; in some cases, companies build brands on top of the QAS to create additional value in the market or avoid direct competition between similar products that use the same label.

However, QAS schemes carry a risk of locking in producers to low productivity systems, which deliver end products according to very specific standards. These standards may limit innovativeness, which would have negative consequences for producers when consumer preferences change. Compared with private brands, QAS are often designed to preserve tradition and therefore are inherently less flexible than private brands.

Some QAS make significant contributions to rural development by supporting higher prices for raw materials and, in some cases, farm incomes. In addition, QAS contribute to environmental objectives by maintaining traditional, low-input agricultural production systems and valuable landscapes (Comté, Dehesa de Extremadura). Possibilities also exist for on-farm processing or agro-tourism.

Only a few QAS measure the organoleptic quality of the end-product directly. In most cases, they assume that specific regional origin and production process characteristics guarantee a superior, or at least different, consumer product. End-product grading can ensure that all producers within a scheme meet the highest standards (collective reputation of the scheme), or it can further differentiate top-quality products.

Market differentiation QAS also have the potential to maintain and improve the position of European agriculture as a high-end producer of exclusive food products to high-income consumers worldwide. Thus far, only one of the studied schemes (Parmigiano-Reggiano) has developed an export strategy to grow beyond the mature European market.

Conclusions

Combing the findings of the stakeholder consultation with the in-depth case studies shows that many market differentiation QAS appear in the EU, but more important in terms of coverage within the EU agri-food chain are the in-chain quality management QAS. Although they are few in number, they represent the focus of most of the stakeholder concerns raised. Regarding the validity of these conclusions, most existing QAS contain elements of both groups, but for clarity, this discussion includes only the archetype.

Market differentiation QAS represent a special form of joint marketing and thus compete with brands. The internal organization therefore must support this joint approach. As the findings of the economic analysis (case studies) show, the most successful market differentiation schemes are those that most closely resemble brand approaches.

In the future, market differentiation QAS will develop and perhaps disappear as consumer demand and preferences shift. The current trend suggests more differentiation, which will increase the numbers of QAS but also create circumstances in which those QAS will remain niche, market-orientated schemes. Only a few QAS can compete on a larger scale and longer term with brands.

The costs of participation in in-chain quality management QAS often seem prohibitive for smaller farms, especially if these small businesses must comply with several schemes across different marketing channels. Yet the benefits are substantial in terms of gaining or maintaining market access and obtaining easier vertical integration along the agri-food chain.

Farmers, traders and food processors all believe that in-chain quality management QAS are turning into 'compulsory' private standards, as a result of increased global trade and the larger share earned by retailers' own brands. Compliance with these standards is therefore often perceived as a prerequisite for supplying products to the global market.

This point highlights the importance of involving all stakeholders in the development and application of in-chain quality management QAS. Some stakeholders voiced their willingness to participate but thus far have been involved to only a limited extent. The role of public authorities could help mediate such involvement.

The parallel development of in-chain quality management QAS ensures considerable overlap, which has prompted expressions of interest in benchmarking and mutual recognition. Great efforts still are needed, especially to ease the international exchange of goods within and beyond the EU. In addition, in-chain quality management QAS force compliance with existing laws regarding food safety, especially through intensified control. An open discussion thus remains about the possibilities of using synergies.

Because fundamental changes seem unlikely, QAS appear to be an integral part of current and future food marketing in the agri-food chain. Different pressures will arise, especially regarding mutual recognition and benchmarking in the field of in-chain quality management QAS, which set the standards for participation in food product marketing. This participation is nearly a prerequisite. Food supply chain participants should use this opportunity to tailor these QAS to their own requirements, which move beyond legal obligations. With regard to market differentiation QAS, no major changes are expected, because they affect specific niche markets but rarely have an overall impact in the agri-food chain. Market differentiation QAS thus provide an option for food supply chain participants to market specific products, other than single-company brands.

RESEARCH LIMITATIONS AND FURTHER RESEARCH NEEDS

In the framework of the stakeholder consultation, the unclear definition of QAS appeared to create a major obstacle during the discussion. The multiple functions of QAS also lead to different perceptions. In addition, no consumer organizations participated in the stakeholder hearing, though their input would have been valuable. The economic analysis of the case studies revealed the difficulty of gathering sufficient data to conduct a quantitative analysis. For the mostly qualitative analysis herein, the best information available relates to long-established and well-functioning QAS, which may bias the conclusions. Analyzing intermediate sections of the food chain made these limitations even more severe, which indicates that these interactions should be a focus of further research that attempts to analyze how QAS might influence the integration of the food supply chain and the marketing of quality foods.

References

1. European Commission (2006a), *Economics of Food Quality Assurance and Certification Schemes Managed within an Integrated Supply Chain*. Final Report; to be published on http://foodqualityschemes.jrc.ec.europa.eu/en/index.html.
2. Project Web site, http://foodqualityschemes.jrc.ec.europa.eu/en/index.html (accessed May 2008).

3.   Grunert, K.G. (2005), 'Food quality and safety: consumer perception and demand', *European Review of Agricultural Economics*, Vol. 32, No. 3, pp. 369–391.

4.   Henson, S. & Reardon, T. (2005), 'Private agri-food standards: Implications for food policy and the agri-food system', *Food Policy*, Vol. 30, No. 3, pp. 241–253.

5.   Bredahl, M. E., Northen, J. R., Boecker, A. & Normille, M. A. (2001), 'Consumer demand sparks the growth of quality assurance schemes in the European food sector', in Regmi, A. (ed.), *Changing Structure of the Global Food Consumption and Trade*. Market and Trade Economics Division, Economic Research Service, U.S. Department of Agriculture, Agriculture and Trade Report. WRS-01-1, pp. 90–102.

6.   Fulponi, L. (2006), 'Private voluntary standards in the food system: The perspective of major food retailers in OECD countries', *Food Policy*, Vol. 31, No. 1, pp. 1–13, see p. 11.

7.   Codron, J.-M., Giraud-Héraud, E. & Soler, L.-G. (2005), 'Minimum quality standards, premium private labels, and European meat and fresh produce retailing', *Food Policy*, Vol. 30, No. 3, pp. 270–283.

8.   WTO (2007), *Private Standards and the SPS Agreement*. Note by the Secretariat, Committee on Sanitary and Phytosanitary Measures, G/SPS/GEN/746, 24 January 2007; http://docsonline.wto.org/DDFDocuments/t/G/SPS/GEN746.doc (accessed May 2008).

9.   Jaffee, S. & Henson, S. (2004), *Standards and Agro-Food Exports from Developing Countries: Rebalancing the Debate*. The World Bank, Policy Research Working Paper Series No. 3348, Washington; http://www-wds.worldbank.org/servlet/WDSContentServer/WDSP/IB/2004/07/22/000112742_20040722152604/Rendered/PDF/wps3348.pdf (accessed May 2008).

10. Van der Lans, I. A., Van Ittersum, K., De Cicco, A. & Loseby, M. (2001), 'The role of the region of origin and EU certificates of origin in consumer evaluation of food products', *European Review of Agricultural Economics*, Vol. 28, No. 4, pp. 451–477; Van Ittersum, K., Meulenberg, M.T.G., van Trijp, H.C.M. & Candel, M.J.J.M. (2007), 'Consumers' appreciation of regional certification labels: a pan-European study', *Journal of Agricultural Economics*, Vol. 58, No. 1, pp. 1–23.

11. Own presentation, based on http://ec.europa.eu/agriculture/foodqual/quali1_en.htm (accessed February 2008).

12. http://foodqualityschemes.jrc.ec.europa.eu/en/Hearing_en.htm, op. cit.

13. European Commission (2006b), *Food Quality Assurance and Certification Schemes*. Stakeholder Hearing, 11/12 May 2006, Background Paper; http://foodqualityschemes.jrc.ec.europa.eu/en/documents/Backgroundpaper_formatted_final.pdf (accessed May 2008).

14. European Commission (2006c), *Report on the Stakeholder Hearing*. Stakeholder Hearing, 11/12 May 2006, Brussels; http://foodqualityschemes.jrc.ec.europa.eu/en/documents/ReportSTKHHearing_final.pdf (accessed May 2008).

15. Ibid.

16. Ibid.

17. EurepGAP has been re-established and transformed itself into GLOBALGAP since September 2007. See http://www.globalgap.org.

18. European Commission, 2006c, op. cit.

19. European Commission, 2006a, op. cit.

7 *Organic versus Conventional Farming: A Marketing Survey on Wine Production*

BY FEDERICA CISILINO[*] AND LUCA CESARO[†]

Keywords

organic farming; grapes and wine production, marketing strategies.

Abstract

In this chapter, we will first describe the main farm characteristics, production, trade channels and sales of a sample of organic grapes producers. Second, we explore their main marketing strategies. Third, we examine the productive potential and market regulatory mechanisms of these producers. Fourth, and finally, we provide evidence of the motivations for organic choice and product marketing

Introduction

In recent decades, a considerable proportion of Italian agriculture has been devoted specifically at high-quality production. As part of this desire for improvement, two phenomena are worth noting: the very rapid growth of organic farming in recent years, and the spread of recognition of quality assurance schemes and certifications for products – for which, in terms of numbers recognized at the European level, Italy is second only to France.[‡] Italy ranks third in the world as a producer of organic goods, after Australia

---

[*]   Dr Federica Cisilino, INEA (National Institute of Agricultural Economics), Friuli Venezia Giulia Regional Office, via F. Crispi, 53 – 33100 Udine, Italy. E-mail: cisilino@inea.it. Telephone: + 39 0432 502 064.

[†]   Dr Luca Cesaro, INEA (National Institute of Agricultural Economics), Friuli Venezia Giulia Regional Office, via F. Crispi, 53 – 33100 Udine, Italy. E-mail: cesaro@inea.it. Telephone: + 39 0432 502 064.

[‡]   Reg. EC 510/06, which reformed Reg. 2081/92 on the subject of a protection system of geographical food designations, conforms to the World Trade Organization rules for the registration of applications, equivalence and

and Argentina, and it is first in Europe in terms of the number of farms (about 37 000); it represents approximately 26 per cent of organic farms in the EU.[1] Approximately 1 million hectares are organically farmed, which is 6.2 per cent of the national cultivated areas.[2] National Law 164/92 regulates the designations of origin and typical geographical indications of wines.[3] Nationally, 341 wines have designations of origin, of which 34 are controlled and guaranteed.[§] Most this production (56 per cent) is in northern Italy, with more than 8 million hectolitres. Within this context, organic viticulture is a tiny niche. Among high-quality food products, organic wine (an improper nomenclature, because it actually is wine obtained from organic grapes) represents a small market that has only recently gained the interest of the large producers.

Literature Review

According to Santucci, studies of the economic and social aspects of organic farming remain rather rare.[4] In an evaluation of the business choices linked to the conversion to organic farming, traditional short-term indicators of income appear poorly suited to evaluating the financial and economic advantages of this choice. If organic farming instead is a long-term choice, then the evaluation techniques should be adopted accordingly to analyze the initial costs (costs of conversion) and future earnings and benefits (higher earnings).

The increasing dispersion of organic farming in Europe during the past decade has stimulated the interest of many economists, in terms of both trade dynamics with related market strategies and farm production and revenue performance. In the medium and long term, organic farming can achieve acceptable profit and efficiency levels.[5] Yet the most common research approach has been to compare organic and conventional farms. According to this branch of research, an analysis of the two different production systems can offer important information from both a micro-economic (for example, evaluating economic chances to convert) and macro-economic (for example, evaluating specific policies) perspectives.[6] However, comparative analysis introduces some methodological problems, including the questionable effectiveness of the comparison itself, because it consists of two systems with very different production techniques, different technical-productive patterns (though it is possible to define a specific pattern for each group) and heterogeneous groups, mostly because conventional farming mixes agronomic techniques, including perhaps organic ones. With respect to this last issue, conventional farming might be considered the most widespread agricultural system in a given territory or it could been imagined as everything but organic techniques and methods.[7] The risk of taking non-homogenous systems into account is very high, from both technological and managerial points of view.

RESEARCH QUESTIONS

Another approach to compare the two productive systems defines conventional farms as an approximation, or how an organic farm would be if it were conventional. The

---

control. Italy holds the record for specific recognitions: 155 products are registered, representing 21.5 per cent of the entire EU basket.

§   In Friuli Venezia Giulia, 13 types of wine belong to the three categories, of which two received the label of designation of controlled and guaranteed origin.

similarity between the two kinds of enterprise, which operate in the same context, depends on the same levels of potential production and available resources. Therefore, the resulting hypothesis posits that technological homogeneity exists between the two production systems. However, this approach also introduces many problems, including the following:[8]

- The selected variables may depend on the system/context, whether organic or conventional farming.
- More innovative farms often reveal a greater inclination to convert.
- A self-selection bias, because if all farms had the same information about maximizing profits, there would be no reason for the comparison, because every farm would adopt the most rewarding production technique.

With regard to organic and conventional farming, the best solution would be to consider a constant sample of farms, such as a panel analyzed over a specified time. This approach would enable evaluations of the conversion period, including the most important impacts on farm economic performance and market behaviour. A temporal analysis is preferable (if possible) because it allows both a *within-* and *between-*farm analysis.[9] This concept represents one of our purposes for further analysis.

Other recent studies instead favour the application of a spatial approach, which analyzes farms' structural and economic characteristics. This method fails to take into account the possible effects of a change in business management, nor does it acknowledge the effective advantage of converting (cost-opportunity evaluation). What emerges from this literature review is that in the recent past, attention focuses mostly on the demand side of organic products, without touching on production aspects, such as the motivation for conversion, technical and economic issues, or marketing strategies.[10] The need for further studies on organic farming, including those that do not adopt a strictly economic or income-related point of view, becomes clear from the lack of such analyses in existing literature. Organic farming has been subject to specific statistical studies only recently and unfortunately has produced not very encouraging results.[11] Even with large data sets, institutional sources point to significant differences, sometimes greater than 20 per cent.[12] Some statistical sources that gather general agricultural data also might provide detailed information about organic farming. For example, recently the FADN databank has been used to compare traditional and organic farms, despite the difficulties associated mainly with the size of the organic sub-sample.[13] A review of the analysis methods and the results of some analyses and comparisons appear Scardera and Zanoli's study, which also analyzes the availability of data and some study cases.[14]

Although this study examines a particular organic product (wine), it has much wider origins, which invite reflection on the following questions:

- How advantageous is organic farming, from a farm performance point of view, compared with conventional crops? Does a comparison between organic and conventional crops in terms of production efficiency have any meaning?[15]
- To what extent is producing in a sustainable way recognized as an added value for a product that is placed on the market at a higher average price than its conventional equivalent? Are the presumably lower profits of organic farms compensated for by a higher market price for organic goods?[16]

- What effects do government funds to support organic farming have on income, and what advantage over conventional farming derives from this support?
- How is it possible to consider socio-environmental factors in an organic versus conventional comparison?

Attempting to provide answers to these open questions cannot be the goal of this empirical study, which starts with survey results. Rather, this study attempts to provide evidence of the contrast between organic and conventional farming, starting from a behavioural analysis of a small sample of organic farms (involved in wine production). To obtain the necessary information, we developed a questionnaire with which we can identify farm characteristics, productive potential and the main market regulatory mechanisms. Furthermore, this study provides evidence of the motivations behind an organic choice, product marketing choices and trade channels and sales.

Context Analysis

According to the last agriculture census, there are approximately 790 000 wine producers in Italy, with vineyards accounting for approximately 5.4 pe cent of the national cultivated areas. Italian viticulture as a whole is progressively adapting to the needs of the market, with the number of farms reducing to the detriment of their size.[17] At the same time, the number of farms orientated towards high-quality standard wines¶ continues to grow, even if there are some differences between north and south.[18] As Table 7.1 shows, the variations in the Friuli Venezia Giulia region mostly confirm the national trends. During the 10-year period addressed in the last two Census data,[19] the

Table 7.1 Viticulture in Italy and Friuli Venezia Giulia: cultivated surface/land area (hectares) and number of farms[20]

	Farms			Vineyards		
	N.	% Var.	Land (he)	% Var.	Average per farm (he)	
	2000	2000/1990	2000	2000/1990	2000	1990
Italy						
Permanent crops	1 858535	-12.3	2 457 993	-11.8	1.32	1.32
of which:						
Vines high-quality standard grapes	108 711	17.4	233 522	22.4	2.15	2.06
Vines for other wines	694 894	-36.2	442 057	-34.2	0.64	0.62
Total	2 551 822	-14.2	13 212 652	-12.2	5.18	5.06

¶ These productions receive the Designation of Controlled Origin (DOC) and Designation of Controlled and Guaranteed Origin (DOCG).

Table 7.1 *Concluded*

	Farms			Vineyards		
	N.	% Var.	Land (he)	% Var.	Average per farm (he)	
	2000	2000/1990	2000	2000/1990	2000	1990
Friuli Venezia Giulia						
Permanent crops	13 343	-43.5	22 753	–	1.71	0.6
of which:						
Vines high-quality standard grapes	3195	-2.7	12 935	14.2	4.05	3.45
Vines for other wines	9808	-50.2	4819	-31.2	0.49	0.36
Total	34 304	-39.1	238 806	-7.0	6.96	4.56

number of farms belonging to the whole permanent crops sector has slightly declined at the national level (-12.8 per cent), but it has dramatically decreased in Friuli Venezia Giulia (-43.5 per cent). Among farms producing wines from high-quality standard grapes, the decrease at the regional level occurs in terms of the number of farms (-2.7 per cent), with a substantial increase at the national level (+17.4 per cent). However, the importance of high-quality viticulture in Friuli Venezia Giulia appears confirmed by the average extension of vineyards (4.05 he), which in 2000 approximately doubled the national level (2.15 he).

Viticulture in the Friuli Venezia Giulia region, though unimportant in quantitative terms, demonstrates a marked specialization in quality compared with the national average, especially in the areas of *Collio Goriziano* and *Colli Orientali del Friuli*, on the border with Slovenia.

Organic Viticulture: A Survey in Friuli Venezia Giulia

STRUCTURE AND IMPLEMENTATION

This study focuses on some problems related to the prospects and marketing strategies of a group of organic producers, in a scenario characterized by strong entry barriers as well as growing market opportunities. Regional organic farming is characterized by a few farms of modest dimensions, which are active in various productive sectors and therefore indicate little incidence on the market of single products. In Friuli Venezia Giulia, there are approximately 350 organic farms,** which represent 2 per cent of the total, mostly concentrated on hills and high plains. Production involves about 2800 hectares, and the prevailing cultivations are field crops (35 per cent), grazing livestock (21 per cent),

---

** Of the 352 registered organic farms in Friuli Venezia Giulia, 257 are producers, while the remainder are manufacturers, gatherers or importers.[21]

viticulture (11 per cent) and horticulture (11 per cent). After a strong expansion during 2000–2001, the sector is now substantially stable.[22]

A survey, designed to clarify some strategic and market aspects, is based on the administration of an *ad hoc* questionnaire to a sample of producers of organic grapes.[23] The data were collected during spring and summer 2003 and updated in April and May 2007. Three farming types are considered: organic farming (57.1 per cent of farms), conversion farming (28.6 per cent) and conventional farming (14.3 per cent),[††] though the results focus on the organic farming respondents. The survey addressed a small group of organic wine producers (14 farms) as a subsample of the regional FADN database. The investigation was conducted on the basis of the willingness of the farmers to provide data and information. The sample represents nearly the 50 per cent of the total organic wine producers located in Friuli Venezia Giulia (in 2007). The results, though they describe a specific and locally important production system, do not aspire to being statistically representative for the sector as a whole.

The questionnaire consists of three parts: The first investigates farm characteristics through a brief production analysis; the second examines quality, product sales strategies and commercialization; and the third part contains information related to farm size (turnover/sales revenue, employees) and structure.[24] The replies provided by the participants made it possible to analyze the following:

- Main farm characteristics, including size and sales revenue.
- Main motivations that prompted the farmers to start organic production.
- Main obstacles to the development of organic crops.
- Marketing plans, promotion and advertising tools.
- Perception of the competitors.
- Commercial channels used to promote the product.
- How to improve the current market share of Friuli wines.

MAIN CHARACTERISTICS OF SURVEYED FARMS

Location, size and revenues

The farmers who took part in the survey are located in typical wine-growing areas of the region, with a heavy concentration in *Collio Goriziano*. The business types are very simple, mainly one-person businesses (57.1 per cent), limited partnerships (21.4 per cent) and limited companies (14.3 per cent). In addition, 85.7 per cent of the respondents do not belong to any syndicate or cooperative. Big wineries exhibit a general lack of interest in producing organic wine; even if the grapes produced organically may be favoured and sometimes sought out, they get made into a product sold as a high-quality wine, not as

---

[††]   Organic farming is a system that avoids or largely excludes the use of synthetic inputs (for example, fertilizers, pesticides, hormones, feed additives) and, to the maximum extent feasible, relies on crop rotations, crop residues, animal manures, off-farm organic waste, mineral-grade rock additives and biological systems of nutrient mobilization and plant protection, according to the EU Commission. Organic agriculture also may be defined as a unique production management system that promotes and enhances agro-ecosystem health, including biodiversity, biological cycles and soil biological activity, accomplished by using on-farm agronomic, biological and mechanical methods to the exclusion of synthetic off-farm inputs. An organic farming approach follows the conversion of land from conventional management to organic, according to specific standards (conversion period); conventional farms must exceed this conversion period before receiving organic certification.

'organic wine'. The survey also reveals that the prevalent size is very small, with 64.3 per cent of the farmers interviewed declaring annual sales revenue of less than 100 000 euro, and 92.9 per cent of farms employ fewer than five people. Only 7.14 per cent have more than 10 employees. In terms of sales revenues, in 2006, 42.5 per cent of the surveyed farmers indicated sales revenues between €51 000 and €100 000, and 21.4 per cent indicated lower (€0–50 000) or average (€101–249 000) revenues. Only 7.1 per cent achieve the highest sales revenue band (€500 000–2.4 million). Expected sales revenues for 2007 generally confirm the trend of the previous year, with the exception of farmers in the sales revenue bands of €101–249 000 and €250–499 000. The former predicted a diminution of sales revenues; the latter anticipated an increase. These declarations of expected sales revenues seem to indicate that the medium-sized farms are more dynamic and foresee consistent increases in their sales revenues for the coming years.

With regard to production, 57.1 per cent of farms have been organic for more than 5 years, 28.6 per cent for more than 2 years, and only 14.3 per cent began their organic efforts just 6 months prior to the survey. Fifty per cent of the organic producers make their own wine. About 21.4 per cent deliver grapes to wineries (14.1 per cent) or other farms (7.1 per cent) for vinification.

UNDERSTANDING ORGANIC FARMING RELATIONSHIP MARKETING

Main motivations and obstacles

The reasons for converting from conventional to organic farming are important for justifying a production choice that often has no clear economic or income-related explanation. The principal motivations leading to the conversion to organic farming are reported in Table 7.2.

The results show the high inclination toward ethical motivations, followed by the aim of raising product quality and the chance to fill a market niche. For some farms, especially in past years, the conversion to organic production was aided by EU and regional grants, which partly supported their ethical and ideological motivations. Many farmers have begun to consider the organic choice attractive from an economic point of view, irrespective of public aids, because this type of product, characterized by high quality standards, has started to achieve strong commercial results. Until a few years ago, these farmers did not certify their product and, in some cases, sold them with no indication on the organic origin of the grapes. Other farmers have begun their conversion in response to growing market demand. Their motivations to produce organically essentially can be attributed to their attempt to acquire commercial advantages deriving from the sales of wine made from organic grapes. These are the most competitive farms on the market, with clear marketing strategies in other sectors as well (for example, mid-priced quality wine).

Two of the main obstacles for producers of wine from organic grapes are high production costs and limited market demand (Table 7.2). The former are higher than those for conventional crops because of the greater need for skilled personnel. The lack of demand suggests that organic wine has yet to be fully recognized, even if organic farming gradually has expanded its market. Although the grants have stimulated many organic farms and prompted a transition from a tiny niche market to an ever-expanding market,[25] the success of organic wine is still questionable. The complexity and cost of

Table 7.2   Principal motivations that prompted organic production[a] (%) and main obstacles to the growth of organic crops[b] (%)

MOTIVATIONS	First choice	Second choice	Third choice
Economic (financial)	7.1	0.0	7.1
Ethical	67.8	7.1	0.0
Quality and innovation	42.8	14.3	0.0
Diversification	7.1	7.1	7.1
Market demand	14.3	0.0	0.0
Regional, national, EU funds	7.1	7.1	14.3
OBSTACLES		**First choice**	**Second choice**
High production costs		35.7	7.14
Certification costs		21.4	21.43
High price compared to conventional		7.1	7.14
Lack of market demand		28.5	0.00

[a]   The total does not correspond to 100 because they are replies to multiple choice questions. The table reports the first three choices.

[b]   The total does not correspond to 100 because they are replies to multiple choice questions. The table reports the first two choices.

certification are major obstacles to organic conversion, even if certification provides a useful tool for business organization and improved quality of the products.[‡‡] Because no international, legal definition of organic wine exists, few farms can sell their produce as organic, and even a label touting 'wine from organic grapes' is rarely used. In the majority of cases, the farmers declared that they sold their organic products through direct sales, so establishing a trust-based relationship with buyers apparently is more rewarding than any label. All farmers have organic grape certification, but few of them put this information on their wine bottle labels. Many farmers sell organic products through direct sales and establish valuable affiliations with customers (58 per cent). They believe that high quality is perceived as the most important (if not the only) information that consumers look for when making a purchase (at least currently). Furthermore, no clarification exists regarding which terminology is allowed. Although the EU excludes processed products (Directive 2092/91) and concedes the definition of organic only to raw materials, some standards for organic wine have been adopted voluntarily by the certifiers and producers. These standards impose limits on the use of sulphur dioxide, sugar and yeast. The debate about sulphur dioxide remains ongoing, but this lack of definition affects promotion strategies and sales.[27]

[‡‡]   In 2005, thanks to the CAP reforms and the reopening of the Rural Development Plan for organic farming in some regions, a more than 16 per cent increase occurred in organic productions.[26]

Advertising and promotions: perception of the market and competitors

A considerable percentage of farmers (35.7 per cent) indicated that planning their marketing was an important or quite important innovation (7.1 per cent) for their farm organization and management. The strategies adopted appear in Table 7.3.

Quality (product) is the first factor considered by producers of organic wine, though an efficient advertising campaign and improvements in the relationship between the producer and distribution also are perceived as necessary. Price is considered a primary strategy, a tool that not only covers production costs but also makes the product competitive. The replies provided by the conventional enterprises generally are homogeneous, according to the 50 per cent distribution of the replies across all suggested options. A generally very low quota of sales revenue gets devoted to advertising (less than 5 per cent). None of the farmers interviewed has a proper marketing plan, mainly because of their farms' sizes. The size factor also tends to prohibit the use of skilled labour, other than that of the farmer or relatives. Management needs are met internally, especially for vineyard cultivation and vinification of the grapes, as are the administrative duties, which are often entrusted to family members. Only harvesting and the sparkling process (if present) require external (seasonal) labour or contractors. Yet marketing activities exist, even if they may seems somewhat primitive, coherent with the farm sizes and characteristics.

Table 7.3   Marketing strategies and promotional tools adopted by organic farms (%) and perceptions of main competitors (%)

MARKETING STRATEGIES	Yes	No	Missing	Total
Product	73.2	10.1	16.7	100.0
Price	36.9	38.1	25.0	100.0
Advertising & promotion	51.1	32.2	16.7	100.0
Distribution	28.0	47.0	25.0	100.0
Other	16.7	58.3	25.0	100.0
PROMOTIONS	**Yes**	**No**	**Missing**	**Total**
National and international events	60.7	14.3	25.0	100.0
Regional events	43.5	31.5	25.0	100.0
Events organized by local groups	28.0	47.0	25.0	100.0
Promotions at retailers	42.2	32.8	25.0	100.0
Sporting and cultural events	15.5	51.2	33.3	100.0
Exhibitions and competitions	35.1	39.9	25.0	100.0
COMPETITORS	**Organic farms**	**Conventional farms**	**Missing**	**Total**
Co-regional	14.3	50.0	35.7	100.0
Other Italian regions	21.4	14.3	64.3	100.0
Other countries	0.0	14.3	85.7	100.0

Despite their low cost, the types of promotion actions are both interesting and efficient (Table 7.3). Approximately 11 per cent of the farmers regard themselves as leaders in the production of organic wine. Moreover, neither conventional nor organic farmers perceive other organic farms as competitors. Both categories state that the competition comes mainly from other conventional farms, as illustrated in the last section of Table 7.3.

Sales and market share

The share of production sold directly to consumers is significant (22.8 per cent), with an average of 42.5 per cent of total sales (see Table 7.4). Wholesalers represent the second supply channel, with a share of 19.2 per cent, followed by restaurants (10.9 per cent) and specialist shops (7.6 per cent). The sales share of the large-scale retail trade is still paltry (2.1 per cent) and is either regional (1.3 per cent) or national (0.8 per cent). The large-scale retail trade is absent in other EU and extra-EU countries, where wholesalers divide up the monopoly with direct sales.

Farmers' perceptions of the level of competitiveness of Friuli organic wine also is reported in Table 7.4. These data show that the product is perceived as highly or quite competitive not only in Italy but also in Europe and the United States. The market of Eastern European countries is not important for 52.4 per cent of the respondents. They

Table 7.4   Average values of organic farming sales by main market areas (%)[a] and perceptions of regional wines in the major markets (%)

TRADE CHANNELS	FVG	Italy	EU	Extra-EU	Total
Direct sales	22.8	3.8	15.9	0.0	42.5
Wholesalers	0.0	3.3	12.9	2.9	19.2
Large-scale retail trade	1.3	0.8	0.0	0.0	2.1
Ordinary shops	0.0	0.0	0.0	0.0	0.0
Specialist shops	1.3	3.8	2.5	0.0	7.6
Restaurants	5.3	4.4	1.3	0.0	10.9
Total	30.7	16.1	32.6	2.9	82.3
PERCEPTION – COMPETITORS		Italy	EU	EEC*	USA
Significant		30.9	19.9	15.5	26.8
Quite significant		60.6	63.7	23.8	35.1
Not significant		0.2	8.1	52.4	29.8
Missing values		8.3	8.3	8.3	8.3
Total		100.0	100.0	100.0	100.0

[a]   The total is less than 100 because of missing values.

*   East European countries.

are aware that they produce high-quality wines, which are limited by both product promotions and low production volumes. They cannot compare to the capacity of other Italian regions, such as Piedmont, Tuscany, Veneto or Emilia-Romagna, much less the French wine-growing regions and other European and non-European countries. The enterprises examined generally are open to foreign markets and do not apply specific marketing strategies, perhaps with the exception of their policy of pricing or product range. Because exports are mainly through direct sales or wholesalers, the price is strongly influenced by the buyer, the volume of exports and competitors' prices. The widening of the distribution channels, together with ongoing promotion activities, could improve the market share of Friuli wines. Most of the farmers interviewed are convinced that the wine produced in the region is competitive in terms of quality, but they believe the image needs to better promotion. Regarding the quality-to-price ratio, many farmers feel at a disadvantage compared with other Italian regions and suggest that the creation of a single regional trademark could lead to wider recognition and product enhancement. Although the policy of a regional trademark is constrained by EU restrictions, a policy for promoting the local wines could be effective, if well organized.

Conclusions

This analysis highlights some characteristics and strong and weak points of organic grape production. The study findings are influenced by (1) limited data collection resources, (b) the characteristics of the organic wine sector (for example, small size) and (3) the need to consider organic farms that belong to the regional FADN databank.

The businesses examined are small or very small, with an average turnover of less than 100 000 euro and fewer than five employees. The market for these farms is mainly local and national, while the European market is only marginal. The results can be summarized in some key areas: competition and innovation, labels and marketing tools, and motivations and promotions.

COMPETITION AND INNOVATION

Both technically and for conversion management, organic production methods necessitate a high level of innovation. Growers who convert to organic methods generally are more open to innovations. The strategies adopted by organic enterprises to remain competitive mainly involve promotion of the trademark, but no organizational model is based on a production chain. The few initiatives promoted by wine co-operatives appear to encounter major business and sales difficulties. These obstacles may become more challenging if the market expands. Distribution channels play a more important role in organic wine production than in conventional markets, though they are still underdeveloped. Natural channels such as direct sales are currently the best established, whereas distribution through the large-scale retail trade is almost inexistent.

LABELS AND MARKETING TOOLS

The survey suggests that labelling is not a matter of how to tell if a wine is organic but rather, how to avoid this information. Declaring that the wine is organic does not seem to be a strategic communication element, even if some consumers base their purchases entirely on the appeal of the labels. Trying to obtain a reasonable explanation, we find that understanding wine labels can be difficult and intimidating for consumers who are primarily interested in the taste of the contents. Therefore, giving no information other than an indication of high quality seems to be the choice preferred by the majority of respondents. Furthermore, labels that carry the term 'organic wine' are not allowed, though it would be possible to use the term 'organically grown grapes'.§§ Most important, the farmers involved in the survey stated that direct sales could favour communications with customers, as no other type of trade channel could. The farm's trademark is seen as a guarantee of quality, leading to consumer confidence. This branding slows the development of a common strategy for advertising and marketing. Putting information on the label is perceived as almost inconvenient, as supported by the big wineries' general lack of interest in producing organic wine. Even if they prefer to buy grapes produced organically, and sometimes seek them out, they put the grapes into a product sold as a high-quality wine, not as an organic wine.

As far as marketing tools are concerned, the survey shows that the choice of organic farming methods offers the possibility of a competitive advantage, thanks a wider range of products and limited but increasing entry to foreign markets. Advertising on the Internet is specific to organic farmers, showing that they have begun an innovative process, not just that linked to production but also in terms of business management. A producer's alliance could coordinate marketing efforts, which seem too weak and fragmented to provide an efficient and successful campaign for enhancing the Friuli Venezia Giulia wines at an international level.

MOTIVATIONS AND PROMOTIONS

The organic sector, which experienced a strong expansion between 1996 and 2000, now needs to consolidate. Two critical factors for organic produce are price and information, which actually are closely correlated. The current perception is that either the cost is too high or the value is not sufficiently communicated. Since the appearance of organic wine on the market, communication has probably not been efficient enough, nor has information about the benefits of organic produce as investment for well-being (both health and environment) been adequate. If this assessment is true in general, it is even more so for a product like wine, which from the outset might have expected to benefit from the positive awareness of traditionally stronger organic sectors (for example, fruit, vegetables, breads). Currently, high prices accompanied by diminished purchasing power and the consequent fall in consumption, which significantly affects premium price products, has demonstrated the industry's incapacity to get better organized and exposed organic producers to the contractual might of the large-scale organized distribution and retail trade. Moreover, the lack of an informational campaign to clarify the properties and peculiarities of organic products means the consumer receives no help in purchasing decisions. Many producers believe that institutions should play a more important role in the future. A suitable policy of support could provide an opportunity for promoting products and give the market the possibility of not only consolidating but

---

§§   Organically grown grapes are grown according to strict standards, without the use of chemical or synthetic pesticides or fertilizers and with minimum added sulfites.

also recommencing growth, though perhaps at less feverish rhythms than in the past. Among the foreseeable opportunities for the near future is the possibility of expanding organic produce supply for the tourism business and restaurants. Such as commercial outlet, as the survey indicates, is particularly well suited to a product like wine.

The choice of organic production for ethical reasons predominates among the replies to the survey. Approximately half of the farmers declared that their main reason for converting was ethical. This point is worth stressing, because it testifies to their motivational calibre.

A possible explanation might be a general quest for eco-sustainability by farmers who, in a climate of growing awareness and responsibility, place more importance on the ethics of their production processes. The result is a product that consumers who consider the purchase as part of a series of ethical actions, linked to their more sustainable and equitable lifestyles, may buy. Nevertheless, the results of the survey also demonstrate that if a producer abandons high-quality conventional farming for organic production, they need to be guaranteed at least the same income level.

FURTHER CONSIDERATIONS

These results suggest that future development could be characterized by two major factors: quality and sales and marketing. Another factor to consider is the competitive position of organic wine production, which suffers exposure to both national and international competition. Within this context, it will be necessary to target ancillary services and marketing, not just aimed at the end-consumer but also and especially at the distribution system. The first goal would be to ensure the visibility of added value derived from product certification. This process can initiate only with a new marketing philosophy that places viticulture within a wider scheme, capable of promoting high-quality local products.

This analysis offers some ideas, but further studies would be desirable to obtain more information. A future research development in evaluating the demand side might be to study consumer perceptions of organic wine to determine if they perceive any differences between high-quality and organic wine. These perceived differences might appear not only in terms of intrinsic product quality and organoleptic properties but also with regard to healthiness and hedonism, key features for fruit and vegetable produce. Nonetheless, perhaps because wine produced from organic grapes is undistinguishable from high-quality wine,[¶¶] at least to the average palate, the market does not appear to be destined for major growth in the near future.

References

1.   INEA Istituto Nazionale di Economia Agraria (2006), *L'agricoltura italiana conta*. INEA, Roma; INEA Istituto Nazionale di Economia Agraria and MIPAF Ministero delle Politiche Agricole e Forestali (2002), *Italian agriculture in figures* INEA, Roma.
2.   SINAB (2006), <http://www.sinab.it/> (accessed 25 May 2007).

¶¶   High-quality wine is often produced with organically grown grapes, even if this input is not communicated to the consumer.

3.   MIPAF Ministero delle Politiche Agricole e Forestali (2001) *L'agricoltura Biologica in Italia al 31 dicembre 2001* <http://www.politicheagricole.it/default.html> (accessed 20 April 2007).

4.   Santucci M. (2002), 'Limiti e necessità di comparazione tra biologico e convenzionale' in Scardera, A. & Zanoli, R. (eds) *L'agricoltura biologica in Italia, Metodologie di analisi e risultati dell'utilizzo dei dati RICA*. INEA, Roma.

5.   Offermann, F. & Nieberg, H. (eds) (2000), *Economic performance of organic farms in Europe. Organic farming in Europe. Economics and policy*, Vol. 5. Hohenheim Universität: Hohenheim Press

6.   Scardera, A. & Zanoli, R. (2002), *L'agricoltura biologica in Italia, Metodologie di analisi e risultati dell'utilizzo dei dati RICA*. INEA, Roma.

7.   Offermann and Nieberg, op. cit.

8.   Offermann, F. & Lampkin, N. (2005), *Organic farming in FADNs–comparison issues and analysis*, Proceedings of the Second EISfOM European Seminar, Brussels, November 10–11, 2005. Research Institute of Organic Agriculture FiBL, Frick, Switzerland.

9.   Santucci, op. cit.

10.   Rossetto, L. et al. (2002), 'Le imprese vitivinicole biologiche nel veneto: caratteristiche strutturali e strategie commerciali', in *Il mercato della carne e del vino da agricoltura biologica nel Veneto*, Padova: Veneto Agricoltura Press.

11.   Giovannucci, D. (2003), 'Emerging issues in the market and trade of organic products', in *Organic agriculture, sustainability, market and policies*, Wallingford: OECD Press, CABI Publishing; Hallman, D. (2003), 'The organic market in OECD countries: Past growth, current status and future potential', in *Organic agriculture, sustainability, market and policies*, Wallingford: OECD Press, CABI Publishing.

12.   Cembalo, L. & Cicia, G. (2002), 'Disponibilità di dati ed opportunità di analisi del Database RICA Biologico' in Scardera, A. and Zanoli, R. (eds) *L'agricoltura biologica in Italia, Metodologie di analisi e risultati dell'utilizzo dei dati RICA*, INEA, Roma; Cicia, G. (2000), 'Il biologico in Campania due anni dopo: l'impatto del regolamento CEE 2078/92', in De Stefano F., Cicia G. & Del Giudice F. (eds) *L'economia agrobiologica in Campania: un difficile percorso*, Editrice Scientifica Italiana, Napoli.

13.   Cisilino, F. & Madau, F.A. (2007), *Organic and conventional farming: a comparison analysis through the Italian FADN*, Proceedings of 103rd EAAE Seminar, First Mediterranean Conference of Agro-food Social Scientists, Barcelona, 23–25 April 2007.

14.   Scardera & Zanoli, op. cit.

15.   Offermann & Lampkin, op. cit.

16.   Offermann & Nieberg, op. cit.

17.   Mazzocchi, C. et al. (2000), *Organic viticulture in Italy*, Proceedings of 6th International Congress on Organic Viticulture, Vol. 1, pp. 39–40. Convention Center Basel, Frick, 25-26 August 2000.

18.   Pomarici, E. & Sardone, R. (2001), *Il settore vitivinicolo in Italia – Strutture produttive, Mercati e competitività alla luce della nuova Organizzazione Comune di Mercato*. INEA Roma; Pomarici, E. & Sardone, R. (2000), *Il settore vitivinicolo in Italia*. INEA Roma.

19.   ISTAT Istituto Nazionale di Statistica (1990, 2000), *Censimenti generali dell'Agricoltura*, ISTAT, Roma.

20.   Ibid.

21.   ERSA (Development Agency of Friuli Venezia Giulia Region), <http://www.ersa.fvg.it/tematiche/agricoltura-biologica/la-realta-dellagricoltura-biologica-in-fvg> (accessed June 2008).

22. Ibid.

23. Marbach, G. (1996), *Le ricerche di mercato*, UTET, Torino.

24. De Luca, A. (1995), *Le applicazioni dei metodi statistici alle analisi di mercato*, Milano: Franco Angeli.

25. Compagnoni, A. et al. (2001), *Organic farming in Italy*, Stiftung Omologie & Landau SOL, Baddurkheim, Germany. <http://www.organic-europe.net> (accessed 20 April 2007).

26. SINAB, op. cit.

27. Cozzolino E. (2004) (a cura di), *Viticoltura ed enologia biologica*, CRPV, Bologna: Edagricole.

8 Critical Aspects of Consumption of Genetically Modified Foods in Italy

BY MARCO PLATANIA* AND DONATELLA PRIVITERA†

Keywords

GM products, consumer behaviour, factor analysis.

Abstract

In this chapter, we examine the problems connected with the perception and rejection of genetically modified (GM) foods in Italy, highlighting both external factors and subjective, psychological and social issues that may affect acceptance rates. Using multivariate analysis, we also test whether interview respondents have homogeneous attitudes towards new GM technologies.

Introduction and Background

Modern biotechnologies offer the promise of improving people's living conditions, yet despite these possibilities, we still know little about the long-term effects of interactions between GM or transgenic organisms and the environment, nor are the potential advantages or negative consequences of GM products clear for human health.

The worldwide debate over GM products is heated; in Italy in particular, the issue has created great doubt and perplexity – not least because it contrasts with that country's general strategy to promote traditional, small to medium-sized enterprises that typify the

* Dr Marco Platania, Dipartimento DiSTAfA, 'Mediterannea' University of Reggio Calabria, località Feo di Vito, 89061 Reggio Calabria, Italy. E-mail: marco.platania@unirc.it.

† Dr Donatella Privitera, Dipartimento DiSTAfA, 'Mediterannea' University of Reggio Calabria, località Feo di Vito, 89061 Reggio Calabria, Italy. E-mail: donatella.privitera@unirc.it.

Italian food industry and produce 'genuine', traceable products with profound links to Italian territory, history and culture.

However, the continuing debate and concerns have not hindered worldwide increases in GM crop cultivation. According to the International Service for the Acquisition of Agri-biotech Applications, between 1996 and 2006, the amount of land devoted to transgenic crops increased significantly, from 1.66 to 102 million hectares, especially in the United States (54.6 million hectares, or 53 per cent of the total area), Argentina (18 million hectares), and Brazil (11.5 million hectares). The main crops grown on this land include soybeans (60 per cent of total hectares), maize, cotton and canola,[1] all ingredients frequently used to process food for both human consumption (for example, starch and flour used to create candy, soups) and animal fodder.

Eurobarometer surveys, involving 25 000 people in 25 countries, also highlight extremely positive perceptions of biotechnological applications for medicine (with biotechnology) and the benefits for human health resulting from reduced use of pesticides and minimized environmental impact.[2] Furthermore, people attach importance to the low cost of GM products compared with conventional ones and accept the approval of authorities (36 per cent). The countries that contain the highest percentage of GM rejecters though are Austria, Greece, Hungary, Germany and Latvia; those most accepting of the foods are Malta, the Czech Republic, the Netherlands, Spain, Belgium and Portugal. Overall though, when people perceive sufficient reasons to accept GM foods, past a certain threshold, they appear to be inclined to find additional reasons to accept GM foods.

From a legislative viewpoint, no comprehensive EU regulations pertain to the coexistence of GM and conventional (or organic) crops, despite Europe's acclaimed support of consumer health and environmental protection through information about the nature and characteristics of food products (for example, EU Directive 18/2001, EU Regulation 1829-30/03).[3] Nor do any control mechanisms guarantee a transparent system that provides customers or authorities with reliable information about the international trade in these products.

In academic literature, the debate about biotechnology and its applications in agriculture and food production largely focuses on assessing the real economic benefits for agricultural enterprises and the impact on the food industry, though doubts remain about whether producing transgenic crops actually provides added value for farmers.[4] Other studies show that introducing GM crops leads to variable reductions in costs, though the products do not meet consumer requirements in terms of improved quality or protection.[5]

Experts also address the importance of labelling GM foods, because without detailed label information, consumers likely cannot distinguish among products to support their personal preferences.[6] Both the message contained on GM labels and the source of that message may have significant impacts on the success of the labelling program and on consumers' perceptions of GM foods.[7] Furthermore, consumers' willingness to pay (WTP)[8] often depend on a limited acceptance, developed in a situation of substantial uncertainty and lack of knowledge. Therefore, the central technology, the health risks a person will accept and consumer trust in private and public organizations represent key variables.[9]

According to Pardo et al., European consumers' doubts about GM foods are justifiable, because the new technologies seem to have undesirable effects that may upset the balance of nature.[10] These same authors also note though that an optimistic attitude and positive beliefs about biotechnological applications can precede and filter the reception of new information, such that differences in attitude may depend partly on the level of socio-economic development of a country and the reference culture.[11] Morris and Adley disagree;[12] according to their sample of Irish consumers, knowledge and an understanding of GM technologies does not automatically induce consumers to adopt a more positive attitude but instead polarizes existing attitudes. Trail et al. also note that differences in consumer demand have implications for international trade, because varying attitudes towards GM foods in different countries (for example, United States versus EU) require a different labelling system, which could change consumer attitudes.[13]

Methodology

This research attempted to analyze consumer attitudes towards GM products, including the main reasons they reject GM products. In this context, it seems more valuable to investigate 'GM rejecters' rather than those who accept these products.

To this end, we distributed a survey questionnaire that focuses on consumer information, beliefs and attitudes, their opinions of GM products and their lifestyles. Interviewees indicated their knowledge about certain categories of products and their perceptions of any related food risk by noting their agreement (1 'strongly disagree' to 5 'strongly agree') with a series of statements‡ about GM products. The statements generally referred to the possibility they would buy GM products and covered various food-related and socioeconomic issues, as Table 8.1 shows.

Table 8.1   Reasons for buying and not buying GM foods

GM products...	
Increase public health risks	var1
Are harmful for the environment	var2
Are dangerous for my children	var3
Lead to unemployment in agriculture	var4
Are harmful to animals	var5
Do not improve the quality of food	var6
Do not comply with my ethical and religious beliefs	var7
Contribute to the economic development of my country	var8
Reduce production costs	var9
Reduce the price of conventional products	var10
Promote innovative farming methods	var11
Reduce the use of additives in agriculture	var12
Are certified	var13
Increase profits for national food industries	var14

---

‡   We developed these statements from a review of relevant literature and submitted the derived variables to various stakeholders for evaluation.

The statements easily divided into two groups, namely, positive or negative statements about the consequences of GM food production, in reference to both microeconomic and macroeconomic scenarios.

The first group cited food safety risks (var6), environmental (var2 and var5) and social issues (var1, 3, and 4) and ethical and moral considerations (var7). The second group included statements about consequences for food industries (var14), the development of the national economy (var8 and var11) and cost considerations (var9 and var10). Other statements noted the possibility of certification or guarantees (var13) and possible health advantages (var12).

We included both positive and negative statements to avoid influencing the respondents. Furthermore, the attributes of GM products mentioned in the survey included both scientifically accepted assertions (for example, less use of chemical additives to protect plants) and controversial statements reported in the mass media (for example, consequences for public health). Many consumers likely are already concerned about biotechnological innovations, especially because of recent food-related scares, such as BSE.

In the second part of the survey, to distinguish consumers on the basis of their attitude toward GM products, we asked respondents to consider a shopping scenario in which they could opt for a GM tomato product containing fewer chemical additives, when the price of 'conventional' tomatoes is 1.50 €/kg. They could choose two possible answers: interested or not interested. Thus, we could distinguish initially between those willing to buy GM products and those unwilling, albeit in a purely hypothetical scenario.[§] In turn, we identified non-consumers (or rejecters) and focused our analysis on this group to gain a better understanding of the reasons for their rejection.

Finally, in the third section of the survey, we collected general information about the socioeconomic characteristics of our respondents, including their individual inclination toward risk-taking (for example, engage in smoking, betting, adventure sports, high-speed driving, health checks).

To code and analyze the data, we first turned to descriptive analysis and employed multivariate analysis techniques, especially factor analysis, to obtain homogeneous segments of consumers. Factor analysis demands elaborate techniques for analyzing the interrelationships within a group of variables and identifying those factors that appear to offer fundamental information about the observed structure. This methodology thus attempted to explain correlations between observed variables as a function of unobserved factors, known as 'components', 'dimensions' or 'latent constructs'. Furthermore, factor analysis can transform aggregated observations into a simple structure that provides as much information as the initial structure.

Of the various multivariate analysis techniques, factor analysis holds the greatest interest because of its possible application in the business sphere, particularly for market research.[14] Within the scope of demand segmentation, it summarizes a series of appraisals of specific characteristics of the products analyzed. Namely, this methodology can provide a concise explanation of the relationships identified by a market survey or condensing and reducing the data to lose the smallest amount of relevant information.[15] Specifically, we applied factor analysis to a matrix of values referring to GM products; subsequently, on

---

§   Of the total sample of 1011 consumers, 736 rejected the shopping scenario, 255 accepted and 20 refused to respond.

the basis of the components extracted, we calculated the factor score for each interviewee and created segments based on the aggregations of two scores. Each group therefore contains interviewees who registered the strongest links with the extracted components, according to a method tested in previous research.[16]

The survey was carried out by means of direct interviews, during which 1250 questionnaires were administered in supermarkets in the five main towns of the Calabria region. Sampling in the supermarkets was random; systematic sampling was used to select the interviewees.[17]

Analysis of Data and Main Results

SOCIOECONOMIC FEATURES OF THE SAMPLE

During the survey administration, 1011 valid questionnaires were completed; the socioeconomic features of the sample are listed in Table 8.2.

More than half of the interviewees were women (53.1 per cent of the total), and the main age range was 25 to 44 years (44.8 per cent). There was a significant presence

Table 8.2 Sample features

Variables	Number	%	Variables	Number	%
Sex			**Occupation**		
Male	475	46.9	Executive	91	9.0
Female	536	53.1	Teacher-clerical worker	248	24.5
Total	1011	100.0	Tradesman	83	8.2
			Manual worker	88	8.7
Age			Student	140	13.8
14-24	164	16.2	Housewife	144	14.2
25-44	453	44.8	Pensioner	91	9.0
45-64	318	31.5	Unemployed	68	6.7
Over 64	76	7.5	Other	58	5.9
Total	1011	100.0	Total	1011	100.0
Marital status			**Education**		
Single	360	35.6	Primary school	67	6.8
Married	562	55.5	Middle school	247	24.4
Widow/er	45	4.5	Secondary school	486	48.0
Divorced-separated	44	4.4	University degree	211	20.8
Total	1011	100.0	Total	1011	100.0

Table 8.2 *Concluded*

Variables	Number	%	Variables	Number	%
Place of residence			**Number of family members**		
Town	624	61.7	1	70	6.9
Country	387	38.3	2	128	12.6
Total	1011	100.0	3	223	22.1
			4	347	34.3
			More than 4	243	24.1
			Total	1011	100.0

of consumers with a medium to high level of education (that is, secondary school and university graduates together account for over 60 per cent of the sample); in terms of employment, the greatest percentage were teachers and clerical workers (24.5 per cent), students (13.8 per cent) and housewives (14.2 per cent). Most of the interviewees live in towns (61.7 per cent) with families of more than four members (34.3 per cent).

An interesting element regarding the relationship between consumers and biotechnology is the degree of awareness: Almost 60 per cent of those interviewed stated that they were familiar, while only 10 per cent were uncertain about them, which confirms that consumers can respond to questions about problems connected with food (Table 8.3).

As further confirmation, 63.6 per cent of respondents stated that they were fully aware of what GM foods were (verified by control questions). Virtually all the respondents (95.5 per cent) wanted to be informed of the presence of GM products by means of clear label because they feel their right to an informed choice is being threatened. Furthermore, a majority of consumers (55.4 per cent) believed their supermarket sold GM products, even without any additional information, especially products whose processing requires greater safety measures and uses 'uncertain' raw materials (for example, pork products, meat).

The survey also assessed consumer perceptions of the risks connected with food by asking respondents to identify the foodstuff that poses the greatest health risk (see Table 8.4), which enabled us to evaluate consumer opinion regarding institutions and responsibility for safeguarding health through commercially available food products.

The product considered to pose the greatest risk is fish (17.3 per cent), followed by pork (13.3 per cent) and vegetables (13.2 per cent). Two findings were of particular interest: The perception of risk linked more closely to food products of animal origin than of vegetable origin, and the percentage of respondents who did not perceive any elements of risk in their food choices was very low.

CONSUMER REASONS FOR AND AGAINST GM FOODS

In Figure 8.1, we plot the mean values obtained for each variable from the entire sample of consumers. Of the negative statements, only var1 (public health risks), var6 (do not improve food quality) and var7 (do not comply with my ethical and

Table 8.3   Consumer awareness of GM technology

	Number	%
1. Have you ever heard of GMOs (genetically modified organisms)?		
Yes	584	57.8
Uncertain	108	10.7
Not at all	319	31.5
2. Do you think any of the products sold in this supermarket contain GMOs?		
Yes, definitely	560	55.4
No, definitely not	82	8.1
Don't know	369	36.5
3. Would you like to be informed by clear indications on a label?		
Yes, definitely	966	95.5
No, definitely not	14	1.4
Don't know	31	3.1
4. To what extent are you influenced in your attitude towards GM products by your family habits?		
Very much	147	14.5
Not at all	864	85.5

Table 8.4   Categories of food products and subjective perceptions of risk

Product	Number	%
Fish	175	17.3
Pork	135	13.4
Vegetables	134	13.3
Fruit	79	7.8
Wheat, flour, and so on	58	5.7
Rabbit	57	5.6
Lamb, mutton	43	4.3
Milk	26	2.6
Other	231	22.8
No answer	73	7.2
Total	1011	100.0

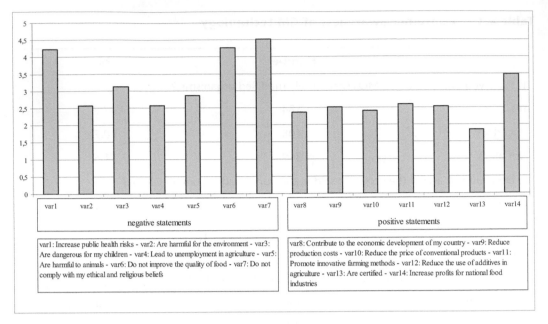

Figure 8.1   Mean judgements about GM foods (1–5 scale)

religious values) received very high scores, such that consumers strongly agreed with these statements. The other statements, though negative, do not appear convincing to consumers; the mean scores for variables related to environmental dangers (var2 and var5) and social risks (var3 and var4) range between 2.5 and 3, which corresponds to a mid-range opinion. For the positive statements, the mean score is approximately 2.5, which suggests uncertainty on the part of the consumer. The only two variables with notably different results are var13 (certified), with a mean score of less than 2, and var14 (increase profits for food industries), for which consumers exhibited a high level of agreement.

To gain a clearer understanding of the results of this preliminary analysis, we next assessed them separately according to the classification of negative versus positive judgements.

The former, ordered by level of agreement, showed that the variables with the highest scores are var7 (do not comply with my ethical and religious beliefs), var1 (public health risk) and var6 (do not improve food quality), which supported our analysis based on the mean scores (Table 8.5). However, the results obtained for the remaining negative statements (var2–5) differ, such that consumers express heterogeneous opinions, with scores spread across the five levels of agreement.

Among the positive statements, again ordered by level of disagreement, we found no uniformity of opinion. For example, the variable that prompts the strongest disagreement, var13 (certified), reaches no higher than 50 per cent, and the level of disagreement drops even further for the next variable, var8 (promote the economic development of my country), to 28 per cent. In addition, from the second variable downward, the most common response is no longer 'I strongly disagree' but 'I agree in part' (Table 8.6). As Table 8.6 further shows, for the third to the fifth ranked variable, consumers express midrange opinions.

Table 8.5   Classification of negative statements by level of agreement (%)

	Strongly disagree	Disagree	Agree in part	Agree	Strongly agree
var7 – Do not comply with my ethical and religious beliefs	2.8	2.2	8.0	14.3	72.7
var1 – Increase public health risks	2.5	4.7	15.8	21.8	55.1
var6 – Do not improve the quality of food	1.5	4.1	14.4	25.6	54.3
var3 – Are dangerous for my children	18.5	12.6	27.7	20.8	20.4
var5 – Are harmful to animals	21.5	17.5	28.0	19.8	13.2
var2 – Are harmful for the environment	26.4	22.5	28.6	12.8	9.7
var4 – Lead to unemployment in agriculture	21.6	28.3	29.4	13.1	7.6

Table 8.6   Classification of positive statements by level of disagreement (%)

	Strongly disagree	Disagree	Agree in part	Agree	Strongly agree
var13 – Are certified	47.8	29.0	16.4	4.7	2.2
var8 – Contribute to the economic development of my country	28.0	30.2	25.4	11.0	5.5
var11 – Promote innovative farming methods	26.5	18.8	30.6	16.5	7.6
var10 – Reduce the price of conventional products	25.3	28.1	30.8	12.3	3.5
var12 – Reduce the use of additives in agriculture	21.1	29.8	30.2	12.5	6.3
var9 – Reduce production costs	19.2	33.5	30.3	11.1	5.9
var14 – Increase profits for national food industries	8.5	11.6	29.2	25.2	25.6

The analysis of the tables therefore confirms the results we obtained by calculating the mean scores. Whereas consumers seem more confident in expressing their opinions in response to the negative statements, they appear perplexed by the positive claims, provoking an uncertain rather than a negative attitude toward GM products.

Because with this research, we also hoped to gain insight into the reasons for rejecting GM products, we next compared the two sets of consumers. Using the same methodology applied to analyze the total sample, we assessed the mean scores and note the low level of agreement with the positive statements about GM products. For negative statements, the scores are not homogeneous and revealed different degrees of intensity (Figure 8.3).

As Figure 8.2 reveals, the opinions expressed by the two segments essentially coincide (or differ only slightly) with regard to the potential to save money due to the introduction of GM products (var9 and 10), the benefits for the economy (var8 and var14) and the use

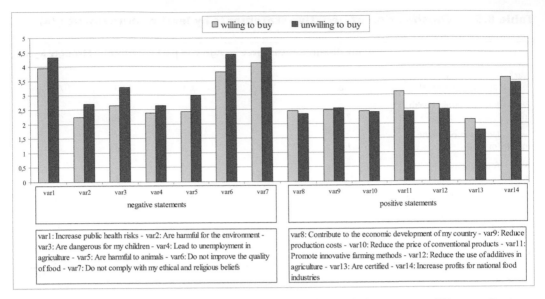

Figure 8.2   Comparison of opinions of consumers willing or unwilling to buy GM products (mean scores)

of fewer harmful chemicals in agriculture (var12) – all positive statements. The negative variables instead display slight disagreements about the possible consequences of GM innovations for farm workers (var4) and public health risks (var1).

The greatest disagreements pertain to food (var6) and agriculture (var11) quality, noncompliance with ethical and religious beliefs (var7), risks for future generations (var3) and environmental risks (var2 and var5). Consumers unwilling to buy GM products appear to give greater weight to these considerations than do consumers who are willing to purchase them.

Again, we analyzed the classification of statements separately for consumers willing and unwilling to buy GM products and distinguish between negative and positive statements. Table 8.7 reveals a clear level of agreement among 'rejecters' regarding the negative attributes: noncompliance with ethical and religious beliefs (var7), lack of improvement in food quality (var6) and public health risks (var1). On these reasons, most rejecters display the highest level of agreement.

The hierarchy of statements changes when we examine the attitudes of consumers willing to buy GM products. As we show in Table 8.8, var1 comes second and var2 last, even though the order followed by the first group changes only slightly. We further noted the difference in the intensity of the agreement expressed, which suggests greater uncertainty, according to the intermediate range of opinions.

If we analyze the positive attributes, ordered by level of disagreement, we find that the rejecters seem more perplexed by certifying GM products (var13) and the possibility that introducing GM farming promotes innovations in agriculture (var11) (Table 8.9). At the bottom of the list, we find var14, which refers to a possible increase in profits for food industries.

Again, for the attitudes of consumers willing to buy GM products (Table 8.10), the order of classification changes. A surprising item is var11, which refers to innovation and

Table 8.7   Classification of negative statements by consumers unwilling to buy GM products, by level of agreement (%)

	Strongly disagree	Disagree	Agree in part	Agree	Strongly agree
var7 – Do not comply with my ethical and religious beliefs	1.5	1.9	5.2	12.7	78.8
var6 – Do not improve the quality of food	0.8	3.5	10.2	22.8	62.6
var1 – Increase public health risks	2.6	3.7	13.6	20.0	60.2
var3 – Are dangerous for my children	17.0	8.6	28.3	22.6	23.5
var5 – Are harmful to animals	18.6	15.3	29.6	21.3	15.3
var2 – Are harmful for the environment	23.1	22.0	29.6	13.6	11.6
var4 – Lead to unemployment in agriculture	21.6	25.0	30.3	14.6	8.5

Table 8.8   Classification of negative statements by consumers willing to buy GM products, by level of agreement (%)

	Strongly disagree	Disagree	Agree in part	Agree	Strongly agree
var7 – Do not comply with my ethical and religious beliefs	6.7	3.2	16.1	18.9	55.1
var1 – Increase public health risks	2.4	7.8	22.4	27.1	40.4
var6 – Do not improve the quality of food	3.5	5.9	26.7	33.7	30.2
var3 – Are dangerous for my children	22.6	24.	26.2	15.4	11.5
var5 – Are harmful to animals	29.8	24.0	23.5	15.7	7.7
var4 – Lead to unemployment in agriculture	21.7	37.8	26.8	8.7	5.1
var2 – Are harmful for the environment	35.8	24.0	25.6	10.2	4.3

drops from the second place it earned among the rejecter group to sixth place for the accepter group.

The opinions of this latter group also appear much more uncertain than those of the first group (which confirms the result from our analysis of the negative statements). For example, if we consider var11, we note a much higher percentage (25 per cent) of respondents who mark 'I strongly agree' than in the first group (12 per cent).

CHARACTERISTICS OF CONSUMERS UNWILLING TO BUY GM PRODUCTS

To perform a more detailed analysis, we apply factor analysis to the matrix of opinions about GM products expressed by consumers unwilling to buy them, which reduces the

Table 8.9   Classification of positive statements by consumers unwilling to buy GM products, by level of agreement (%)

	Strongly disagree	Disagree	Agree in part	Agree	Strongly agree
var13 – Are certified	50.8	29.0	14.6	4.0	1.5
var11 – Promote innovative farming methods	32.1	19.7	28.7	13.7	5.8
var8 – Contribute to the economic development of my country	29.3	29.2	25.6	10.1	5.8
var10 – Reduce the price of conventional products	25.8	28.9	28.7	13.1	3.6
var12 – Reduce the use of additives in agriculture	22.7	29.9	29.4	12.4	5.6
var9 – Reduce production costs	18.9	33.6	29.5	11.8	6.1
var14 – Increase profits for national food industries	9.9	12.4	28.6	23.0	26.1

Table 8.10   Classification of positive statements by consumers willing to buy GM products, by level of agreement (%)

	Strongly disagree	Disagree	Agree in part	Agree	Strongly agree
var13 – Are certified	39.0	28.7	21.3	6.7	4.3
var8 – Contribute to the economic development of my country	24.3	32.9	24.7	13.3	4.7
var10 – Reduce the price of conventional products	23.7	25.7	37.2	9.9	3.6
var9 – Reduce production costs	20.2	33.2	32.4	9.1	5.1
var12 – Reduce the use of additives in agriculture	16.5	29.8	32.6	12.9	8.2
var11 – Promote innovative farming methods	10.6	16.1	36.1	24.7	12.6
var14 – Increase profits for national food industries	4.3	9.4	31.0	31.4	23.9

reasons to a limited number of categories. Thus, we identify five components with an overall extracted variance of approximately 62 per cent (Figure 8.3).¶

The first component (15.4 per cent of variance extracted) consists of variables related to the negative social and environmental effects of GM products. The second component

¶   We follow traditional procedures to identify common factors. After verifying the statistical significance of the data with a correlation matrix, partial correlation, KMO and the Bartlett test – which suggest reducing the initial matrix through a measure of sampling adequacy analysis[18] – we extract factors from the correlation matrix using the principal components method. The criteria for determining the number of factors to account for the correlation between variables are an eigenvalue greater than 1 and scree plots. These methods identify five components with an overall extracted variance of 64.24 per cent, though they are unclear and not uniformly described. We apply an orthogonal rotation using the Varimax method, which facilitates reading the matrix of extracted components.[19]

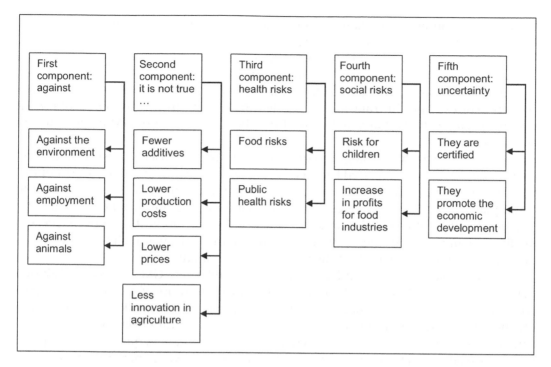

Figure 8.3   Structure of the components extracted

accounts for 12.8 per cent of the variance extracted and comprises positive effects that perplex consumers, such as the possibility of saving money for producers and consumers and potential changes in agriculture, including less pollution and a greater degree of innovation.

The third component, which accounts for 12.4 per cent of the variance, relates to health risks, including the dangerous effects consumers believe GM products might have on food and public health (var6 and var1). Consumers express a high level of agreement with both these statements and demonstrate high awareness of these issues.

The fourth component (11.0 per cent of the variance extracted) relates to social dangers, such as risks for future generations (var3) and advantages for food industries (var14). In this case, the factor analysis groups together a negative effect (var3) and a positive one, which indicates that consumers' opinions of possible wealth increases for industries that manufacture GM products should not be interpreted as positive.

Finally, the last component (9.9 per cent of variance) comprises variables related to other (hypothetical) advantages: GM products might be certified by third parties who would provide guarantees of food quality (var13), and they might improve the national economy (var8).

After identifying the factors linking homogeneous groups of purchasers, we next examine the purchasers themselves, namely, those consumers who show strong ties to each of the extracted factors.** The resulting segmentation, which features distinct agglomerated

---

** In practice, after calculating the factor score, we hierarchically order the 736 basic questionnaires, then consider only those cases that have strong ties with the extracted components. A particularly strong tie appears when the value is 1.00 or greater; all cases with lower or negative values are discarded.[22]

preferences,[20] appears to correlate with an 'advantages pursued' segmentation that focuses on differences between systems of values.[21]

Even though all the consumers analyzed in this section expressed a negative attitude toward buying GM products, and thus should be considered hostile to the set of attributes, we find distinctions with regard to the variables over which they express the greatest concern. For example, in some cases a segment may show greater awareness of certain possible positive effects that could derive from the introduction of GM crops to Italian agriculture.

The first segment, 'the ecologists', consists of consumers with a high level of education (that is, university degrees), aged between 25 and 44 years, whose families consist of two or three members. They are very well informed and use their gathered information to make purchase decisions. They consider information about the nutritional value of a product, its place of production and the presence of marks indicating provenance and quality very important in their buying decisions. In line with their inclination to opt for quality, they consider the price of the product less important.

The ecologists also are well informed about issues related to GM products and already have developed specific beliefs about them. They are more aware of the possible negative consequences that the introduction and consumption of transgenic crops may have on the environment and animals. They also perceive harmful effects on the national agricultural system, which in Italy focuses more on quality than on quantity. They appear to lead a relatively safe lifestyle; the variables measuring willingness to take risks are almost all negative. Finally, they devote significant time to reading, listening to music, surfing the Internet and reading newspapers.

We refer to the second segment as 'the savers', who have attained a medium to high level of education and often have large families. Their beliefs about GM products are based mainly on the possible reduction in production costs and the prices of conventional products. That is, though they reject GM products, they demonstrate some confusion regarding the possibility of saving money. An analysis of their lifestyle suggests a greater willingness to take risks (for example, they often lay bets).

The third group, defined as 'the uninformed', consists of consumers with medium to low levels of education and large families. This group is the only segment that expresses a low opinion of the possibility that introducing GM products might affect food quality and pose a public health risk. They pay little attention to the value of information and adopt a very sober lifestyle: They do not smoke, tend not to use cars, do not travel frequently and do not go to the cinema.

Consumers aged 46 to 64 years, mostly living in the country, from families that greatly influence their opinions of GM products, constitute the fourth group, 'the pessimists'. When choosing a product, they pay more attention to some information but not other; for example, they look closely at ingredients but do not consider knowledge of nutritional values important. Their beliefs demonstrate their concern about potential social risks in the near future for GM crops, such as increased profits for food industries and the possible risks for future generations. Their lifestyle does not exhibit a willingness to take risks.

Finally, 'the uncertain' segment contains male consumers who do not live alone, live in the country, belong to a medium to low social class (mainly manual workers with a medium to low level of education) and have small families. They pay little attention to formal characteristics when choosing products; they are not interested in being informed by a label about the goods they are buying, they are uninterested in the ingredients in a product and they do not attach importance to trademarks or the absence of preservatives. Price, however, is of great importance to this segment of consumers.

The lifestyle of this final group aligns with their system of choices. They have regular medical checkups, play sports but do not often go to the cinema, do not buy books and do not listen to the radio. Their system of beliefs suggests that the development of GM products would be a guarantee of greater economic development for the country and that these products would offer better food quality guarantees.

Conclusions and Implications

Despite its limits, this study contributes to a better understanding of consumers who express uncertainty about or reject GM products. By verifying and analyzing consumers' knowledge about GM products and the reasons for their rejection, we offer better insight into the nature of such attitudes. Consumers exhibit knowledge and a certain degree of familiarity with the issue (almost 60 per cent of those interviewed are aware and only slightly more than 10 per cent are unaware). In particular, this study clearly confirms consumer interest in label information; more than 95 per cent of the sample note their preference for such information, apparently not to resolve a lack of information but rather because of their uncertainty and diffidence toward both the media and food industries. This concern has various effects, including the belief among consumers that supermarkets sell GM products not labelled as such.

The lack of trust in food industries represents a primary reason consumers reject GM products, albeit in a hypothetical shopping scenario. Other considerations include the fear that these products may pose a health risk to the consumers and their children and that the introduction of transgenic products will not improve food quality. Ethical and religious beliefs also play a significant role.

Comparisons between consumers who accept and those who reject GM products also yield interesting results. Rejecters indicate greater awareness of the risks associated with their consumption. Among the positive considerations, the only variable that distinguishes these groups relates to the possible advantages for agriculture, and opinions on this topic varied greatly. In addition, the classifications of the negative considerations reveal great behavioural differences between the two groups.

Our methodology further enables us to investigate the profiles of rejecters and identify five groups. Of these, the ecologist segment seems most strongly convinced of the risks of GM foods, and the social characteristics of this group strengthen the idea that information provides the main tool consumers use to reject these products.

In contrast, other groups exhibit controversial attitudes. The uninformed group seems uncertain about some of the effects, such as food quality and public health risks, whereas the uncertain group believes GM products guarantee greater economic development and greater food safety guarantees if certified. These segments therefore contain consumers who, despite their rejection, are close to accepting GM food.

The results therefore reveal the extent to which the value of information may increase in critical situations, not only future food scares but also in cases of uncertainty, such as is the current situation for GM products. The fear and mistrust shown by some segments of consumers should make public policymakers think carefully. Accurate, impartial information would enable consumers to make informed choices and reduce information asymmetry, which in turn would decrease consumer uncertainty and the costs needed to ensure that the purchases are safe.

Finally, our segmentation result suggests several possible future scenarios for GM foods. The ongoing debate about GM products and their effects has increased awareness, yet uncertainties remain. Primarily, these information asymmetries relate to the amount of risk involved in the purchase of GM food.

Although economic interests likely will prompt multinational corporations to press for the authorization of GM foods, public policymakers must adopt a transparent attitude and ensure the presence of GM ingredients is evident. In Italy currently, GM crop cultivation is prohibited by a Ministerial Decree (n.72 of March 29, 2005), and similarly, many areas in Europe conform to the 'Charter of GM Free Regions'. This mostly economic decision attempts to safeguard agricultural enterprises and respond to requirements for quality, tradition and safety – attributes Italian consumers increasingly look for in their food products.

Overall, our results undoubtedly confirm the topical nature of the issue, as well as the fundamental role of information for influencing consumers' opinions.

Acknowledgments

This paper was developed within the auspices of the 'Exploitation and diffusion of legume no GM' project, performed in collaboration with the Calabria Region (Italy). It is the result of collaboration between the authors. Marco Platania wrote the 'Methodology' and 'Characteristics of consumers unwilling to buy GM products' sections, and Donatella Privitera wrote 'Introduction and Background', 'Socioeconomic features of the sample' and 'Consumer reasons for and against GM foods' sections. The 'Conclusions and implications' are a joint contribution.

References

1.   ISAAA (2007), *ISAAA Brief 35-2006: Executive Summary*, http://www.isaa.org (accessed January 2007).
2.   Eurobarometer (2006), *Europeans and Biotechnology in 2005: Patterns and Trends*, report to the European Commission's Directorate-General for Research, May.
3.   EU-COM (2006), *Report on the Implementation of National Measures on the Coexistence of Genetically Modified Crops with Conventional and Organic Farming*, OJ C/2006/104.
4.   Mauro, L. & Prestamburgo, M. (2002), 'L'introduzione degli organismi geneticamente modificati (OGM) nella produzione agro-alimentare italiana: un'analisi dei possibili effetti economici', *Politica Agricola Internazionale*, Vol. 2, pp. 7–19; De Stefano, F., Cicia, G., Verneau, F., & Cembalo, L. (2002), 'L'impatto economico derivante al sistema agroalimentare italiano dalla liberalizzazione delle colture transgeniche', *Politica Agricola Internazionale*, Vol. 2, pp. 21–52; Vieri, S. (2002), 'Organismi transgenici e sistema agroalimentare italiano', *Economia Agroalimentare*, Vol. 2, pp. 7–32; Vieri, S. (2006), 'Prospettive di introduzione delle coltivazioni transgeniche nel sistema produttivo agricolo italiano', *Rivista di Economia Agraria*, Vol. 1, pp. 29–57; Esposti, R. & Sorrentino, A. (2003), 'Biotecnologie in Agricoltura: principi, politiche e conflitti tra USA ed UE', *La Questione Agraria*, Vol. 1, pp. 73–08.

5. Price, G.K., Lin, W., & Falck-Zepeda, J.B. (2001), *The Distribution of Benefits from the Adoption of Agricultural Biotechnology*, Proceedings of 5th International Conference Biotechnology, Science and Modern Agriculture, Ravello, June 15–18.

6. Golan, E. & Kuchler, F. (2000), *Labelling Biotech Foods: Implications for Consumer Welfare and Trade*, International Agriculture Trade Research Consortium. Symposium, Montreal, Canada, June 26–27; Teisl, M.F. & Caswell, J.A. (2003), 'Costi e benefici dell'informazione sugli OGM', *La Questione Agraria*, Vol. 4, pp. 23–47; Nayga, R.M., Fisher, M.G. & Onyango B. (2006), 'Acceptance of genetically modified food: comparing consumer perspectives in the United States and South Korea', *Agricultural Economics*, Vol. 34, pp. 331–341.

7. Roe, B. & Teisl, M.F. (2007), 'Genetically modified food labelling: the impacts of message and messenger on consumer perceptions of labels and products', *Food Policy*, Vol. 32, pp. 49–66.

8. Boccaletti, S. & Moro, D. (2000), 'Consumer willingness-to-pay for GM food products in Italy', *AgBioForum*, Vol.3, No. 4, pp. 259–267; Lusk, J., Daniel, M.S., Mark, D. & Lusk, C. (2001), 'Alternative calibration and auction institutions for predicting consumer willingness to pay for nongenetically modified corn chips', *Journal of Agriculture and Resource Economics*, Vol. 26, pp. 40–57; Cembalo, L., Cicia, G., & Verneau, F. (2001), *Prodotti transgenici e consumatori: il ruolo della conoscenza e dell'attitudine al rischio*, Proceedings of XXXVII Convegno SIDEA, Catania; Grunert, K.G., Lahteenmaki, L., Nielsen, N.A., Poulsen, J.B., Ueland, O., & Astrom, A. (2001), 'Consumer perceptions of foods products involving genetic modification – results of a qualitative study in four Nordic countries', *Food Quality and Preference*, Vol. 12, pp. 527–542; Canavari, M., Davoodabady Farahany, F. & Nayga, R.M. (2005), 'Consumer acceptance of traditional and nutritionally enhanced genetically modified food: a probit analysis of public survey data in Italy', in *Food Agriculture and Environment. Economic Issues,* De Francesco, E., Galletto, L. & Thiene, M., (eds) Milano: Franco Angeli.

9. Sparks, P., Shepherd, R., & Frewer, L.J. (1994), 'Gene technology, food production and public opinion: a UK study', *Agriculture and Human Values*, Vol. 11, pp. 19–28; Frewer, L.J., Howards, C., & Shepherd, R. (1996), 'The influence of realistic product exposure to genetic engineering of food', *Food Quality and Preference*, Vol. 7, pp. 61–67; Frewer, L.J., Howards, C., & Shepherd, R. (1997), 'Public concerns in the United Kingdom about general and specific applications of genetic engineering: risk, benefits and ethics, science', *Technology and Human Values*, Vol. 22, pp. 98–124; Frewer, L.J., Scholderer, J., & Bredahl L. (2003), 'Communicating about the risks and benefits of genetically modified food: the mediating role of trust', *Risk Analysis*, Vol. 23, pp. 1117–33; Bredahl, L. (1999), 'Consumers' cognitions with regard to genetically modified foods. Results from a qualitative study in four countries', *Appetite*, Vol. 33, pp. 343–60; Bredahl, L. (2001), 'Determinants of consumers attitudes and purchase intentions with regard to genetically modified food. Results of a cross-national survey', *Journal of Consumer Policy*, Vol. 24, No. 1, pp. 23–61; Saba, A., Rosati, S., & Vassallo, M. (2000), 'Biotechnology in agriculture. Perceived risks, benefits and attitudes in Italy', *British Food Journal*, Vol. 102, No. 2, pp. 114–21.

10. Pardo, R., Midden, C., & Miller, J.D. (2002), 'Attitudes toward biotechnology in the European Union', *Journal of Biotechnology*, Vol. 98, pp. 9–24.

11. Loureiro, M.L. & Hine, S. (2002), 'Discovering niche markets: comparison of consumer willingness to pay for local (Colorado grown) organic and GMO-free products', *Journal of Agricultural and Applied Economics*, Vol. 34, No. 3, pp. 477–87; Caswell, J.A. & Joseph, S. (2006), *Consumers' Food Safety, Environmental, and Animal Concern: Major Determinants for Agricultural and Food Trade in the Future?* IATRC Summer Symposium, Bonn, May 28–30.

12. Morris, S.H. & Adley C.C. (2000), 'Genetically modified food issue: attitudes of Irish university', *British Food Journal*, Vol. 102, pp. 669–682.
13. Trail, W B., Lusk, J.L., House, L.O., Valli, C., Jaegle, S.R., Moore, M. & Morrow, B. (2006), *Considering Consumers' Demands – The Case of GMO Approval and Labelling Systems*, IATRC Summer Symposium, Bonn, May 28–30.
14. Iacobucci, D. (1996), 'Classic factor analysis', in *Principles of Marketing Research*, Bagozzi, R.P., (ed.) Oxford: Blackwell Business; Cool, K. & Henderson, J. (1997), 'Factor and regression analysis, power and profits in supply chains', in *Statistical Models for Strategic Management*, Ghertman, M., Obadia, J. and Arregle J., (eds) The Netherlands: Kluwer Academic Publishers.
15. Molteni, L. (1993), *L'analisi multivariata nelle ricerche di marketing*, EGEA, Milano.
16. Platania, M. & Privitera, D. (2006), 'Typical products and consumer preferences: the 'soppressata' case', *British Food Journal*, Vol. 108, No. 5, pp. 385–395
17. Kirsh, L. (1965), *Survey Sampling*, New York: John Wiley & Sons.
18. Hair, J.F., Anderson, R.E., Tatham, R.L. & Black W.C.(1995), *Multivariate Data Analysis*, Englewood Cliffs, NJ: Prentice Hall.
19. Sharma, S.J. (1996), *Applied Multivariate Techniques*, New York: John Wiley & Sons.
20. Kotler, P. (1991), *Marketing Management. Analysis, Planning, Implementation and Control*, Englewood Cliffs, NJ: Prentice Hall.
21. Lambin, J-J. (1994), *Le marketing strategique*, Paris: Ediscience International.
22. Platania & Privitera, op. cit.

Solving the Controversy Between Functional and Natural Food: Is Agri-food Production Becoming Modular?

BY ROBERTO ESPOSTI[*]

Keywords

modular production, mass-customization, technological innovation.

Abstract

This chapter investigates the complex and evolving relationship between innovation and consumer preferences in food production. The combination of new tastes and tendencies on the demand side with new and partially unprecedented technological opportunities on the supply side makes this relationship particularly relevant in reshaping the agri-food sector. These changes are often considered controversial, as they apparently entail incompatible consumer attitudes. New technologies would seem incongruent with emerging consumer claims. This chapter stresses how this combination can bring about production modularity in the agri-food business, increasing opportunities for targeting production to very specific consumer needs while still maintaining the large-scale (mass) character of most food processes (mass-customization), together with substantial changes to key managerial strategies and practices.

[*]   Dr Roberto Esposti, Department of Economics, Università Politecnica delle Marche, Piazzale Martelli 8, 60121 Ancona, Italy. E-mail: r.esposti@univpm.it. Telephone: + 39 0721 207 119.

Introduction and Background

The complex relationship between innovation and consumer preferences in food production and consumption usually is analyzed according to two basic conventional concepts. First, the agri-food sector is regarded as a low- or medium-tech sector, in which the role of innovation for competition is not as crucial as it is in many other sectors. Second, consumers are supposed to have a risk-adverse and conservative attitude toward food consumption, which sometimes also takes the form of technological scepticism or scare.

This chapter provides a critical review of these two conventional interpretations. On the one hand, increasing evidence indicates the evolution of food consumption, as demonstrated by the success of apparently opposite kinds of products in the contemporary market: organic (or all-natural) food, functional food and convenience food products. The second section analyses how these seemingly contradictory consumption tendencies reveal that an increasing number of consumers show a propensity (depending on age, gender, social status and so forth) to move quite rapidly, autonomously and often unpredictably between products, that is, within an ever-increasing 'food space'.

What makes this growing complexity particularly relevant in reshaping the agri-food sector, on the other hand, is the combination of these new tendencies on the demand side and the new and somewhat unprecedented technological opportunities emerging on the supply side. The third section of this chapter therefore analyzes how the diffusion of three major general purpose technologies (biotechnologies, information and communication technologies [ICT] and nanotechnologies) in agri-food production has given rise to a flow of new food products, as well as a shorter life cycle and higher degree of failure for these novelties.

The combination of new technological opportunities and increasing consumer dynamism creates the conditions for modular production in the agri-food business. The emergence of modularity is subject to analysis in further detail, in terms of the growing opportunity for targeting production to very specific (and even individual) consumer needs while maintaining the large-scale (mass) character of most food production; this idea is the general concept of mass customization applied to the agri-food sector.

A major implication of modular production and mass customization in the agri-food business, however, involves the substantial change it imposes on food firm management, regardless of size or consumer targets, as the fourth section of this chapter notes. First, modular production makes a firm's technological portfolio management more complex and crucial for its competitiveness. Second, the increasing complexity of modular food production, in terms of consumer safety and acceptance, requires a significant extra effort with progressively more articulated liabilities. These managerial implications, together with a new technological environment that allows for modular production, may be regarded as forces driving toward increasing sectoral concentration.

Relationship Between Technological Innovation and Consumer Behaviour

Apparently, it is difficult to interpret the emerging new technologies in agri-food production along an individual trajectory. New technological perspectives of the different components of the agri-food sector (agriculture, food industry, trade) emerge very specifically and sectorally. However, this fragmented and diverse flow of new technological opportunities can draw a unique, innovative horizon, exposing many new perspectives and managerial issues for the agri-food sector and its branches, whose technological innovativeness apparently differs. In his famous taxonomy of innovative trajectories, Pavitt classifies agriculture and trade as 'supplier dominated' sectors and the food industry as a 'scale intensive' sector.[1] Schettino recently updated Pavitt's study and classified the food industry itself as a supplier-dominated sector.[2]

In the former case, it implies a lack of autonomous sectoral innovative strategies, because they ultimately are controlled by science-based or 'specialized' supplier sectors and enterprises. With regard to the food industry, its greater innovative propensity usually is considered to perform less than those of most other manufacturing sectors. For example, in its Community Innovation Survey (CIS), Eurostat distinguishes among high, medium-high, medium, medium-low and low-technology sectors; the food industry belongs to the last group. Furthermore, the food industry is never included among the science-based or high-technology sectors, which represent the major drivers of technological evolution, even the other sectors, by providing a continuous flow of innovations in the form of new intermediate or capital goods.

The limited innovative potential that apparently characterizes the agri-food industry is confirmed by its low R&D intensity (sectoral R&D expenditure/sectoral value added, or sectoral GDP). According to Eurostat (CIS) data, the food sector remains permanently among those industries that lag behind with respect to such indicators. In OECD countries, the research intensity for the food industry averages about 0.3 per cent, whereas it is 11 per cent for the pharmaceutical industry and 2.4 per cent for the aggregate manufacturing sector.

In a recent analysis, Foresti confirms that in most Western countries, the food industry is not an R&D-intensive sector.[3] As Table 9.1, comparing the research intensity of the food sector with a typical high-tech branch, namely, the pharmaceutical industry, when considering wider innovative and technological indicators, the general contribution of innovative activities to sectoral and firm is lower than the average among the manufacturing industries and much lower than that of the most innovative branches. Despite research intensity, as Table 9.2 reports, other indicators confirm that the food industry cannot be included among high-tech sectors.

When agriculture joins the analysis, the general picture does not change. Public R&D (a significant part of agricultural R&D in many countries) often amounts to approximately 1 per cent of the sectoral value added. In the United States, the case is the same, even though a significant private R&D component (for example, university research, not computed as public R&D) is involved. Furthermore, other innovative indicators, such as the inflow–outflow of patents across industries, corroborate that within the agri-food sector, patent inflow largely prevails over patent outflow, which confirms the substantial dependence on the innovative trajectories of other branches.[5]

This evidence should be enough to support the idea of a technologically conservative agri-food system. Consumer behaviour itself may seem consistent with this general idea and represents the major motive, in that food consumption is often repetitive and dependent on established habits and tastes, which slowly and sceptically respond to new proposals and products.

Table 9.1   Research intensity (R&D/value added) of food industry and pharmaceutical industry by firm size in some Western countries, 2004

Size (number of workers)	Food industry				Pharmaceutical industry			
	1–499	500–999	>1000	Total	1–499	500–999	>1000	Total
France	0.7	1.2	2.8	1.0	5.2	4.2	10.5	7.0
Germany	0.2	0.5	2.3	0.5	3.4	4.7	15.3	11.2
Italy	0.2	0.4	1.7	0.4	0.9	1.6	6.1	2.5
UK	0.4	0.4	1.8	0.9	3.2	7.2	11.2	6.0
USA	0.1	0.2	5.9	1.2	2.4	1.6	31.9	8.0

Source: Foresti[4]

Table 9.2   Contribution of innovations to performance of food and pharmaceutical sectors in Italy, 1998–2000

	% Innovative firms	R&D expenditure per worker (thousand €)	% R&D for projects and marketing	% sectoral revenue of innovative firms	% revenue generated by new or improved products
Food Industry	38.2	7.5	11.0	59.2	23.5
Pharmaceutical Industry	54.7	19.9	16.8	75.9	24.6
Total Manufacturing	38.1	9.3	7.6	65.0	29.5

Source: ISTAT

However, is this really a realistic and exhaustive picture of the sector? The first point of potential disagreement is the allegedly conservative attitude toward food consumption. Especially in recent years, food producers have been challenged to expend continuous innovative efforts by increasingly demanding, curious and critical, but also capricious and impulsive, consumers. New attitudes and consumption patterns have generated increasingly differentiated and segmented food demand, as well as a substantial and novel impulse towards a more innovative and creative food sector.

Sectoral research and innovative indicators therefore should be interpreted with caution. The low innovativeness of the food sector may be, at least partially, a statistical illusion for three major reasons. First, these 'low' indicators may be expressions of the great number of small and very small enterprises within the sector that neither undergo any individual research activity nor show any particular innovative strategy. However, if research attention focused on the small number of top companies operating at either national or international levels, a substantially different picture might emerge.

Second, sectoral data could be inappropriate and misleading. Some top companies actually extend their activity into various sectors; these multinational, multisectoral firms attribute their innovative effort and performance to a single sector or country, which is neither appropriate nor informative. This aspect is particularly evident for new kinds of products, which often induce new firms to enter the sector (newcomers), though they still operate in other sectors (sectoral convergence).

Third, being mainly demand-driven, innovation activity within the food sector usually focuses on generating a continuous flow of only incremental or imitative new products whose relevance is largely underestimated by conventional innovative indicators. The creation of new products in the food sector is much more informative about innovative effort and performance than are conventional R&D intensity or patent counts. The share of a firm's revenue attributable to new products (last column of Table 9.2), can itself be a misleading indicator, though it suggests performance by the food sector that is much closer to high-tech industries. Sectors in which this share is generated by a larger number of new products should be considered more innovative, because they indicate a higher intrinsic innovative propensity and dynamism, even when they are incremental or have a limited product life cycle and a high degree of failure.

According to studies carried out by the US Department of Agriculture (USDA) on the basis of data provided by the New Product News service[6] (Table 9.3), during 1990–2007 period, between 13 000 and 22 000 new products were introduced in the domestic market by the agri-food sector. The USDA estimates that in 2004, approximately 320 000 different packaged food products were available to US consumers. On average, a retailer could accommodate 'only' 50 000 products on its shelves. Thus, the huge variety of products was excessive with respect to the available stocking capacity. The trend of new food product introductions was positive until 1995, after which it started declining,[7]

Table 9.3   Yearly new food product introductions in the United States[a]

	1990	1995	2000	2005[*]
Food products	10 301	16 863	9145	NA
Non-Food products (Pet-food, products with cosmetic or medical use, and so on)	2943	5709	7142	NA
Total	13 244	22 572	16 390	18 722
Of which: all natural and organic	NA	4%	12%	NA

[a]   Source: Harris[9]
[*]   Source: Martinez[10]

then stabilized in recent years, though some segments are still growing intensively (for example, non-food, organic, all-natural). Nonetheless, the introduction of new food products has remained massive and continuous. According to Connor, about half of all new product introductions involve food products.[8]

Economists studying the agri-food business specifically have mostly neglected this major innovative effort, and few analyses of the phenomenon have been proposed.[11] One possible explanation for this gap involves the belief that this huge flow of introductions is not very significant or and somehow misleading, because it overstates the sectoral innovative potential. According to the USDA,[12] 95 per cent of new products are considered 'non-innovative'. Few of these introductions are really new and survive in the market for a long time.[13]

A market survey by Ernst&Young reports that most new US food products (78 per cent) actually are limited developments of the products that firms already sell in the market, so-called line extensions, whereas only 22 per cent are really new to the firm, that is, new brands. Among these latter products, many products are simply imitations (me-too products). Eventually, the survey finds that only 3 per cent of this continuous and intense flow of food product introductions is substantially new to the market (one-of-a-kind products). As such, the life cycle of these products is short and brutal. In 1996, the US failure rate – products surviving in the market for less than a year – was 72 per cent for really new products and 55 per cent for line extensions.

Nevertheless, under the surface of an apparently conservative and traditional attitude, firms (especially top companies) engage in major efforts to launch new products, even though they bear the risk of a high failure rate.[14] Incremental innovations incessantly introduced in the market often do not involve the intrinsic or organoleptic characteristics of food products but rather the way they are prepared, conserved, packaged or presented, in an attempt to match increasing or changing consumer needs. To interpret this massive innovative effort, concentrated on seemingly less relevant and incremental properties of food products, this analysis returns to the original driver of the entire process, that is, changing food consumption behaviour.

Consumption Scenarios: The Concept of the Food Space

CONSUMPTION SCENARIOS

Increasingly demanding and diversified consumers are the engines of innovation, at least in the food sector, as confirmed by a careful analysis of the data on food purchasing behaviour. A survey by Federalimentare (Italian food firm association) of Italian market in 2004 indicates that for processed products, 35 per cent of food purchases include 'not-traditional' products that have some degree of relatively recent characterization or improvement:[15] 9 per cent that designate the origin, 1 per cent organic, 8 per cent 'novel food' products and 17 per cent improved traditional products, such as with additional characteristics or services. The remaining 65 per cent of the processed food market consists of traditional or undifferentiated products.

These differentiated products are somehow innovations. Even when the product remains inherently unchanged compared with its traditional or natural properties, the way it gets supplied to consumers is new. Unprecedented packaging, labelling, certification and other kinds of information all are new services with significant and sometimes remarkable technological implications.

The increasing market share of these new products is often interpreted as a consequence of two opposite consumption tendencies, though they actually are two co-existing and somehow interdependent phenomena. According to the survey of the Italian market,[16] we can distinguish two baskets of new products: the quality basket, which includes products with a designation of origin, organic and all-natural products (that is, minimally processed products containing no additives), and the time-saving basket, which features functional food (health-promoting and/or disease-preventing claims, beyond basic nutritional function), convenience food (or commercially prepared food designed for ease of consumption), and so forth. In 2003, the former basket represented 5.8 per cent of the Italian food consumption and the latter 4.8 per cent. However, expenditures are increasing significantly in both cases; during 2000–2004, the increases were 2.3 per cent for the quality basket and 13.9 per cent for the time-saving basket, which also experienced market price increases (+8.5 per cent and +18.4 per cent, respectively). Evidently, the two baskets behave as complements rather than substitutes, though the latter enjoys more encouraging performance.

In this latter, more dynamic segment, the most striking case is the worldwide growth of functional food such as soft and powered drinks and milk products (also called 'dairy functional food'), which together constitute more than 50 per cent of this segment. In 2000, the overall world market value of functional food (or 'phood') was estimated at US$33 billion; in 2004 and just in the US, it already had reached approximately US$25 billion.

The US currently represents more than 50 per cent of the world phood market, which now exceeds 2 per cent of the overall food market. In Europe, the phood market is less developed, with an estimated value of between US$4 and $8 billion, still less than 1 per cent of the overall food market. It is more developed in Central and Northern Europe than in Mediterranean countries. For example, two-thirds of the European market for dairy functional food concentrates in Germany, France, the United Kingdom and the Netherlands. In some of these countries, market share also is significant and increasing for food products with apparently opposite characteristics; for example, Germany constitutes approximately one-third of the overall European market for organic products.

The emergence of these new food consumption patterns suggests a much more fundamental change than that measured by conventional statistical indicators. The growth of the phood market, for example, may be an almost negligible phenomenon, according to most R&D or innovation variables, whereas it indicates some fairly deep transformations within the agri-food sector. The qualitative evidence regarding this major transformation involving food science and technology is quite impressive.

First, consider the great amount of neologisms for this food. Here is just a short and incomplete list: functional food, convenience food, light food (containing at least 50 per cent less fat or 50 per cent fewer calories furnished by fat compared with a similar conventional product), fit food (nutritional superiority and potential health benefits), fortified food (supplemented with essential nutrients, either in quantities greater than those normally present or that are not present naturally), ethical food (production and trade fit some ethical criteria, such as animal welfare, environmental sustainability or fair trade) and nutraceuticals (extracts of foods, usually contained in a medicinal format, that claimed to have a medicinal effect on human health). Many other new terms easily could be added to this list.

Second, the renewed and, in some respects, surprising interest of the scientific community in this topic is noteworthy. By 2004, the EU financed 36 research projects (Fifth and Sixth Framework Programme) dealing with new technologies and food. Again, these programs highlight the great variety of neologisms and bizarre names (for example, Entransfood, Profood, Nutracells, Enosefoodmicrodetect, Qpcrgmfood, Goodfood).

Just as numerous, and with very diverse denominations but a common and convergent background, are new scientific journals (*Innovative Food Science & Emerging Technologies, Trends in Food Science & Technology, The International Review of Food Science and Technology*) and scientific conferences on the subject (as just a few examples: 'Integrating safety and nutrition research along the food chain: the new challenge', 'The invisible frontier: biomedicine, nutraceuticals, nanobiotechnologies', 'Life style challenges to food science & technology', 'Thinking beyond tomorrow: a safe and nutritious food chain for the consumer', 'The economics and policy of diet and health', and 'Economics, policy and obesity').

This scientific ferment suggests an evident and declared multidisciplinary inspiration, in search of a kind of new profession: a food scientist, able to synthesize the different disciplines involved and embody this new scientific and technological horizon.

If consumers take leading roles in shaping the innovation trajectories within the agri-food sector, the common feature that underlies this considerable innovative ferment is the attempt to exploit, on the supply side, these new consumption attitudes. Similar to Harmsen and colleagues, the future of the agri-food sector can be depicted in terms of three possible alternative directions or scenarios[17] depending on how consumers might choose from the following potential technological options:

- Naturalness scenario, expressing consumers' prevailing techno-scepticism.
- Technology-driven health or functionality scenario, expressing consumers' prevailing tech-optimism.
- Tight-spending or convenience (price and time) scenario, expressing consumers' prevailing tech-opportunism.
- The designations of these scenarios in turn express the prevalence of three kinds of food products that embody such consumer attitudes:
  1. Naturalness: organic, all-natural, designation of origin, typical products.
  2. Functionality: functional food.
  3. Convenience: convenience food or tertiary processed foods.

However, these scenarios, as well as respective consumer attitudes and product typologies, should not be regarded as alternative cases. The major novelty in food consumption, and thus in food technology, is the coexistence of all these attitudes. The rise of all these tendencies may be a substitute for traditional, undifferentiated products, those whose characteristics have not been improved, expanded or renewed in any of these three directions. Eventually, all 'new' products, though apparently very different and even opposing, jointly aim to compete for market share with traditional and undifferentiated food products.

In turn, it seems plausible to depict the future not according to the prevalence of one of these three alternative scenarios but rather as their combination as three 'axes' or directions, along which overall food demand is moving. The main challenge for food production thus becomes to find a technological platform that allows for the combination and synthesis of these directions in 'new' forms with respect to traditional production systems.

Conceptualizing the coexistence of these apparently contradictory consumer trends, and the consequent technological opportunities thus generated, remains a challenge. The concept of 'food space' may be helpful.

FOOD SPACE

Conceptually, a consumer places traditional and new needs in the space generated by the three directions or axes. The food space therefore represents a set of technologically affordable combinations of the joint satisfaction of the three needs. Each product can be defined as a point in this space, with technological innovations (new products) positioned on the surface of the food space (*innovative surface*) as various new forms from which consumers can choose, according to their own specific and special circumstances (for example, availability of time, income, access to information) see Figure 9.1.

New Technologies and Modular Production: Agri-food Mass Customization

This conceptual representation attempts to clarify what may appear to be a set of new, unrelated and scattered opportunities; the 'chaotic complexity' of innovative ferment may be interpreted as the expression of a single technological force capable of continually inflating this surface, thus making the three axes coexist and co-evolve over the continuum. This underlying technological force is correlated with increasingly articulated, diverse

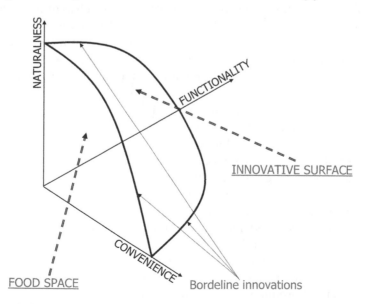

Figure 9.1   Evolving directions of food demand: food space and the innovative surface

and complex consumer attitudes. From this perspective, technological innovations in the agri-food sector are and will continue to be increasingly pervasive, just because they will be capable of being 'modulated' according to consumers' specific needs and thus will move over the 'frontier' of the three directions of food demand.

THE AGE OF MODULARITY

The unifying force that links the current innovative trends is technological modularity, a concept predominant in different production contexts as a means to disclose major new opportunities and issues.[18] In practice, technological modularity means the organization of production according to the following basic elements:

- quicker and cheaper development of new products;
- much larger number of new products (higher variety);
- much greater product complexity (decoupling tasks, design freedom and continuous upgrading all indicate the opportunity to 'design' new products in a much more flexible way).

These aspects may constitute a breakthrough in many production sectors, because they allow for the mass production of goods and services, that is, large production scale and lower unit costs, but still maintain the possibility of 'customizing' products according to the consumer's specific needs and attitudes (mass customization). As a consequence, on-demand business is no longer limited to handicraft contexts but may also pertain to the large scale – even international or global – industrial environment. Somehow paradoxically, the larger the production scale, the greater the advantages of modularity, as described in the next section.

Classical cases of modularity include the production of hardware (IBM), software (3Com, Palm Pilot) and, more generally, ICT (Motorola), as well as more traditional sectors, such as the automotive industry (Chrysler), financial services (Fidelity Investments) and watches (Swatch).[19] However, modularity has never been associated with the food industry or any of its major companies. No concrete cases indicate production has been built on modularity, and two essential prerequisites of modularity rarely are detected in food production.

These prerequisites are the presence of 'critical' consumers who express new and specific needs and the availability of a modular technology, such that it is technologically feasible to engage in low-cost production processes. By combining modules, different elementary technological components can be easily, effectively and efficiently integrated into many forms across a 'logic architecture' (or interface).

For food production, increasing evidence indicates the existence of the first prerequisite, and with respect to the latter, the novelty of modular technology seems to be in agricultural and food production and transformation. In other production sectors, modular production has become affordable, due to the advent and combination of versatile new general purpose technologies (GPT), which are capable of various applications in diverse contexts. These GPT have permeated economic history, as well as agriculture and food production. The internal combustion engine and electricity both represent GPT, because they exposed new possibilities and induced processes and product innovations in all fields of economic activity. The major novelty is not the emergence of the GPT but rather their potential combination and convergence toward specific scopes (converging technologies). The convergence of different GPT makes modular production affordable, even in sectors and production contexts apparently unsuitable to applications of modularity.

A plausible innovative horizon for the agri-food sector seems to emerge from the combination of three technological revolutions currently in effect: biotechnology, nanotechnology and ICT. These GPT and their combination may create new technological paradigms in all three of the key stages of agri-food production: raw material production (namely, agriculture), food processing (including packaging and distribution) and food design.

INNOVATIVE SURFACE AND NEW PARADIGMS

Returning to the food space (Figure 9.1), note that the most important aspect of modular production is its adaptability to a wide array of specific requirements, which creates a continuum of intergrades and variants that can comply with continuous new food consumption needs, combinations of naturalness/tradition, functionality and convenience. Therefore, the idea of modularity, as a major and generalized perspective of agri-food production, should not be confined to novel or functional food. Rather, the most relevant aspect of modularity is the intelligent combination of technological modules to achieve, improve or secure food origin, purity and traceability (that is, naturalness) and provide higher food safety standards (for example, using electronic sensing to ensure the purity and origin of wine).

This analysis may seem very abstract and, though plausible, not sufficiently supported by empirical facts. Nonetheless, clear evidence exists of these new technological opportunities in various segments of the agri-food system, as follows.

Intelligent agriculture (or *linked-systems agriculture*) refers to a series of expressions: (1) precision agriculture or precision farming uses new technologies, such as global positioning, sensors, satellites, aerial images and information management tools to assess and understand in-field variations; (2) intelligent machines use robotics and automation in agricultural production; or (3) molecular agriculture or (bio)farming, which usually indicates the cultivation of plants to produce non-food molecules, such as pharmaceuticals, 'edible' vaccines and biodegradable plastics. Despite their different connotations or meanings, these new terms express new possibilities and the potential of the primary sector to achieve a more targeted, designed or customized form of production. These solutions lead to agricultural supply whose quantitative and qualitative outcomes are less uncertain and more targeted to specific and complex consumption requirements, as well as able to adapt quickly to changing environmental conditions and the specific needs of downstream uses.

Intelligent processing outlines advanced technological and organizational solutions and combinations. Among them, careful processing reflects the design of 'minimum' and 'low input' food processing strategies, which guarantee quality and food safety; mild or fair technologies aim to minimize thermal and chemical damages in food processing and conservation; an intelligent environment, which is equipped with sensors, actuators and computers to enable processing to adapt to changing external conditions; or intelligent packaging and materials, also known as active packaging or functional packaging, which includes functions or materials that switch on and off in response to changing external/

internal conditions and communicate the status of the product to consumers. As an example, an interesting case of active packaging is the highly successful, foam-producing 'widget' in a can of beer. These technologies, in all stages of processing, indicate the increasing capacity for integrating and adapting technological and organizational solutions with greater flexibility to fulfil predetermined functions or objectives.

Nutritional genomics (or *nutrigenomics*) emphasizes the potentials of biotechnology (especially in combination with ICT and nanomaterials) to design and control food production, thanks to increasing in-depth knowledge about the links among gene sequences, proteins, metabolism and specific physiological functions. It represents the effective application of functional genomics to food production. The main perspective of nutritional genomics involves designing customized food and diets, according to the specific, personal, metabolic characteristics of the consumer, and eventually according to therapeutic needs. Concrete applications in the food industry include dietary interventions for individuals at risk of diet-related diseases for which susceptible genes have been identified, such as type-2 diabetes mellitus, obesity, cardiovascular diseases, some autoimmune diseases and some cancers.

Concluding Remarks: Critical Managerial Aspects of Modular Agri-food Production

NEW OPPORTUNITIES

Meeting the growing demands for food naturalness and tradition often is associated with the immutable techniques of traditional production and thus a lack of technological innovation. On the contrary though, new technologies should be regarded as a major opportunity to protect and improve product naturalness, origin and safety. Traditional food processes do not necessarily fulfil consumer requirements in this respect. In particular, 'high-class' food products must protect and secure their origin and their organoleptic quality, which requires continuous and specific technological improvements. These technological solutions should not be viewed as a contamination of the essential characteristics of the product, such as its 'purity' and typicality, but rather as an opportunity for conservation and restoration. In several Western countries, especially those with a long and rich gastronomic traditions and a large variety of high-quality and typical food products (for example, Italy, France, Spain), new opportunities from modular production should relate mainly to this dimension of the food space.

More generally, opportunities abound in the practical expression of the innovative surface in some specific segments of the food market, such as organic and typical products. The needs expressed by consumers in these segments are not conservative and stable, in that they permanently, and sometimes unpredictably, evolve and indicate new tendencies. Modular technology offers a great variety of options for accompanying this continuous evolution and achieving this dynamism in quality, even in contexts previously considered unsuitable in this respect (for example, combination of naturalness and functionality in next-generation organic products).

NEW ISSUES

In addition to these new opportunities, modular production may raise some critical issues that require consequent and consistent new managerial strategies and institutional arrangements. Major issues implied by the advent of modular food production include the following:

- property rights regimes and the problem of anti-commons;
- advent of newcomers and increasing concentration;
- new risk profiles and respective liabilities.

The first concern is a consequence of the basic character of modularity itself, implying a combination of technological components and innovations (modules and/or interfaces). The access to so many modules/interfaces held (in terms of property rights) by so many owners remains a complex issue. The need to coordinate the property rights regimes of the different technologies involved makes technology portfolio management strategies more and more relevant, even for agri-food companies.

Complex property right regimes implied by modular technologies also raise a major risk for society in terms of overall welfare. Specifically, the combination of such property rights could prevent or hinder socially relevant horizontal and non-rival technological developments. This concern is the so-called problem of anti-commons[21] and mainly concerns those minor or 'niche' segments of the agri-food sector in which firms may encounter major difficulties in accessing the modular solutions designed and owned by firms that operate in different market segments or even sectors.

The second issue raised by modular production pertains to sectoral concentration, in that modularity favours concentration for two major reasons. First, it usually induces scale economies. Although conventional production may give rise to decreasing returns to scale due to the more than proportional organizational and managerial costs, modularity can reveal major advantages in terms of greater flexibility in large-scale production organization and planning (Figure 9.2).

Second, the advent of new modular technologies in agri-food typically is associated with the arrival of new companies in the sector. Technologies converging towards different applications give large firms the opportunity to converge in different sectors. Thus, modularity itself attracts newcomers, mostly from high-tech oligopolistic sectors in which the key modules or interfaces were actually produced, which creates higher concentration (and market power), even in entry markets. One typical example is the arrival of large multinational firms from the pharmaceutical sector into the agri-food sector; Novartis Consumer Health, a division of the multinational Novartis (owned by Sygenta, an even larger pharmaceutical multinational firm), operates in the functional food market.

Finally, the third issue involves the implications of modularity in terms of food safety and consequent liabilities. On the one hand, modularity may bring about innovative opportunities, even in terms of food safety, with respect to known and well-established safety risks. On the other hand, new and mostly unknown kinds of food risk might be generated, or suspected, by unprecedented combinations of technologies and modules. At least in principle, the convergence of technologies may imply the convergence of risks and no one can rule out, *a priori*, that new risk profiles will emerge from the combination

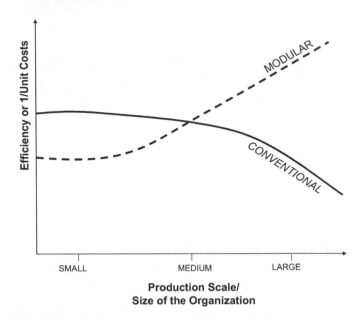

Figure 9.2   Relationship between efficiency and production scale, according to underlying technology

Source: Adaptation from O'Grady[22]

of individually safe technologies. For example, still unknown are the possible effects and risks arising from the combination of nanotechnologies and biotechnologies (that is, nanobiotechnologies) in food production and consumption.[†]

In general terms, the combination and interaction of many modules and interfaces in a new product redefine both accountability and responsibility (or liability, in more appropriate legal terms) to consumers with regard to the properties of the new product. In particular, is the final producer liable for product safety, or is the entity that produces the modules and interfaces or designs the new product? Moreover, what does liability imply in this respect? Is it just a problem of food safety, or should liability extend to all product properties for which it has been purchased? For example, who guarantees that probiotic food really satisfies the function for which it has been advertised, and as such, is also responsible for it? According to these kinds of arguments, liability issues seem likely to become much more complex as modularity increases product complexity.

References

1.   Pavitt, K. (1984), 'Sectoral patterns of technical change: towards a taxonomy and a theory', *Research Policy*, Vol. 13, pp. 343–373.

---

† The exploitation of nanotechnologies in food production is generically designated as *nanofood*. Approximately 200 applications of nanotechnologies to food production appear worldwide (though most are still in the experimental stage, and few already are in commerce). As an example, nanocapsules incorporate tuna fish oil, a source of Ω-3 fatty acids, into bread. In 2003, the estimated size of the nanofood market was about US$2.6 billion, projected to grow to US$7 billion by 2006 and US$20 billion by 2010. These technological developments indicate the huge potential for food production but also raise remarkable objections on ethical and, above all, safety grounds.

2. Schettino, F. (2007), 'US patent citations data and industrial knowledge spillovers', *Economics of Innovation of New Technology*, Vol. 16, No. 8, pp. 595–33.

3. Foresti, G. (2005), 'Specializzazione produttiva e struttura dimensionale delle imprese: come spiegare la limitata attività di ricerca dell'industria italiana', *Rivista di Politica Economica*, Vol. XCV, No. 3–4, pp. 81–122

4. Ibid.

5. Fanfani, R., Lanini, L. & Torroni, S. (1996), 'Invention patents in Italian agrifood industry: analysis of the period 1967–1990', in Galizzi G. & Venturini, L. (eds) *Economics of Innovation: The Case of Food Industry*, Physica-Verlag, Heidelberg, pp. 391–406; Esposti, R. (2002), 'Public agricultural R&D design and technological spill-ins. A dynamic model', *Research Policy*, Vol. 31, No. 5, pp. 693–717.

6. Harris, J.M. (2002), 'Food product introductions continue to decline in 2000', *FoodReview*, Vol. 25, No. 1, pp. 24–27; Lee, C.H. & Schluter, G. (2002), 'Why do food manufacturers introduce new products?', *Journal of Food Distribution Research*, Vol. 33, No. 1, pp. 102–111; Martinez, S.W. (2007), *The U.S. Food Marketing Sysytem: Recent Development, 1997–2006*, USDA–Economic Research Report N. 42, Washington, D.C.

7. Lee & Schluter, op. cit.

8. Connor, J.M. (1981), 'Food product proliferation: a market structure analysis', *American Journal of Agricultural Economics*, Vol. 63, No. 4, pp. 607–617.

9. Harris, op. cit.

10. Martinez, op. cit.

11. Lee & Schluter, op. cit.; Corbett, J.J. (2004), 'A survey of new food product introductions and slotting allowances in the New England marketplace from a food broker's perspective', *Journal of Food Distribution Research*, Vol. 35, No. 1, pp. 44–50.

12. Martinez, op. cit.

13. Connor, op. cit.

14. Ibid.; Corbett, op. cit.

15. Federalimentare (2004), *L'industria agroalimentare in Italia*, 2° Rapporto Federalimentare-ISMEA, Parma. According to the current EU regulation, novel foods are foods and food ingredients that have not been used for human consumption to a significant degree within the Community before 15 May 1997.

16. Ibid.

17. Harmsen, H., Andersen, A.M.S. & Jensen, B.B. (2004), 'Future impact of new technologies: three scenarios, their competence gab and research implications', in Evenson R.E. & Santaniello, V. (eds), *The regulation of Agricultural Biotechnology*, Oxon: CABI Publishing, pp. 213–238.

18. O'Grady, P. (1999), *The Age of Modularity*, Iowa City: Adams and Steele Publishers.

19. Ibid.

20. Harmsen et al., op. cit.

21. Esposti, R. (2004), 'Complementarietà, coordinamento e problemi di anticommons nell'innovazione biotecnologia', *La Questione Agraria*, No. 2, pp. 99–134; Fonte, M. (2004), 'Proprietà intellettuale e dominio pubblico: il caso delle agro biotecnologie', *La Questione Agraria*, No. 3, pp. 129–154.

22. O'Grady, op. cit.

10 Controversies in Managing Competencies: The Case of Development and Launching New Functional Food Products

BY JOFI PUSPA,* TIM VOIGT† AND RAINER KÜHL‡

Keywords

capabilities, corporate responsibility, knowledge, marketing.

Abstract

This chapter proposes four types of capabilities that are identified as the foremost anchored elements for setting up a firm's core competency, relevant for development and launching of a functional food: technology and external knowledge adjustment, market orientation, marketing and networking capabilities. This chapter therefore will discuss the following three issues. First, the functional food field as an attractive business field of the future in the context of the concept of core competencies. Second, the linkage between strategic content and strategic process, identified by extending the framework to the analysis of competencies in new product development. Third, and finally, the development of certain competencies in the processes of new product development (development and launching/marketing phases).

* Dr Jofi Puspa, Chair of Food Economics and Marketing Management, Justus-Liebig University of Giessen, Senckenbergstrasse 3, 35390 Giessen, Germany. E-mail: jofipuspa@yahoo.de.

† Mr Tim Voigt, PhD Student, Chair of Food Economics and Marketing Management, Justus-Liebig University of Giessen. E-mail: tim.A.Voigt@agrar.uni-giessen.de.

‡ Dr Rainer Kühl, Chair of Food Economics and Marketing Management, Justus-Liebig University of Giessen. E-mail: Rainer.Kuehl@agrar.uni-giessen.de.

Introduction

In the era of industrialization, consumers are becoming more knowledgeable and demanding. There is therefore an emerging need for firms to improve their internal capability for exhibiting appropriate responses to consumers' global demand. Successful firms in this international competition require a sustainable stream of organizational and technological innovations to maintain their leadership in their respective market segments. The requirements of successful innovation demand that firms establish a new capability for combining multidisciplinary areas and building global networking for advanced technology and material outsourcing. Newly emerging markets, such as that for functional food, are consumer-driven industries that exist globally. The functional food market provides a concrete example of an innovative product category in the domain of the food and beverage industry. Functional food (FF) is characterized by great value added to the product core that corresponds to changing customer needs, as well as providing a basic supply of nutrients.

The FF industry also possesses a clear, distinctive market structure (consumer, competition, technology and know-how) in comparison with the conventional food and beverage markets. Consumers of functional food vary in terms of their psychological perceptions (for example, awareness, acceptance, attitude, involvement level) of potentially added values.[1] The position of FF, in the grey area between food and medicine products, provides an image of unnatural, high-tech and additive food which may create two different perceptions. In an extreme example, the first pool contains people who have high self-preference and are among the first innovators of new foods. The second group includes consumers who are sceptical and reluctant to accept FF, either because they do not have knowledge of or belief in the functional benefits of FF or because they worry about issues of product safety related to unnatural, added ingredients and the high-technology profiles of FF. Thus, for functional food firms, market sensing and understanding the dynamics of consumers' needs, belief, knowledge and acceptance remain essential drivers of innovation.

To develop a FF product, a firm faces challenges with regard to adopting new technologies and know-how. A firm's own technology and know-how may need to be developed, beyond finding new designs (for example, new taste, packaging or positioning) or simple product modifications, which are the usual paths in the food industry. Other aspects related to investigating, searching for, finding and purifying new components that provide added value and then testing their efficacy and safety are all relevant for the FF market. As a consequence, developing a new FF product requires more capital investment, a longer development process and more managerial capabilities by the firm.

In converging industries such as the functional food market, firms also face unusual challenges associated with adopting the combined knowledge, capabilities and technologies of two or more different markets. Moreover, to be successful in such a converging industry, firms may be required to learn and adopt a new kind of culture, as well as new organizational behaviours and new market structures. Converging industries require firms to create new competencies that enable them to be established in the new market. Because the FF market has only recently emerged, functional food firms must shape the market structure and market behaviour (apply market-driving market orientation) on their own rather than keep the status quo (accept existing consumers, preferences, market structure) or apply market-driven market orientation.[2] Moreover, in the course of the evolution of high-tech products, we often find research and development (R&D) processes separately executed from marketing activities. In contrast, in the field of fast-moving consumer goods, especially with regards to FF, the starting process of research and the ensuing process of development and launching (D&L) often are strongly connected. This significant difference results because the implementation of organizational and technological competencies for employing the necessary resources and activating the corresponding capabilities is not simply transferable from one case to the other.

In the next section, we consider how corporate strategy, in the context of functional food might be influenced by market dynamics (see Figure 10.1). In this regard, we emphasize tendencies toward market convergence as a central driving force. We then attempt to understand the linkage between a firm's general corporate strategy and the innovation processes for new FF products. In the following section, we analyze capabilities and organizational routines, with a primary emphasis on detecting the distinction between early (new product development) and late (product launch and marketing) phases of the innovation process. We conclude the chapter with a discussion and some implications.

Market Convergences

Since the early 1990s, a new converging market, the functional food market, has emerged and developed. This sector merges the food industry, the pharmaceutical industry (including over-the-counter [OTC] and prescription drugs) and the food-chemical industry. According to Functional Food Science in Europe (FUFOSE), a food can be regarded as functional if it is satisfactorily demonstrated to affect beneficially one or more target functions in the body, beyond adequate nutritional effects, in a way that is relevant to either an improved state of health and well-being and/or the reduction of the risk of disease.[3] This chapter does not focus on the issue of genetically modified functional food, because of its profile as outliner, especially in terms of the innovation process, consumers' perceptions and ethical concerns. Instead, most widespread and usually discussed examples of FF are the incorporation of natural substances, such as sitosterol and sitostanol (plant sterols), into conventional food products such as margarine, milk and yogurt, which provides a health benefit by lowering plasma cholesterol levels by 10–15 per cent. This cholesterol-lowering effect is more or less equal to the effect of a low-dose treatment of classic cholesterol-lowering drugs such

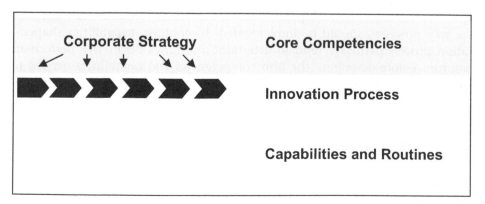

Figure 10.1 Discussion flow

as statins. FF offers a concrete example of a dynamic, innovative product category in the food and beverage industry domain. Dynamics on the product level also can affect the dynamics of markets and industries – or to be more precise, they may induce market convergences. The magnitude and acceleration of market convergences in the market are typical indicators for assessments of market dynamics. That is, market convergences are initiated and pushed by product and process innovations, which effect a change in market boundaries. In this context, the emergence of new products or technologies allows two or more economically separated markets to converge.[4]

In terms of convergence processes in high-tech industries, firms reinforce their efforts to find linkages between their own focal knowledge and external knowledge sources from other markets. In the case of the FF industry, firms face unusual challenges not only in terms of establishing a continuous market orientation concept but also in creating new competencies that will enable the firm's establishment in the new market. This convergence process between the food and pharmaceutical markets includes the increasing importance of the application of cooperative strategies for gaining the necessary market knowledge. Level of competition likely will increase in the future, caused by the new substitutive relationship of products from rival companies from both industries. Hence, the food industry displays the highly dynamic development process and potential in special submarkets, such as those for FF.

Both food- and pharmaceutical-based firms have basically the same opportunities for entering the FF market. However, different organizational adjustments are required for each kind of firms. For food-based firms, capabilities in terms of fast learning and understanding of the dynamics of consumers' needs and competitors' activities must shift into technological- and research-oriented capabilities, which usually are developed as a core competency in the pharmaceutical industry. The opposite adaptation process must take place in the pharmaceutical industry.[5] Rigidity in competence adjustments on either side hamper the fast and successful development of innovative functional food products. In a nutshell, in the functional food industry, where the boundaries of the food, pharmaceutical and food-chemical markets are merging, a strategic concept that combines capabilities and assets, usually developed and required by either food or pharmaceutical industries, is the salient basic requirement for corporate management.[6]

Core competence is not only a root system that provides nourishment, sustenance, and business stability but also the engine for new business development. Moreover, core competence provides a guide and pattern for the firm's diversification and market entry strategy.[7] The efficient firm's corporate core competencies should chart the direction and the way processes should be implemented. In contrast, capabilities shaped along innovation process pathways tend to determine the firm's distinctive performance in the long run. Before describing the firm competencies and capabilities needed to deal with the product development and launch/marketing phases, we address the innovation process in the area of functional food.

Innovation Process

When defining the requirements for the FF industry, we might underline at least two types of relevant innovation: product and process innovation. Product innovation refers to changes in the product/service an organization offers, whereas process innovation means changes in the ways in which a product/service is created and delivered.[8] These two types of innovation represent the main application domains of a FF firm's corporate strategic content. The product (innovation) development process consists of a distinct pathway of how to align the product development efforts of a food item and a drug, especially when they pertain to the development of a functional food product that will be marketed with a distinct health claim (see Figure 10.2). Basically, we argue that the product development process of a FF product represents a mixture of the development processes of all three integrated items, namely, food-chemical, packaged finished food and pharmaceutical products. Remarkably, these processes may run simultaneously or in parallel. The functional ingredient might be developed and manufactured within the food firm, by an external food-chemical firm, or by a different research centre. The development process – from mixing the functional ingredients to creating the finished food product to the complete manufacturing of the FF food product – follows the development processes of the food industry. Furthermore, to market this new product with a certain health claim, the firm must undergo a process similar to the development process for a drug. According to European Commission Regulation Nr. 1924/2006 (26 and 29), health claims must be based on generally accepted scientific evidence. Furthermore, health claims based on newly developed scientific evidence must undergo a different type of assessment and authorization. According to EFSA (Q-2007-066), substantiation of a health claim requires data from human studies that address the relationship between consuming the food/constituent product and the claimed effect.[9] That is, to use a new health claim for product marketing purposes, firms must conduct different studies in humans, similar to the design of phase II or III clinical studies in the development process of pharmaceutical products.

Process innovation principally consists of the technology and knowledge used to manufacture and deliver the firm's products or services. The innovation knowledge lifecycle, describing the use and creation of knowledge in the innovation process, includes the firm's focal (existing knowledge) and acquired knowledge. With regard to the innovation process of the convergence of the FF market, there are at least three salient types of knowledge cycles: community knowledge (for example, about consumers), organizational knowledge (for example, managing a team, motivating people) and working knowledge (for example, concrete working or task context). Furthermore, taking into account the nature of the convergence market, it is crucial for a firm to enable effective knowledge acquisition from both the original markets, food and pharmaceutical. It also is necessary to consider the adoption of new technologies, especially for food firms, which usually work with basic technology applications. Possible new technological applications deal more or less with the extraction processes for functional ingredients, purification of bioactive components used as raw materials and genomic approaches that may be reinforced in the pharmaceutical R&D and manufacturing processes.

Two potential processes running simultaneously can define innovation strategies: deliberate and the emergent processes.[10] In the pharmaceutical industry, innovative strategies dealing with new substance development are centralized with top management and characterized by top-down implementation processes. In that case, deliberate strategic processes follow. The deliberate strategy-making process is conscious, analytical and based on a rigorous set of data. In contrast, food industry firms, especially those concerned with product innovation, often use the emergent process to develop new products. This strategy bubbles up from within the organization, the cumulative effect of day-to-day

A. Phase of Drug Development Process in USA

B. Phase of Food Development Process

C. Phase of Functional Food Development Process

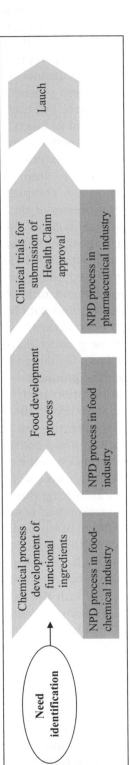

Figure 10.2 New product development processes[11]

prioritizations and investment decisions made by middle managers. With regard to the FF innovation development processes, it seems that firms may need to adopt a different development process strategy that combines both deliberate and emergent processes. In the beginning, during the early phase of the decision-making process regarding entering the new FF industry, a deliberate strategic innovation process should be applied to generate the firm's pre-eminent strategic concept, such as formulating its core competencies, which may include some considerations of new technologies to be applied, areas of research strategy, the core concept of a possible business enlargement strategy and so forth. In the strategic implementation phase, the second, emergent process may be more relevant. The proactive involvement of the middle managers in this stage can benefits in the form of flexibility and appropriateness with regard to the dynamics of market conditions, because middle managers work more closely with external stakeholders, such as consumers, suppliers, retailers and so forth, and the strategy will influence all these groups of persons.

Capabilities Relevant for Developing, Launching and Marketing a Functional Food

The development of the FF industry is quite challenging because of the difficult tasks associated with functional ingredient development (through internal development or external suppliers) and consumer relations, regulations and the marketing and selling of products. With regard to developing core competencies in the FF industry, we must distinguish between a competence for developing an innovative product and a competence for launching or marketing such innovative products. In the product development phase, market orientation, technology and knowledge adoption, as well as some networking principals, are relevant issues for discussion. For the product launch and marketing phase though, marketing knowledge/skills, marketing assets and, again, networking abilities with external institutions are important features. As mentioned previously, the FF industry accommodates the boundaries of both the food and pharmaceutical industries. Therefore, a core competence can be based on the convergent capabilities from areas of either industry (see Figure 10.3).

COMPETENCIES FOR DEVELOPING FUNCTIONAL FOOD

The following discussion focuses on salient capabilities that have a strong strategic impact on the meta level. The findings from previous new product development (NPD) literature that considers the strategic content dimension of core competencies are highlighted. Accordingly, we identify four types of (meta-) competencies that play key roles in the NPD processes for functional food.

Combining different component competencies (architectural competence)

A strong connection exists between platform development and leverage opportunities for a core competence as an architectural competence. Leveraging is the basic principle on which the platform concept is based.[12] As Kogut and Kim show,[13] strategic platform concepts get

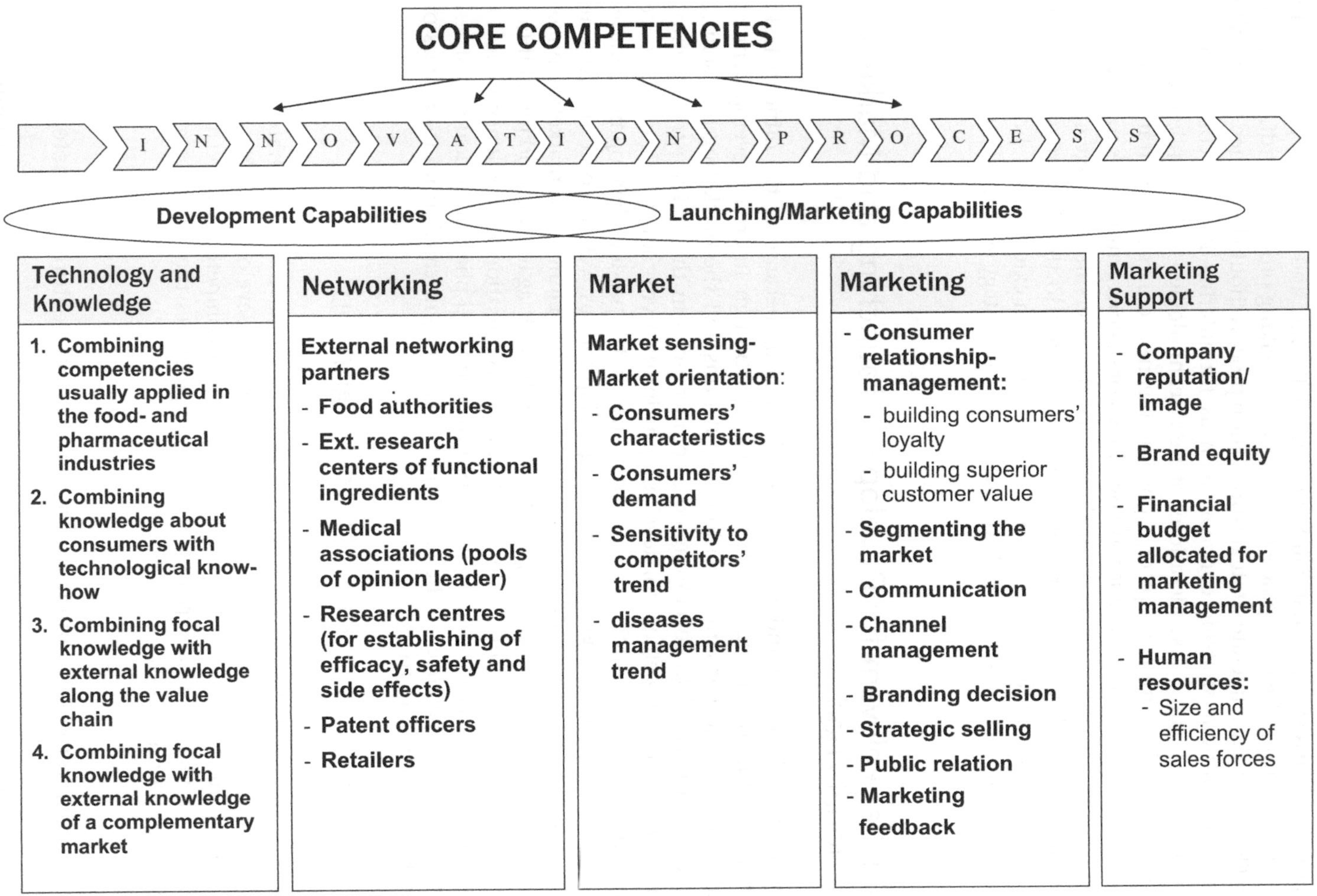

CORE COMPETENCIES

INNOVATION PROCESS

Development Capabilities

Launching/Marketing Capabilities

Technology and Knowledge	Networking	Market	Marketing	Marketing Support
1. Combining competencies usually applied in the food- and pharmaceutical industries	**External networking partners**	**Market sensing-Market orientation:**	- **Consumer relationship-management:**	- **Company reputation/ image**
2. Combining knowledge about consumers with technological know-how	- Food authorities - Ext. research centers of functional ingredients	- Consumers' characteristics - Consumers' demand	- building consumers' loyalty - building superior customer value	- **Brand equity** - **Financial budget allocated for marketing management**
3. Combining focal knowledge with external knowledge along the value chain	- Medical associations (pools of opinion leader) - Research centres (for establishing of efficacy, safety and side effects)	- Sensitivity to competitors' trend - diseases management trend	- **Segmenting the market** - **Communication** - **Channel management**	- **Human resources:** - Size and efficiency of sales forces
4. Combining focal knowledge with external knowledge of a complementary market	- Patent officers - Retailers		- **Branding decision** - **Strategic selling** - **Public relation** - **Marketing feedback**	

Figure 10.3 Development of core competencies for the functional food industry

applied in different industries, though their work has not been applied previously to the food industry. In the case of functional food, the components include nutrition solutions on the one hand and pharmaceutical solutions on the other hand. Both are required for the establishment of stable core competencies in the long run. Employing a platform strategy that provides a complex architecture of component products will become a critical success factor for leveraging core products into multiple end products. In addition, decisions about a firm's platform strategy must be aligned and coordinated with the cooperation strategy.

Combining knowledge about consumers with technological know-how

The development of new FF products requires extensive R&D efforts, combining technological knowledge (for example, new technologies required to explore a new ingredient, new interactions of multiple ingredients, testing metabolism patterns, safety and efficacy) on the one hand and customers' expectations about added value and on new product development requirements on the other hand. The great challenge is to combine both competencies. A simple, efficient way to leverage such knowledge and competencies involves a process of identifying further deployments for the application of existing products.[14] Often, alternative uses of a product are not discovered by the firms themselves but rather by the firms' customers.[15] Identifying alternative applications of a technology or competencies largely depends on a firm's absorptive capacity and ability to tap customers' absorptive capacity.[16]

Combining focal knowledge with external knowledge along the value chain

In general, firms deploy a whole spectrum of external technology acquisition possibilities, such that each mode of transfer has specific advantages and disadvantages.[17] The food processing and pharmaceutical industries employ very complex value chains, so they must ensure close and simultaneously flexible cooperation to establish focal knowledge.

Combining focal knowledge with external knowledge of a complementary market

Convergence processes enable firms to reinforce their efforts to find linkages with external knowledge sources from other markets. Cockburn and Henderson find that such linkages (gatekeepers) provide crucial success factors for knowledge creation in the pharmaceutical industry.[18] Powell and colleagues also find significant alliance relationships in the R&D used by firms in the biotech sector.[19] Food processing companies that intend to establish a core competence in the field of FF must connect their focal knowledge with that of the complementary market. In this regard, the analysis of market pulses has great importance for long-term R&D activities, because it builds experience in a sophisticated field of research. Furthermore, complementary assets exist in terms of market convergences between actors from the pharmacy industry and those from the food market.

To adopt the external knowledge related to knowledge of the supply chain and the complementary market, both pharmaceutical and food industry firms can use various strategic alternatives. External knowledge can be adopted both through non-institutionalized and institutionalized sources. Institutionalized forms of obtaining access to external knowledge include licensing, technology buying, contract R&D and cooperation forms such as joint ventures, virtual corporations, mergers, acquisitions and strategic alliance.[20] Knowledge also might be adopted by establishing a loose knowledge sharing community, such as through networking with other parties along the supply chain. Moreover, external knowledge can be obtained from non-institutionalized access, such as recruiting personnel directly from other companies or even competitors.

COMPETENCIES FOR LAUNCHING/MARKETING FUNCTIONAL FOOD PRODUCTS

The development of marketing capabilities and competencies provides a primary way for firms to achieve a competitive advantage.[21] Marketing competencies in a broader sense cover all aspects relevant for actualizing exchanges in the market for the purpose of satisfying human needs and wants. Marketing competencies also include customer relationship management skills, supply chain management and similar market-based assets.[22] Vorhies and Morgan compile at least eight distinct marketing capabilities that contribute to business performance:[23] product development, the ability to establish an appropriate pricing strategy, abilities to establish channel management, marketing communications, selling or the process by which firm acquires customer orders, market information management, marketing planning and implementation.

To fulfil these tasks, either the food or pharmaceutical company usually must develop a distinct marketing competency that combines capabilities and assets . This requirement is all the more important because of the distinctiveness of the marketing cultures of each industry (see Figure 10.4).

	Food Industry	Pharmaceutical (OTC and prescription drugs) Industry
Consumers	Familiar with product usage. Non-loyal, usually variety seeking or inertia buying process. Low involved consumers. Price sensitive.	Familiarity with and knowledge about the product vary among consumers. Loyal to certain product classes. Medium to highly involved consumers. Less sensitive to price (no bargaining power).
Marketing strategy	Distributed widely or mass market distribution via retailers. No specific segmentation is applied. Product communication is concerned with affective issues (taste, newness, originality) and end-consumer oriented. Communication is usually done via mass communication media and low involved media such as TV or radio spots. Consumers' need orientation and competitors' trend are based on developing a new product. Firm applies a product differentiation strategy.	Limited distribution channel, only via pharmacies and OTC drugstores. Segmentation is mainly based on product's indication. Communication is based on scientific findings or cognitive issues. Communication through personal communication or direct face-to-face promotion by firm's medical representatives. New products developed based on diseases management orientation. Product innovativeness is central issue for a NPD process.

Figure 10.4 Marketing strategies of food and pharmaceutical products

Having delineated the scope of marketing capabilities defined by previous authors, we also acknowledge that the discussion of the boundaries of marketing competencies should cover broader marketing terminology. 'Marketing' comprises not only the 4P strategy but also understanding of market potential, relationship management with stakeholders and product development. In this case, the broader scope of marketing consists of a mix of 'Kotlerian' marketing paradigms and the Nordic networking marketing concept[24] (see Figure 10.3).

The Kotlerian marketing concept for functional food

A firm's marketing capabilities should be targeted toward building and improving consumer loyalty, such that competencies dealing with launching and marketing a functional food product must focus on the development of knowledge and skills related to markets and marketing (in a narrow sense of the term). Market capability includes the firm's ability to understand the dynamics of consumers' psychological sets and thus develop an appropriate and efficient marketing strategy. The firm needs the ability to capture, react to and satisfy consumers' needs, wants and demands for healthy food and nutrition-based prevention. For the FF industry, it is also relevant to have an idea of the dynamic trends of disease management in the medical community. Understanding such trends will enable the firm to adapt its strategic planning appropriately. Moreover, understanding a competitor's innovative moves will enable the firm to develop its own strategic planning efficiently.

Marketing theories also emphasize that superior customer value orientation, as a major goal for establishing the firm's (marketing) capabilities and competencies, can be achieved though four core dimensions of customer value: attribute, benefits, attitude and network effects/communality.[25] We suggest that a firm's intelligent communication capability, as often used to market medicine and food products, is a marketing core capability necessary for successful marketing of a product in general. In this case, a superior communication platform related to cognitive attributes (scientific issues) and (health) benefits can be established without ignoring affective issues (for example. taste, convenience, frequently used as primary messages for food products). Communication messages should not just follow the medical communication model. Rather, the firm's capability to create a simple, easy-to-understand communication campaign based on scientific medical findings represents a salient strategic communication capability.[26] This combination platform also will be valid for media selection. For most consumers, sophisticated education will be more efficient for establishing awareness, knowledge and acceptance toward healthy foods, so television, radio or magazine advertising may not be sufficient to create trust in functional food (with its real and proved health benefits).[27] A long-term education program, as is often performed to promote pharmaceutical products, instead might be adopted. This strategy will be more or less relevant when the firm enters the mass market (which also includes unaware consumers).

Consumers of FF also have different levels of awareness and acceptance regarding a healthy life style and disease prevention, unlike their common familiarity with normal food products. Therefore, it will be more efficient to market FF to selected customer segments. A firm's ability to identify, select and develop such a market segment is the hallmark of the initial success of launching and marketing FF products.[28] Examples of stepwise segments include groups of patients (small segment but usually fully aware and motivated), consumers with high risk factors, consumers who have awareness of disease prevention, consumers who demand better health in general and finally all consumers (including the unaware group).[29] This capability should be combined with the ability to place the newly developed FF product appropriately within the existing product portfolio, especially by food firms, which often have a very complex product assortment. Such a capability in turn will influence the firm's strategy for branding, channel and positioning strategies and thus eventually the success of the product launch.

Nordic networking concept for functional food

A firm's (marketing) competencies for entering the FF industry should be based on its capability to establish networking with external stakeholders, on top of its market and marketing capabilities. From a marketing perspective, (Nordic) networking with others not only provides a possibility of intra-institutional knowledge exchange but also enables the implementation of marketing tasks. In this context, it seems important to establish a working relationship with external parties, such as members of the medical community who will act as opinion leaders. Medical associations often provide guidelines and suggestions about disease management and nutrition-related topics. Furthermore, establishing a good partnership with food authority bodies can be beneficial, especially for new product submission and the usage of health claims as a unique selling proposition.

These mentioned capabilities for launching and marketing an innovative FF product on the corporate level span and support multiple lines of business, combined with other firm-specific tangible assets such as marketing support (see Figure 10.3). They provide the baseline for establishing the firm's core competencies. Robust marketing support resources will accelerate the development of a firm's distinctive capabilities and make them superior and resistant to imitation.[30] We argue that such marketing supports are also relevant to ensure the success launch of a functional food product in the market.

Conclusions

The converging functional food industry merges the areas of food, pharmaceuticals (including OTC and prescription drugs) and the food-chemical industry. In these branches, firms face unusual challenges in terms of not only adopting their combined knowledge, capabilities and technologies but also creating new competencies that will enable the new firm to persist in the converging market. We note that both deliberate and emergent innovation processes can be applied appropriately within the FF industry. The deliberate strategic process is suitable for initiating the business, whereas the emergent process is more relevant for strategic implementation. When considering the requirements for a firm's capability and organization routines, we identify at least four types of (meta-) competencies that play key roles during NPD processes. During the launch and marketing phase, a firm should explore its capabilities beyond the 4P strategy. The firm's ability to establish networks with external stakeholders also is important for successful entry into the FF industry.

In terms of practical implications, this chapter suggests that entering a converging market requires a firm to evaluate its readiness (tangible and intangible) to adopt a new culture, new technology applications and new knowledge from other industrial branches. Pharmaceutical firms that intend to enter this market should adopt new market cognizance (that is, market sensing, including consumer orientation, competitors and internal infrastructure), which is usually indigenous to the food industry. In contrast, entering the FF market requires food industry participants to adopt advanced technology and scientifically-based knowledge regarding the nutrient–disease relationship, which is the main domain of the pharmaceutical industry. In the early innovation process, a firm should implement a deliberate process to build a comprehensive, grand entrance strategy that include determining its core competencies, business strategies and platforms for strategy familiarization. However, the participation of middle managers cannot be neglected, because they are the ones who work closely with and have direct access to consumers. Therefore, their involvement in the emergent process of strategy implementation is required to ascertain the flexibility and responsiveness to the dynamics of consumer demand. During innovative development phases, we suggest that to develop core competencies, firms should combine market orientation with technological know-how and focal knowledge with external knowledge along horizontal and vertical chains. The knowledge adoption process can be maximized by establishing sustainable networking channels with important stakeholders. Because of the different marketing cultures applied in both industries, accomplishing successful marketing tasks within a distinct marketing competency concept that combines the capabilities and assets usually developed by either food or pharmaceutical industries is a salient basic requirement. Marketing competencies should be based on a broader definition of marketing that moves beyond the 4Ps to include knowledge, market orientation and relationship management.

In addition, improving the capability to shape the market or apply market-driving market orientation remains an important strategic thinking pattern for entrance into the market because of the nature of the FF industry, which may force firms to shape their own market structure and market behaviour rather than retain the status quo. Based on the Kotlerian concept, we suggest that to launch a FF product, a food firm must develop internal competencies related to communication and segmentation early. It requires a different marketing approach than the one it would use to market conventional food products. Finally, we argue that in addition to establishing Kotlerian capabilities, a food firm needs to network with external players who influence or contribute to the success of a new, innovative, functional food product. By establishing this networking system, a food firm can integrate new, external knowledge. The concept of meta-competencies analysis, which combines the firm's focal knowledge and this newly adopted general knowledge, can improve the firm's internal capabilities.

References

1. Puspa, J. (2005), *Marketing Strategy for Newly Developed Functional Food Products: A Consumer-Company Based Analysis*. Germany: Dissertation.de. Verlag im Internet.
2. Berghman, L., Matthyssens, P. & Vandenbempt, K. (2006), 'Building competences for new customers value creation: An exploratory study', *Industrial Marketing Management*, Vol. 35, No. 8, pp. 961–973.

3. Margaret, A. (2002), *Concepts of Functional Foods*. Belgium: ILSI Europe Concise Monograph Series.

4. Stieglitz, N. & Heine, K. (2007), 'Innovations and the role of complementarities in a strategic theory of the firm', *Strategic Management Journal*, Vol. 28, No. 1, pp. 1–15.

5. Voigt, T. & Puspa, J. (2007), '*Competencies in new product development: The case of functional food*', in Atkan, C.C. & Balta, S. (eds), Perspectives on Business and Management. Selected Proceedings of the 3rd International Conference on Business, Management and Economics, Turkey.

6. Puspa, J., Voigt, T. & Kühl, R. (2007), *Marketing Competencies in the Era of Globalization: The Case of Functional Food*, Proceedings, MIC Conference, Slowenia.

7. Prahalad, C. K. & Hamel, G. (1990), 'The core competence of the corporation', *Harvard Business Review*, Vol. 68, No. 3, pp. 79–91.

8. Tidd, J., Bessant, J. & Pavitt, K. (2005), *Managing Innovation: Integrating Technological, Market and Organizational Change*. 3d edn. London: John Wiley & Sons.

9. EFSA (2007), 'Scientific and technical guidance for the preparation and presentation of the application for authorisation of a heath claim', EFSA-Q-2007-066. *The EFSA Journal*, Vol. 530, pp. 1–44.

10. Christensen, C.M. & Raynor, M.E. (2003), *The Innovator Solution: Creating and Sustaining Successful Growth*. Boston: Harvard Business School Press.

11. Wierenga, D.E. & Eaton, C.R. (2000), 'Phases of product development', Alliance Pharmaceutical Corp, http://www.allp.com/drug_dev.htm. 25.08.2007; McDevitt, C.A. (2002), *Product Development in the Industry/Chemical Process Industry*. Cambridge, MA: The System Design and Management Program, Massachusetts Institute of Technology.

12. Koruna, S. (2004), 'Leveraging knowledge assets: Combinative capabilities-theory and practice', *R&D Management*, Vol. 34, No. 3, pp. 505–516.

13. Kogut, B.W. & Kim, D. (1996), 'Technological platforms and diversification', *Organization Science*, Vol. 7, No. 3, pp. 283–301.

14. Koruna, op. cit.

15. Leonard, D. & Swap, W. (1999), *When Sparks Fly: Igniting Creativity in Groups*. Boston: Harvard Business School Press.

16. Cohen, W. & Levinthal, D. (1990), 'Absorptive capacity: A new perspective on learning and innovation', *Administrative Science Quarterly*, Vol. 35, No. 1, pp.128–152; von Hippel, E. (1986), 'Lead users: A source of novel product concepts', *Management Science*, Vol. 32, No. 7, pp. 791–805.

17. Barabaschi, S. (1992), 'Managing the growth of technical information', in Rosenberg, N., Landau, R., & Mowery, D.C. (eds), *Technology and the Wealth of Nations*. Palo Alto, CA: Stanford University Press.

18. Cockburn, I. & Henderson, R. (1997), *Public-Private Interaction and the Productivity of Pharmaceutical Research*. Cambridge, MA: NBER.

19. Powell, W.W., Koput, K.W. & Smith-Doerr, L. (1996), 'Interorganizational collaboration and the locus of innovation: Networks of learning in biotechnology', *Administrative Science Quarterly*, Vol. 41, No. 1, pp. 79–91.

20. Koruna, op. cit.

21. Day, G.S. (1994), 'The capabilities of market-driven organizations', *Journal of Marketing*, Vol. 58, No. 4, pp. 37–51; Srivastava, R.K., Fahey, L. & Christensen, H.K. (2001), 'The resource-based view and marketing: The role of market-based assets in gaining competitive advantage', *Journal of Management*, Vol. 27, No. 6, pp. 777–802; Srivastava, R.K., Shervani, R. &

Fahey, L. (1998), 'Market-based assets and shareholder value: A framework for analysis', *Journal of Marketing*, Vol. 62, No. 1, pp. 2–18.

22. Srivastava et al., 1998, 2001, op. cit.
23. Vorhies, D.W. & Morgan, N.A. (2005), 'Benchmarking marketing capabilities for sustainable competitive advantage', *Journal of Marketing*, Vol. 69, No. 1, pp. 80–94.
24. O'Driscoll, A., Carson, D. & Gilmore, A. (2000), 'Developing marketing competence and managing in networks: A strategic perspective', *Journal of Strategic Marketing*, Vol. 8, No. 2, pp. 183–196.
25. Keller, K.L. (1993), 'Conceptualizing, measuring, and managing customer-based brand equity', *Journal of Marketing*, Vol. 57, No. 1, pp. 1–22; Kotler, P. (2004), *Marketing Management: Analysis, Planning, Implementation, and Control*, 8th edn. London: Prentice-Hall International.
26. Puspa, op. cit.
27. Heasman, M. & Mellentin, J. (2001), *Functional Food Revolution*. London: Earthscan Publications Ltd.
28. Puspa, op. cit.
29. Ibid.
30. Day, op. cit.

11 Is There a Real Health versus Taste or Price Controversy in Food Marketing? The Case of Functional Foods

BY ATHANASIOS KRYSTALLIS,* MICHALIS LINARDAKIS†
AND SPYRIDON MAMALIS‡

Keywords

functional foods, responsible food marketing, children.

Abstract

Following well-designed research methodologies and adopting robust statistical analysis techniques, this chapter offers insights into some interrelated criteria for the acceptance of functional food (FF) by consumers (that is, healthiness, taste and price). Such criteria often confront one another to shape consumers' decision making, which highlights one of the most common controversies in food marketing: healthiness versus taste and price. The results indicate that FF should deliver health benefits above and beyond standard (high) quality (for example, nice taste for the 'right' price) required by consumers for any common food product. However, careful tailoring of a successful marketing mix must consider that for specific healthiness–taste–price combinations, there are no winning or losing product attributes; in these contexts, real controversies do not seem to develop in consumers' perceptions.

*   Dr Athanasios Krystallis, Senior Researcher, Food Marketing and Consumer Behaviour, Institute of Agricultural Economics and Policy (IGEKE), National Agricultural Research Foundation of Greece (NAGREF), 5, Parthenonos str, GR-141 21, Athens, Greece. E-mail: ATKR@asb.dk. Telephone: + 30 697 366 2975.

†   Dr Michalis Linardakis, Department of Agricultural Economics and Development, Laboratory of Agribusiness, Agricultural University of Athens, Greece. E-mail: mlinard@otenet.gr. Telephone: + 30 778 9114.

‡   Dr Spyridon Mamalis, Department of Economics, Division of Business Administration, Aristotle University of Thessaloniki, Greece. E-mail: mamalis@econ.auth.gr. Telephone: + 30 697 443 7621.

Introduction

For foods to be connected with healthy nutrition, they usually must be related to certain food elements, such as concentrations of fat, salt, dietary fibres or vitamins.[1] In addition to this conventional conceptualization of healthy foods and healthy nutrition, some products claim to have special beneficial effects on the human organism; they usually are referred to as nutraceuticals, pharma foods or functional foods, as well as nutritional foods, medical foods, designer foods or super foods.[2] The basic idea behind functional foods (FF) can summarized in Hippocrates's words: 'Let your food be your medicine, and your medicine be your food.'[3]

Despite the variety of terms used, an internationally acknowledged definition of FF remains lacking. According to the widely acknowledged European project FUFOSE (Functional Food Science in Europe): 'A food can be regarded as functional if it is satisfactorily demonstrated to affect beneficially one or more target functions in the body, beyond adequate nutritional effects, in a way that is relevant to either an improved state of health and well-being and/or a reduction of risk of disease,' and furthermore, FF must maintain their nutritional nature and easily become part of daily diets: 'Functional foods must remain foods, and they must demonstrate their effects in amounts that can normally be expected to be consumed in the diet. They are not pills or capsules, but part of a normal food pattern.'[4]

The lack of an official definition is one of the inhibitory factors for the growth, analysis and monitoring of FF markets.[5] However, consumers rarely define a food according to business or scientific terms related to health or nutrition information. On the contrary, FFs are regarded as a category that respond to consumers' ideas and beliefs about healthiness in food products.[6] According to Schmidt,[7] a lot of consumers receive the FF message; they feel that they have control over their health and know that certain ingredients can help reduce the risk of a disease or improve their health. As a result, this particular category seems to have many development prospects.

The category also has high diversity. From consumers' point of view, it is not a homogenous category, so consumer attitudes affect purchasing intentions for different functional products differently.[8] Among the most important criteria for FF acceptance are consumer health-related attitudes, such as personal motivations to engage in health preservation behaviours and confidence in health-related information such as health claims on food labels, and consumer-perceived characteristics of FFs, such as enhanced healthy images compared with the mediocre hedonic feelings resulting from consumption or the (often) higher price demanded for their purchase.[9]

The latter acceptance criterion constitutes the focal point of this chapter. By following well-designed research methodologies and adopting robust statistical analysis techniques, this chapter attempts to offer insights into consumers' perceived acceptance criteria for FFs' healthiness, taste and price. These criteria often conflict, forming opposing couplets such as healthiness versus taste or healthiness versus price.

Healthiness and Taste as Acceptance Criteria

Evidence in international literature related to these criteria (that is, consumer health-related attitudes and consumer perceptions of characteristics of FFs) is often contradictory. European consumers consider healthiness an important factor affecting their overall dietary choices, which aim to protect their health and avoid diseases.[10] One of the most obvious motives for consuming FFs is the preservation of good health, and research by van Kleef et al. (2005) shows that the relevance of a consumer's health condition in a product's health claim affects the intention to buy functional products.[11] Verbeke reports that a family member with a specific health problem affects the acceptance of FFs positively.[12] Furthermore, Jong et al. show that 40 per cent of the consumers in their sample consider the consumption of particular FFs an easy way to preserve their health, though in subsequent research, they find that 19 per cent of participants with health issues use drugs, 11 per cent use FF and only 5 per cent combine both therapies.[13] These findings show that consumer health issues do not necessarily support the consumption of FFs.

Information about the effects of diet on health and the means of communication constitute additional important factors of FF success.[14] Information about the health benefits of a food can increase the likelihood of its consumption and liking for it, though the results differ according to the effects on health; for example, claims related to improved physical and mental alertness are more likely to motivate consumption of a functional fruit juice.[15] Moreover, confidence in FFs, and consequently their acceptance by consumers, is largely the result of knowledge about the issue. The purchase likelihood of a FF increases when consumers combine the functional characteristic with the consequences of its consumption.[16] That is, consequence-related knowledge (for example, the product contributes to the preservation of health) increases the likelihood of consumption more than does attribute-related knowledge (for example, increased calcium level). In addition to trust in the health claims and communication means, trust in a FF depends on the type of the base product, such that the base product contributes to whether consumers perceive the FFs as healthy by allowing potential buyers to trust the health claims more when the basic carrier has a positive overall image and a history rich in health claims (for example, yoghurt).[17]

Organoleptic attributes of food, especially taste, dominate food choice.[18] The development of FFs usually demands modern technology to modify the composition, which may have negative consequences on taste and other organoleptic attributes and thus would relate inversely to their acceptance.[19] Even if the greater functionality does not affect their organoleptic attributes, specific levels of saltiness, bitterness or sourness result from enhancements of foods with bioactive or natural ingredients.[20] Consumers may express discomfort about the mediocre taste performance by such foods.[21] For example, the severity of a functional juice's bad flavour decreased the probability of its consumption, despite persuasive claims about its advantages.

Thus, FFs cannot be inferior in taste compared with their conventional counterparts. Consumers seem to believe that perceived health benefits and pleasure are not by definition inversely related. This point lies at the centre of the healthiness versus taste controversy for functional foods and constitutes a great challenge for food marketing, especially when the cost-related concerns that accompany any (food) choice also get taken into consideration.

Aims and Objectives

The success of FFs depends on how well they satisfy various, frequently contradictory consumer needs. Using this claim as a point of departure, this chapter attempts to offer some consumer behaviour-related insights of more psychometric nature in relation to the healthiness versus taste or price controversy (that is, what motivates consumers to buy FF, and how consumers connect their perceived knowledge about FFs to knowledge about themselves). Such 'abstract' insights can be particularly useful as a starting point for designing more targeted consumer research within the wider marketing and new product development strategy of a food firm.

In addition, this research attempts to illustrate how such 'abstract' psychometric knowledge can illuminate the healthiness versus taste or price controversy from a more pragmatic, marketing-oriented point of view, to provide concrete suggestions for marketers and new product developers. The functional children snacks category (chips[§] and croissants) provides a suitable context for this task, because the difference between the person who consumes (that is, children) and the person who usually purchases the product (that is, parents) imposes further challenges on food marketers. This study therefore examines how the technical translation of healthiness into different types of functionality (or 'levels') might influence parents' selection;[¶] investigates where functionality types, as opposed to taste or price variations, lie along a classification of parents' selection criteria; and determines if the dilemma between healthiness and other criteria really exists in consumers' perceptions.

In line with this twofold aim, this study uses means–end chains (MEC) and discrete choice (DC) methodology. The research therefore consists of two phases, a qualitative MEC (phase I) and a quantitative DC (phase II).

Research phase I attempts to identify the full range of FF attributes deemed important by consumers during their selection process; investigate the motives behind FF selection and how strongly some of these motives relate to consumers' health-related behaviour; and explore consumers' overall cognitive structures by designating the links that connect important FF attributes, consumers' knowledge about FFs' consumption benefits and consumers' knowledge about themselves.

Building on the findings from phase I, phase II undertakes an investigation of the exact meaning of healthiness by measuring the utility derived by various types of functionality in consumers' selection process, in contrast with the utility delivered by various levels of the selection criteria under examination (for example, taste and price). It also measures consumers' willingness to pay (WTP) for specific combinations of functionality and taste and thereby compares healthiness with taste and price in a ranking of consumer selection criteria importance.

Qualitative Means-end Chain (MEC) Analysis (Research Phase I)

METHODOLOGY

A MEC model seeks to explain how a product or service selection facilitates the achievement of consumers' desired end states. Means are objects (product or service attributes) with which people engage, and ends are valued states of being, such as happiness, security

---

§   In the UK and much of Europe, 'chips' are more commonly known as 'crisps'.

¶   Intrinsically, snacks are not healthy, and their supplementation with calcium or vitamins does not make them healthy in a traditional sense. From a nutrition point of view, it might be more appropriate to consider supplemented food. However, this experimental consumer marketing research aims to incorporate two opposite dimensions in the same experiment: the intrinsic unhealthiness of snacks and the extrinsic healthiness delivered by supplementation with functionality attributes. This inherent opposition fits the wider objective of the book, as well as the specific objectives of this chapter.

or accomplishment. In turn, MECs are hierarchical cognitive structures that relate consumers' product knowledge to their self-knowledge. Consumers regard products as more self-relevant or involving to the extent that their product knowledge pertaining to attributes and functional consequences connect through MECs to their self-knowledge with regard to desirable psychosocial consequences and values.[23] Laddering, the qualitative research method that accompanies MECs, refers to an in-depth, one-on-one interviewing technique for developing an understanding of how consumers translate the attributes of products into meaningful associations with respect to themselves. Laddering uses a tailored interviewing format, primarily a series of directed probes, typified by 'why is this important to you?' questions, with the express goal of determining sets of linkages between the key conceptual elements across the range of attributes (A), consequences (C) and values (V).[24]

To achieve the study objectives, the qualitative phase used a homogeneous sample of younger, well-educated consumers, aged between 35 and 44 years, responsible for food buying in their household. Vannoppen et al. claim that convenience samples are acceptable for laddering research.[25] One criterion for sample inclusion was that respondents should be able to 'speak out' knowledgeably about the product. A pilot questionnaire provided data that served as the basis for designing the MEC application.

Pilot study

The pilot study used a homogeneous sample of 60 consumers who had attained high educational levels (university graduate or higher) and were of early middle age (35 and 44 years). These respondents must have purchased a FF during the month prior to the pilot study (January 2006) and actively participated in food buying decisions for their household. These respondents completed a short pilot questionnaire in the course of a personal interview that lasted no more than 10 minutes. The pilot questionnaire featured a master list of 58 FF attributes, covering seven concrete and abstract categories of a hypothetical FF's marketing mix (Table 11.1). This master list was based on previous literature related to FF acceptance criteria, as well as past MEC applications for food. The participants evaluated each attribute on a five-point Likert type importance scale, anchored at 1, 'not important at all', and 5, 'very important'. Attributes that earned scores of 4 or 5 for at least two-thirds of the participants constitute the short list of attributes used for the subsequent evaluations in the MEC study.

MEC study

The MEC study in March 2006 used a homogeneous sample of 40 consumers; the requirements for participation are similar to those for the pilot sample. The laddering interview process began with an introduction to inform the respondent of the purpose of the interview and ensure understanding of the term 'functional' through reference to the FF definition described at the beginning of this chapter. In the consequences and values elicitation phase, respondents first considered buying a FF, then received the short list of the most important FF attributes identified in the pilot study (Table 11.1). Next, the respondent evaluated each attribute in terms of its importance when buying a FF,

Table 11.1 Attributes list, qualitative MEC phase

Type of attributes[25]	Pilot sample, n = 60	Main sample, n = 40
Perceived quality attributes		
Pure product	96.7%*	45%**
Safe food	100	15
Trust in brand name	83.3	45
Economical in use	56.7	-
Quality product	96.7	40
Part of daily nutrition	76.7	20
Healthy product	100	50
High technology product	50	-
Natural product	96.7	15
Appearance and package-related attributes		
Package size	23.3	-
Environmental friendly package	60	-
Nice package	33.3	-
Practical package	70	10
Different package than the conventional product	20	-
Aluminium can package	23.3	-
Package made of glass	33.3	-
Plastic package	20	-
Paper package	26.7	-
Organoleptic attributes		
Strong aroma	13.3	-
Neutral aroma	20	-
Light aroma	23.3	-
Nice taste	90	40
Neutral taste	10	-
Specific texture	40	-
Specific colour	20	-
Label information attributes		
Information about health/functionality claims	86.7	15
Nutritional value	86.7	20
Quality assurance (for example, ISO/ HACCP)	90	20
Best before date	100	70
Packaging date	80	15
Country of origin	86.7	60
Functionality-related attributes		
Antioxidant ingredients	63.3	15
Removed dangerous ingredients	86.7	15
Fortified ingredients	63.3	0
Added calcium	80	15
Added vitamins and minerals	63.3	15
Added fibre	80	15
Added phosphor	60	0
Added functional ingredients	43.3	0
Low cholesterol level	86.7	55
Low saturated fatty acids content	70	30
Necessary for personal well being	83.3	15
Enforces body defence	96.7	55
Reduces cardiovascular disease risk	83.3	60
Provides more energy	70	15
Provides proved health claims	90	20
Contains probiotics	60	15
Contributes to digestion improvement	86.7	40
Contributes to vision improvement	53.3	5
Contributes to good physical condition	96.7	60
Contributes to osteoporosis prevention	86.7	40

Table 11.1 *Concluded*

Type of sttributes	Pilot sample, n = 60	Main sample, n = 40
Price-related attributes		
Value for money	93.3	15
Same price with conventional products	73.3	5
Price higher than the conventional product	56.7	-
Price lower than the conventional product	70	0
Brand name attributes		
Promotion campaign	43.3	-
Known producing company	76.7	15
Familiarity with the brand name	76.7	20

*   Attributes deemed important when at least 66.7 per cent of the sample scores it as a 4 or 5 in the pilot study.

**   Attributes deemed important when at least one participant (5 per cent) scores it as a 4 or 5 in the main study.

using the same importance scale from the pilot phase. Finally, the attributes deemed most important served as the starting points for the laddering technique.

Through these attributes and interviews, each respondent subconsciously connected product attributes with consequences and/or personal values. The time necessary to complete this task varied between 40 to 60 minutes, depending on the respondent's ability to express themself and their conscious involvement in the functional food purchasing process.

ANALYSIS

The responses of the 40 consumers first were coded into common A-C-V categories. In terms of consequences and values, 17 consequence codes and 9 value codes emerged. The coding process was based on codes from prior FF and food-related MEC studies (Table 11.2). The average number of codes elicited per consumer was 48, forming 10.6 ladders on average. After coding, the analysis continued with the use of the MEC Analyst software, which provides an interactive system of data importation, inserting multiple A-C-V ladders per participant in the form of the relevant codes. After each ladder was input, the data analysis created a tree diagram (hierarchical value map, or HVM) that illustrated the sample's cognitive structure for FFs. The HVM represents 76 per cent of the direct A-C-V links reported by at least 5 persons during the laddering interviews (Figure 11.1).

Quantitative Discrete Choice (DC) (Research Phase II)

METHODOLOGY

In a DC experimental task, respondents indicate their preference for experimentally varied products or product profiles. This preference is expressed in the form of a choice between two, three or more product alternatives, for a predefined number of rounds imposed by the research design. The choice is statistically represented with a multinomial dependent variable, such as [1 0 0] to represent the first of three alternatives being chosen by a respondent. Each product alternative consists of pre-selected product attributes that serve

Table 11.2 List of consequences – values codes elicited from the main sample (n = 40) and relative literature examples, qualitative MEC phase

Consequences	Number of times mentioned by at least one respondent	Literature examples
Functional		
1. Body needs	< 5	1. Body needs[26]
2. Easy to choose – use	< 5	2. Easy to buy/use[27]
3. Eating – living healthy	**20**	3. Eating healthy[28]
4. Physical appearance	< 5	4. Physical appearance[29]
5. Physical health improvement	13	5. Physical well-being[30]
6. Product choice – consumption	11	6. Will buy/ use[31]
7. Bones' health promotion	8	7. –
8. Health promotion	**20**	8. Promotes health[32]
Psychological		
9. Eating enjoyment	**14**	9. Eating enjoyment[33]
10. Economic efficiency	< 5	10. Monetary considerations[34]
11. Feeling good	**17**	11. Feel good[35]
12. Performance improvement	**15**	12. Improved performance[36]
13. Knows what to get (price/ quality)	< 5 < 5	13. Knows what to get[37] (price-quality)
14. Need – desire satisfaction	< 5	15. Quality of life[38]
15. Quality of life	< 5	16. Time saving[39]
16. Time saving	**16**	17. Trust[40]
17. Trust		
Values		
Instrumental		
18. Tradition	<5	18. Tradition[41]
Terminal		
19. Belonging	<5	19. Belonging[42]
20. Good health and long life	<5	20. Good health and long life[43]
21. Inner harmony	**11**	21. Inner harmony[44]
22. Pleasure	**14**	22. Pleasure[45]/Hedonism[46]
23. Psychological satisfaction	9	23. Psychological
24. Security	**16**	24. Security[48]
25. Self-confidence	<5	25. Self-esteem[49]
26. Self-fulfilment	<5	Achievement[50]
		26. Self-fulfilment[51]

as independent variables in the multinomial model. Such modelling has been used in diverse consumer behaviour fields, including shopping behaviour, housing choices, travel mode choices and health care service preferences,[52] as well as food-related research.[53]

To achieve the research objectives, the DC experimental methodology focuses on the Greek snack market, specifically, two different snacks for children: potato chips and croissants. The rationale behind this selection centres on the perception (by children) of taste versus (parents) healthiness of the product. New product development would want to take advantage of the characteristics of the market, namely, that children's snacks are

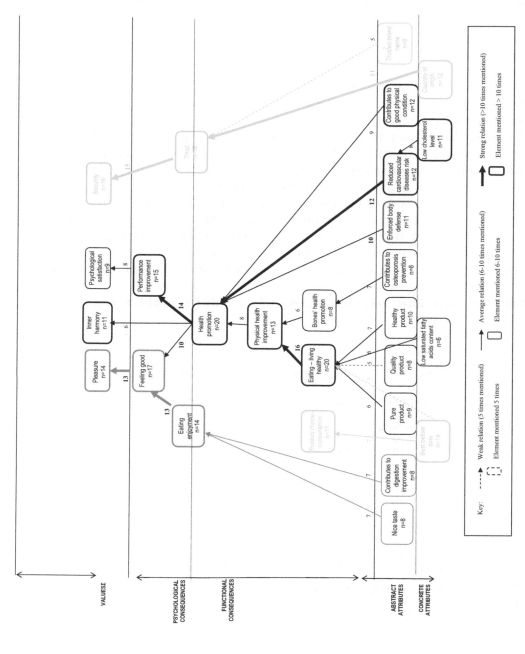

Figure 11.1 Healthiness vs hedonism HVM (cut-off level: 5, n = 40), qualitative MEC phase

consumed and purchased by different persons in the household, who have varying needs and wants; that parents are concerned about their children's health; and that parents are susceptible to their children's wants. Parents likely choose a snack whose taste their children prefer, so to what extent should this preferred snack also be healthy? Moreover, what is a fair price for the 'right' combination of healthiness and taste?

Survey design

In the DC task, respondents expressed their preferences for several experimentally developed snack profiles. The functional product attributes selected for inclusion in the experimental design come from the results of the MEC phase and reflect concrete elements of the snacks' marketing mix. Moreover, the experimental attributes reflect a health versus taste or price controversy: Pleasure is expressed as flavour variations, healthiness as various functionality types and price (premium above standard price) is the expected value of each flavour–functionality combination (Table 11.3).

Consumer preferences are expressed in the form of a choice between two alternatives of the same snack. This choice is statistically represented with a binomial dependent variable, such as [1 0] if the first of two alternatives is chosen. Each pair of alternative snacks appears in visual form on a card (Figure 11.2), with coloured pictures of the two alternatives, to increase the realism of the choice task. The number of cards was defined

Table 11.3 Attribute and attribute levels for products' profiles, quantitative DC phase

	Chips		
	COST	**PLEASURE**	**HEALTHINESS**
	Price, €/unit	**Flavour**	**Functionality**
Usual	0.90 1.00 1.10	Classic Oregano Barbeque	None
Functional	1.20 1.30 1.40	Classic Oregano Barbeque	Enriched with Ca Enriched with vitamins Enriched with Ω-3 fatty acids Enriched with fibre

	Croissants		
	COST	**PLEASURE**	**HEALTHINESS**
	Price, €/unit	**Flavour**	**Functionality**
Usual	0.40 0.50 0.60	Chocolate Cream Marmalade	None
Functional	0.70 0.80 0.90	Chocolate Cream Marmalade	Enriched with Ca Enriched with vitamins Enriched with Ω-3 fatty acids Enriched with fibre

by the orthogonal design facility in SPSS v.13.0. In total, 16 cards were designed per snack, equal to the number of attributes under examination. The order of the cards shown was randomized per respondent to decease any response bias effect (Table 11.4).

On each card, alternative A is always a common snack with no functionality, already available in the market, and alternative B is always the functional snack, a hypothetical product not available in the snack market yet. The three price levels for each alternative A are real prices in various food outlets, whereas the price levels for each functional alternative B are hypothetical prices calculated as price premiums imposed beyond the price of the common snacks (+30 per cent for chips, +60 per cent for croissants). The flavour levels also represent existing product alternatives in the market.

Card type 1: Common versus functional croissants. Type of functionality: added calcium

Card type 2: Common versus functional potato chips. Type of functionality: added vitamins

Figure 11.2 Examples of visual stimuli for data collection in quantitative DC phase

Table 11.4 **Examples of card designs shown to consumers for data collection, quantitative DC phase**

Chips A	Chips B
Cost €0.90 Plain flavour	Enriched with fibbers Cost €1.40 With oregano flavour
☐	☐
None of the above ☐	

Croissant A	Croissant B
With marmalade flavour Cost €0.60	Enriched with Ω-3 fatty acids Chocolate flavour Cost €0.90
☐	☐
None of the above ☐	

Data collection, sample and questionnaire

The sample selected for the quantitative phase of the survey consists of 120 parents with children of an age suitable for snack consumption (parents' mean age: 39.2 years, SD = 6.7), who had purchased at least one children's snack during the month prior to the survey (May 2006) and were aware of the meaning of 'functionality' in a food context. Each respondent evaluated four chip cards and four croissant cards. The total number of observations per product alternative thus is 480 (120 × 4) for both snacks. Most of the respondents are women, well-educated, and of high to average income levels. Overall, their profile resembles that of the sample used in the MEC phase. To collect background information, a structured questionnaire accompanies the cards each respondent evaluated (Table 11.5). Before completing the DC task, the respondents received short explanations of the health benefits of each of the four functionality levels included in the experiment, as well as detailed instructions about how to proceed in the DC task.

Analysis

The analysis of the binomial logit model uses Limdep v.7.0 (Table 11.6), with a significance level of $\alpha = 0.05$. The numbers in parentheses in Table 11.6 show the standard errors of the respective coefficients. The fit of the three models can be evaluated by the percentage of correct predictions for the actual choices of the respondents. In the present analysis, these percentages are 53.3 per cent and 60.4 per cent for chips and croissants, respectively, which indicate that the models capture at least the average (50 per cent) preferences of respondents.

Table 11.5 Respondents' purchasing behaviour, purchasing criteria, involvement and perceptions of children's snacks

1. I generally buy snacks for my children regularly...

Yes	No
68.3	31.7

2. I usually buy snacks for my children...

< 1/week	1/week	> 1/week	Exceptionally – almost never
36.7	23.3	8.3	31.7

3. I buy for my children...

	Often	Sometimes	Never
1. Chips	24.2	35.8	8.3
2. Biscuits	35.0	28.3	5.0
3. Croissants	24.2	31.7	2.5
4. Savoury puffs	5.8	30.8	31.7
5. Crisps	26.7	27.5	14.2
6. Pop-corn	10.0	29.2	29.2
7. Chocolate bars	31.7	31.7	5.0

4. I buy snacks for my children from...

	Often	Some times	Never
1. Corner shop	14.2	30.0	24.2
2. Supermarket	30.0	20.0	18.3
3. Hypermarket	28.3	23.3	16.7
4. Street kiosk	10.8	23.3	34.2

5. When I buy snacks for my children, I pay attention to...

	Strongly agree	Agree	Neither nor...	Disagree	Strongly disagree
1. Taste nice	19.2	32.5	10.8	4.2	1.7
2. Be innovative	1.7	10.0	23.3	23.3	10.0
3. Have seen them on TV	3.3	10.8	21.7	24.2	8.3
4. Be less fatty	17.5	27.5	15.0	6.7	1.7

Table 11.5   *Concluded*

	Strongly agree	Agree	Neither nor...	Disagree	Strongly disagree
5. Have nice appearance	2.5	11.7	21.7	25.8	6.7
6. Be as healthy as possible	35.8	25.0	5.8	0.8	0.8
7. Be as cheap as possible	0.8	10.8	16.7	21.7	18.3
8. Be what children ask for	12.5	25.0	18.3	7.5	5.0

6. When I buy snacks for my children...

	Strongly agree	Agree	Neither nor...	Disagree	Strongly disagree
1. I look at the ingredients written on the label	26.7	18.3	15.0	3.3	5.0
2. I compare them with other snacks before buying	11.7	18.3	21.7	12.5	4.2
3. There are specific brand names that I prefer	15.8	20.0	16.7	10.8	5.0

7. In general, I consider the children snacks that are in the market today to be...

	Strongly agree	Agree	Neither nor...	Disagree	Strongly disagree
1. Nutritious	4.2	12.5	21.7	26.7	3.3
2. A convenient solution	12.5	30.0	16.7	7.5	1.7
3. Unhealthy	19.2	31.7	15.0	1.7	0.8

Results and Discussion

The sample of respondents (early middle-aged, educated consumers) in the MEC phase considers perceived quality-related attributes (see Table 11.1), such as the product being pure, safe, healthy and high-quality, extremely important for their FF buying choices. The MEC sample also attaches great importance to label-related attributes, such as the best before and packaging dates, as well as the type of health or functionality claims,

Table 11.6   Estimation of the parameters of the binomial logit models

	Chips		Croissants	
Valid responses	(370/480), 77.7%		(374/480), 77.9%	
Correct predictions, %	53.3%		60.4%	
Attributes	Coefficient	p-value	Coefficient	p-value
Price	-1.748 (0.663)	**0.008**	-0.402 (1.010)	0.690
Enriched with Ca	0.919 (0.255)	**0.001**	0.909 (0.403)	**0.024**
Enriched with vitamins	0.860 (0.249)	**0.001**	1.079 (0.384)	**0.005**
Enriched with Ω-3 fatty acids	0.894 (0.249)	**0.001**	0.894 (0.386)	**0.021**
Enriched with fibre	1.215 (0.257)	**0.001**	1.125 (0.384)	**0.003**
Oregano	-0.224 (0.133)	0.092		
Barbeque	-0.270 (0.129)	**0.036**		
Cream			-1.059 (0.218)	**0.001**
Marmalade			-0.667 (0.212)	**0.002**

quality assurances and nutritional value. They also consider price attributes (value for money and price the same as or lower than conventional products), brand name attributes (familiarity with the brand and known producing company) and nice taste.

The significance of these attribute categories highlights that even if FFs are perceived as a special category of foods, they remain foods. Consumers must be certain about the FFs' quality and have at their disposal all relevant information to judge the food before they proceed to evaluate its functional benefit, which complements the product's (high) expected quality.

Perhaps the most interesting result from this study relates to the category of functionality-related attributes, in which consumers of the MEC phase exhibited substantial interest, including enforcement of the body's defences, contributions to good physical conditions, the provision of health claims and the removal of dangerous food ingredients. The same consumers assign importance to more concrete functionality attributes, such as low cholesterol levels, contributions to improve digestion, reduced cardiovascular disease risk and low saturated fat content. Overall, these results indicate that consumers in their 40s emphasize disease prevention attributes, such as lower cholesterol and cardiovascular disease risk reduction – a key finding of this research.

To obtain insights related to the FF-related buying motives of consumers and the way FF consumption-relevant knowledge gets stored and organized in their cognition, the analysis also focuses on the main outcome of the study, the HVM.

The most important cognitive construct of the HVM consists of consumers' concern about preventing risks to their health (143 direct links, or 66.8 per cent of the links in the HVM; see Figure 11.2). This construct consists of several health promotion-related benefits from the consumption of a FF, such as health promotion, physical health improvement, and eating and living healthy. Consumers between 35 and 44 years of age take advantage of a variety of product attributes related to their health status preservation, in the sense of risk prevention (low saturated fat content), but also in terms of perceived quality health-related attributes (pure) and label information health-related attributes (best before date). In this sense, the direct link between the cardiovascular disease risk reduction attribute and the health promotion functional benefit is among the most powerful. Satisfying this particular need drives both the feeling good and, predominantly, the improved performance psychological consequences, which form the core link of consumers' cognition about their preference for health-related food attributes, including functionality, and more abstract personality constructs, such as the values of pleasure, inner harmony and psychological satisfaction.

In addition to the health-related cognitive construct, the HVM includes eating hedonism (27 direct connections, 12.6 per cent of links), which centres around the link between eating enjoyment and feeling good benefits, which itself results not only from the consumption of a food with a nice taste but also a food that improves digestion.

Thus, through FFs, consumers try to satisfy mainly their need to improve or preserve their health status. Educated consumers in their 40s rely not only on attributes typical of common food products (for example, practical use, fair price, nice taste, high perceived quality, trusted brand name) but also on more targeted health-related attributes. Thus, they perceive FF products as a hybrid of food and medicine. The results of the MEC study further indicate that these consumers are not willing to lose taste or convenience or risk trust in unknown brands to purchase foods with functional characteristics.

More than two in three parents purchase snacks for their children once per week or less (see Table 11.5). The most important criteria for children snacks' selection for parents are as healthy as possible, taste nice, less fatty and what the children want. Among parents who purchase snacks for their children regularly, most pay attention to the ingredients written on the label; however, familiarity with snack purchasing is weaker, because only half of the regular buyers have preferable snack brand names, and less than half compare different snacks before buying. Overall, the majority of parents perceive snacks as an unhealthy but convenient solution.

The predictive ability of the binomial logit model is very satisfactory (see Table 11.6), which supports the robustness of the DC design. In the case of functionality, the common product serves as a reference category. Hence, the four coefficients (per functionality type) indicate the difference in the utility compared with that of the common product. The four components of functionality are all statistically significant for both chips and croissants. In addition, the signs of the coefficients are positive in all cases, indicating that functionality increases the utility of a snack in comparison with the common product alternative. The results further show that enriching chips with fibre or calcium delivers higher utility than enriching them with vitamins or Ω-3 fatty acids. For croissants, the higher utility comes from enrichment with fibre or vitamins.

The coefficients of price also are statistically significant for chips, but they reveal a negative sign, indicating that more expensive snacks lead to lower utility. However, for croissants, price is not statistically significant.

In the case of flavour, the classic flavour (plain) is the reference category for chips. Hence, the two coefficients (oregano and barbeque) indicate the difference in utility compared with that of the classic flavour. The barbeque flavour reduces the utility significantly, whereas the oregano flavour does not influence utility. Finally, the classic flavour is the reference category for croissants; the coefficients of the remaining two categories are statistically significant with a negative sign, indicating that the chocolate flavour offers greater higher utility in croissants, whereas the flavour with the lowest utility is cream.

These results indicate that the majority of the parents in the sample understand children's snack purchasing. The conclusions from this assessment of the criteria that parents use to purchase snacks for their children merit special attention (see Table 11.5). Several of these observations are encouraging with regard to the market potential of functional children's snack, including the unhealthy image of existing children's snacks, which prompt parents to want a new snack that is as healthy as possible, for which they would pay substantial attention to the ingredients on the label. What the children want and cost considerations are less important purchasing motives than the snack's healthiness, precluding latent demand for health-enhanced children's snacks.

Functionality is a statistically significant attribute for both snacks, which has substantial marketing implications for the children's snacks market. The functional snack alternative is always perceived as different and of greater utility than its common counterparts. At the individual snack level, fibre enrichment is the most important for both chips and croissants.

This information should be coupled with the finding that price is statistically significant only for chips. Thus, when functionality is coupled with price, the greatest market potential pertains to functional croissants, for which a 60 per cent price premium is not enough to activate consumers' price sensitivity. This result may be because croissants are lower priced than chips – apparently too low to stimulate price sensitivity effects. For chips, the 30 per cent premium seems to exceed what consumers are willing to pay for the functional version of an already high-priced snack.

Thus, (early middle-aged, well-educated) parents who want to purchase functional snacks for their children would be willing to pay no more than 30 per cent above standard price for functional chips but even more than 60 per cent above the standard price for functional croissants. The most preferred type of functionality is enrichment with fibre, then vitamins or calcium. Finally, the plain flavour in chips and the chocolate flavour in croissants is the most preferred variation for parents. Assuming that, in terms of taste, parents' choice reflects their children's preferences, functional snack developers can assume that flavour variations in chips and croissants do not substantially alter product choice; rather, such variations in the marketing mix can have negative results on consumer preferences.

To contrast healthiness with the other selection criteria, the binomial logit model can be constrained in terms of flavour (plain for chips and chocolate for croissants) and type of functionality (0 = no functionality, 1 = any type of functionality). In other words, functionality, irrespective of its type, contrasts with the flavour variation with the highest utility per snack type for the given price levels in the DC experimental design.

As the results of the model for chips show (Table 11.7), all the selection criteria used in the experimental design are statistically significant, indicating that no classification rank can be established for chips. In the case of croissants, both healthiness and taste criteria are statistically significant, and only price is indifferent. Overall, the results of the MEC study thus are justified again: consumers do not compromise on taste for healthiness, even when they must decide about the health of their children. It is thus possible that specific product attribute combinations do not provoke the marketing controversies such as healthiness versus taste or healthiness versus price.

Conclusions and Managerial Implications

The aim of this chapter has been to offer insights into interrelated criteria (that is, healthiness, taste and price) for the acceptance of FFs by consumers of a certain socio-demographic profile (that is, early middle-aged, well-educated). Such criteria often compete to shape consumers' decision making, highlighting the much talked-about healthiness versus other controversy in food marketing.

The qualitative study in research phase I offers consumer behaviour-related insights of a more psychometric nature, related to the concept of healthiness versus taste or price, with the FF selection process as an example. Phase I employs MEC analysis to define the most important FF attributes that affect consumers' purchasing decisions. The study further aims to obtain insights into the FF-related buying motives of consumers and design a hierarchy of consumption-relevant cognitive structures to explain FF-related purchasing behaviour.

The FF attributes with the greatest importance for (well-educated, early middle-aged) consumers in the MEC sample are those that typically affect the choice of a common food

Table 11.7 Estimation of the parameters of the constrained binomial logit model

	Chips		Croissants	
Valid responses	(365/480), 76.0%		(374/480), 77.9%	
Correct predictions	53.3%		59.8%	
Attributes	Coefficient	p-value	Coefficient	p-value
Price	-1.717 (0.659)	**0.009**	-0.386 (0.991)	0.696
Enrichment	0.961 (0.213)	**0.000**	0.984 (0.325)	**0.002**
Plain flavour	0.243 (0.105)	**0.020**		
Chocolate flavour			0.857 (0.176)	**0.000**

product, and the most important fuel the healthiness versus taste or price controversy, such as functionality-related information, nice taste, right price and overall healthiness. The most important FF purchasing motive is consumers' interest in their health status. Nevertheless, consumers of this profile do not appear willing to compromise taste and overall eating enjoyment for the health benefit of FFs, irrespective of the type.

The results of the MEC analysis should promote some creative solutions to product positioning problems. Each HVM orientation could be seen as a potential FF market positioning strategy. The HVM presents many alternatives for developing of a strategic placement, like increase or decrease the importance of an element on the map, create or delete a connection among the elements, or strengthen or weaken a connection or create a new element.

The provided HVM suggests that, in contrast with the other six functionality-related attributes associated with the health risk prevention area of the map, digestion improvement pertains to eating hedonism (see Figure 11.2). This finding is particularly useful in the case of children's snacks, for which fibre enrichment (which improves digestion) is the functionality type with the highest utility for consumers. The mere appearance of the eating hedonism area in the HVM is a hint to the food industry that consumers will not accept products that could have a positive effect on their health but present worse organoleptic attributes compared with conventional foods. Digestion improvement and nice taste are equally strongly connected to higher-order constructs of the hierarchy, such as eating enjoyment; therefore, a potential market positioning for snacks enriched with fibre would highlight that the functional product can simultaneously satisfy the need for healthiness AND the pursuit of hedonism. The result would be a strategic placement that could communicate to consumers that they do not have to compromise taste for healthiness.

Building on the insights from MEC phase I, phase II illustrates how to examine the controversy from a marketing-oriented point of view and provides concrete suggestions for new product developers. The DC methodology offers several experimentally developed functional children's snacks, and the binomial logit analysis identifies the exact levels of functionality, taste and price that deliver the highest utility to consumers.

Functionality is a statistically significant attribute for both snacks in the experimental design, which indicates the market potential of functional children's snacks. The functional snack alternative is always perceived as different and of greater utility for respondents compared with its common counterpart. This type of directly usable marketing information has special value in the technical new product development process. Moreover, more concrete information in terms of marketing value emerges with regard to the real potential of other elements of the marketing mix, such as the most and least preferred flavours for each snack and the right price levels for the combinations of healthiness and pleasure per snack, which enables estimations of consumers' willingness to pay.

Finally, sceptical readers should keep in mind that generalizations such as 'healthiness always wins' are unacceptable; careful tailoring of detailed marketing mixes for new product developers must realize that there are no winning or losing attributes for specific healthiness–taste–price combinations; rather, consumers just want it all. In such cases, the technical R&D process must to keep all options open, because real controversies do not seem to develop in consumers' perceptions.

In a similar vein, generalizations of the marketing implications of this study should be avoided, mainly because the samples used in both phases are not representative of the wider population. Nevertheless, the choice of this specific socio-demographic stratum helps satisfy the research design, in the sense that early middle-aged consumers should be the focus of a recruitment process to target parents of children of snack-consuming ages; moreover, a certain (high) level of education ensured the respondents could understand the terminology used in relation to functionality attributes.

The success of FFs depends on how well they satisfy various and frequently complementary consumer needs. The successful development of new functional products should take into account the opportunities presented through consumer research. Food manufacturers should carefully monitor consumer behaviour toward FFs to ensure that these products match consumer needs and expectations and that health claims get promoted by disseminating honest information in an attractive way.

References

1.   Urala, N. (2005), *Functional foods in Finland: Consumers' views, attitudes and willingness to use*. Finland: VTT Publications.
2.   Childs, N.M. & Poryzees, G.H. (1997), 'Foods that help prevent disease: consumer attitudes and public policy implications', *British Food Journal*, Vol. 9, pp. 419–426.
3.   Jonas, M.S. & Beckmann, S.C. (1998), 'Functional foods: consumer perceptions in Denmark and England', Working Paper 55, MAPP, Aarhus School of Business, Denmark. See p. 11.
4.   Margaret A. (2002), *Concepts of functional foods*. Belgium: ILSI Europe Concise Monograph Series, p. 5.
5.   Menrad, K. (2003), 'Market and marketing of functional food in Europe', *Journal of Food Engineering*, Vol. 56, pp. 181-8; Castellini, A., Canavari, M., & Pirazzoli, C. (2002), *Functional foods in the European Union: An overview of the sector's main issues*, 8th Joint Conference on Food, Agriculture and the Environment. Working Paper no 02-12.
6.   Lähteenmäki, L. (2004), Consumers and health: getting the probiotic message across', *Microbiological Ecology in Health and Disease*, Vol. 16, pp. 145–149.
7.   Schmidt, D. (2000), *Consumer attitudes toward functional foods in the 21st century*, IAMA Agribusiness Forum, Chicago, IL.
8.   Urala, N., & Lähteenmäki, L. (2004), 'Attitudes behind consumers' willingness to use functional foods', *Food Quality and Preference*, Vol. 15, pp. 793–803; Urala, N. & Lähteenmäki, L. (2007), 'Consumer's changing attitudes towards functional foods', *Food Quality and Preference*, Vol. 18, pp. 1–12.
9.   Urala, op. cit.
10.   Lappalainen, R., Kearney, J., & Gibney, M. (1998), 'A pan European survey of consumer attitudes to food, nutrition, and health: an overview', *Food Quality and Preference*, Vol. 9, pp. 467–478.
11.   Urala, op. cit.; van Kleef, W., van Trijp, H.C.M., & Luning, P. (2005), 'Functional foods: health claim-food product compatibility and the impact of health claim framing on consumer evaluation', *Appetite*, Vol. 44, pp. 299–308.
12.   Verbeke, W. (2005), 'Consumer acceptance of functional foods: socio-demographic, cognitive and attitudinal determinants', *Food Quality and Preference*, Vol. 16, pp. 45–57.

13. De Jong, N., Ocké, M.C., Branderhorst, H.A.C., & Friele, R. (2003), 'Demographic and lifestyle characteristics of functional food consumers and dietary supplement users', *British Journal of Nutrition*, Vol. 89, pp. 273–281; De Jong, N., Simojoki, M., Laatikainen, T., Tapanainen, H., Valsta, L., Lahti-Koski, M., Uutela, A., & Vartianen, E. (2004), 'The combined use of cholesterol-lowering drugs and cholesterol lowering bread spreads: health behaviour data from Finland', *Preventive Medicine*, Vol. 39, pp. 849–855.

14. Urala, op. cit.

15. Tuorila, H., & Cardello, A.V. (2002), 'Consumer responses to an off-flavor in juice in the presence of specific health claims', *Food Quality and Preference*, Vol. 13, pp. 561–569.

16. Wansink, B., Westgren, R.E., & Cheney, M.M. (2005), 'Hierarchy of nutritional knowledge that relates to the consumption of a functional food', *Nutrition*, Vol. 21, pp. 264–268.

17. Bech-Larsen, T., & Grunert, K.G. (2003), 'The perceived healthiness of functional foods: A conjoint study of Danish, Finnish and American consumers perception of functional foods', *Appetite*, Vol 40, pp. 9–14; Urala, op. cit.

18. Urala, op. cit.; Verbeke, op. cit.

19. Tuorila & Cardello, op. cit.

20. Verbeke, op. cit.

21. Jonas & Beckmann, op. cit.

22. Tuorila & Cardello, op. cit.

23. Walker, B.A. & Olson, J.C. (1991), 'Means-end chains: connecting products with self', *Journal of Business Research*, Vol. 22, pp. 111–118.

24. Bech-Larsen, T., Nielsen, N.A. Grunert, K.G., & Sorensen, E. (1997), 'Attributes of low involvement products – a comparison of five elicitation techniques and a test of their nomological validity', Working Paper 43, Aarhus School of Business, Denmark; Grunert, K.G., Sorensen, E. Johansen, L.B. & Nielsen, N.A. (1995), 'Analysing food choice from a means-end perspective', *European Advances in Consumer Research*, Vol. 2, pp. 366–371.

25. Bech-Larsen, T., Nielsen, N.A., Grunert, K.G., & Sorensen, E. (1996), *Means-end chains for low involvement products – a study of Danish consumers' cognitions regarding different applications of vegetable oil*, Working Paper 41, Aarhus School of Business, Denmark; Fotopoulos, Ch., Krystallis, A., & Ness, M. (2003), 'Wine produced by organic grapes in Greece: using means-end chains analysis to reveal organic buyers' purchasing motives in comparison to the non buyers', *Food Quality and Preference*, Vol. 14, No. 7, pp. 549–566; Krystallis, A. & Ness, M. (2003), 'Motivational and cognitive structures of Greek consumers in the purchase of quality food products', *Journal of International Consumer Marketing*, Vol. 16, No. 2, pp. 7–36; Morris, D., McCarthy, M., & O' Reilly, S. (2004), *Customer perceptions of calcium enriched orange juice*, Agribusiness Discussion Paper 42, Department of Food Economics, National University of Ireland, Cork; Padel, S., & Foster, C. (2005), 'Exploring the gap between attitudes and behaviour: understanding why consumers buy or do not buy organic food', *British Food Journal*, Vol. 107, No. 8, pp. 606–625; Sorenson, D., & Bogue, J. (2005a), 'A conjoint-based approach to concept optimization: probiotics beverages', *British Food Journal*, Vol. 107, pp. 870–883; Urala, N. & Lähteenmäki, L. (2003), 'Reasons behind functional food choices', *Nutrition and Food Science*, Vol. 4, pp. 148–158; Vannoppen J., Verbeke, W., Van Huylenbroeck, G., & Viaene J., (2001), 'Motivational structures towards purchasing labelled beef and cheese in Belgium', *Journal of International Food and Agribusiness Marketing*, Vol. 12, No. 2, pp. 1–29; Jonas & Bechmann, op. cit.

26. Urala & Lähteenmäki, 2003, op. cit.

27. Jonas & Beckmann, op. cit.

28. Padel & Foster, op. cit.

29. Morris et al., op. cit.

30. Jonas & Beckmann, op. cit.

31. Ibid.

32. Urala & Lähteenmäki, 2003, op. cit.

33. Morris et al., op. cit.

34. Ibid.

35. Padel & Foster, op. cit.

36. Urala & Lähteenmäki, 2003, op. cit.

37. Ibid.

38. Krystallis & Ness, op. cit.

39. Jonas & Beckmann, op. cit.

40. Morris et al., op. cit.

41. Schwartz, S.H. (1992), 'Universals in the content and structure of values: theoretical advances and empirical tests in 20 countries', *Advances in Experimental Social Psychology*, Vol. 20, pp. 1–65.

42. Gutman, J. (1982), 'A means-end chain model based on consumer categorization processes', *Journal of Marketing*, Vol. 46, No. 1, pp. 60–72; Kahle, L.R., Beatty, S.E., & Homer, P. (1986), 'Alternative measurement approaches to consumer values: The List of Values (LOV) and Values and Lifestyles (VALS)', *Journal of Consumer Research*, Vol. 13, pp. 405–409.

43. Jonas & Bechmann, op. cit.

44. Zanoli, R., & Naspetti, S. (2001), *Consumer motivation in the purchase of organic food: A means-end approach*, Paper presented at the 72nd EAAE Seminar, Organic Food Marketing Trends, Chania, Greece.

45. Krystallis & Ness, op. cit.; Fotopoulos et al., op. cit.

46. Schwartz, op. cit.

47. Krystallis & Ness, op. cit.

48. Schwartz, op. cit.

49. Gutman, op. cit.

50. Schwartz, op. cit.

51. Zanoli & Naspetti, op. cit.

52. Allenby, G. M. & Rossi, P.E. (1999), 'Marketing models of consumer heterogeneity', *Journal of Econometrics*, Vol. 89, pp. 57–78; Hall, J., Viney, R., Haas, M., & Louviere, J. (2004), 'Using stated preference discrete choice modelling to evaluate health care programs', *Journal of Business Research*, Vol. 57, pp. 1026–1032; Hensher, D.A., Stopher, P.R., & Louviere, J.J. (2001), 'An exploratory analysis of the effect of numbers of choice sets in designed choice experiments: an airline choice application', *Journal of Air Transport Management*, Vol. 7, pp. 373–379; Linardakis, M. & Dellaportas, P. (2003), 'Assessment of Athens' metro passenger behaviour via a multiranked probit model', *Journal of the Royal Statistical Society, Series C (Applied Statistics)*, Vol. 52, pp. 185–200; Louviere, J.J. (1984), 'Using discrete choice experiments and multinomial logit choice models to forecast trial in a competitive retail environment: a fast food restaurant illustration', *Journal of Retailing*, Vol. 60, pp. 81–107; Louviere, J.J., Hensher, D.A., & Swait, J.D. (2001), *Stated Choice Methods: Analysis and Applications in Marketing, Transportation and Environmental Valuation*. Cambridge: Cambridge University Press; McFadden, D. (1978), 'Modelling the choice of residential location', *Transportation Research*, Vol. 673, pp. 72–77; McIntosh, E., & Ryan, M. (2002), 'Using discrete choice experiments to derive welfare estimates for the provision of elective surgery: implications of discontinuous preferences', *Journal of*

Economic Psychology, Vol. 23, pp. 367–382; Scott, A. (2002), 'Identifying and analysing dominant preferences in discrete choice experiments: an application in health care', *Journal of Economic Psychology*, Vol. 23, pp. 383–398; Train, K. (1986), *Qualitative Choice Analysis: Theory Econometrics and Application to Automobile Demand.* Boston, MA: MIT Press.

53. Lochshin, L., Jarvis, W., d'Hauteville, F., & Perrouty, J-P. (2006), 'Using simulations from discrete choice experiments to measure consumer sensitivity to brand, region, price and awards in wine choice', *Food Quality and Preference*, Vol. 17, pp. 166–178; Nganje, W.E., Kaitibie, S., & Taban, T. (2005), 'Multinomial logit models comparing consumers' and producers' risk perception of specialty meat', *Agribusiness*, Vol. 21, pp. 375–390; Scarpa, R., Philipidis, G., & Spalatro, F. (2005). 'Product-country images and preference heterogeneity for Mediterranean food products: a discrete choice framework', *Agribusiness*, Vol. 21, pp. 329–349; Steiner, B., Unterschultz, J., & Gao, F. (2006), *New meats from the Wild West: consumers' perceptions towards alternative meats*, Paper presented at the 98th EAAE, Seminar Marketing Dynamics within the Global Trading System: New Perspectives, Chania, Greece, July.

The Consumer View

PART III

The Consumer View

12 Controversies in Food and Agricultural Marketing: The Consumer's View

BY KEITH WALLEY,* PAUL CUSTANCE† AND STEPHEN PARSONS‡

Keywords

consumers' attitudes, farming practices, food scares, pollution, food miles, animal welfare.

Abstract

Controversies in food and agricultural marketing have a consumer dimension, especially when they involve issues such as food scares, pollution, food miles and animal welfare. These effects become manifest in consumers' attitudes, which provide a valuable means for evaluating past events and predicting future behaviour. This chapter uses a longitudinal data set related to consumer attitudes about various contentious issues in UK food and agricultural marketing to offer some insights and managerial implications. The key results reveal that consumers perceive several livestock and poultry products as too expensive though, conversely, many consumers are willing to pay a premium for products that ensure animal welfare. Despite EU policy encouraging more environmental responsibility, many consumers believe farmers use too much pesticide and contribute to pollution.

* Dr Keith E. Walley, Harper Adams University College, Edgmond, Newport TF10 8NB, UK. E-mail: kewalley@harper-adams.ac.uk. Telephone: +44 1952 820 280.

† Dr Paul R. Custance, Harper Adams University College, Edgmond, Newport TF10 8NB, UK. E-mail: prcustance@harper-adams.ac.uk. Telephone +44 1952 820 280.

‡ Mr Stephen T. Parsons, Harper Adams University College, Edgmond, Newport TF10 8NB, UK. E-mail: stparsons@harper-adams.ac.uk. Telephone +44 1952 820 280.

Introduction

One of the basic tenets of a market is that the driving force is consumer demand. This concept of consumer power and its role within market economies is often referred to by economists as 'consumer sovereignty'.[1] In turn, one of the most important indicators of consumer sovereignty is the attitudes that consumers hold.

According to Malhotra an attitude is a 'summary evaluation of an object or thought'.[2] Although attitudes often are rooted in some aspect of reality, they are perceptual in nature, which means that they may not be an accurate reflection of reality. Duffy, Fearne and Healing believe that consumers receive information in an extremely fragmented form; in the current context, this explanation may account for the discrepancies between attitudes towards the price of selected commodities and their actual price.[3]

Although attitudes are not necessarily a veridical measure of an object, they remain an important concept because they serve as a useful means of not only evaluating past events but also predicting future action. Unfortunately, however, publications such as the Curry Report (published by the Policy Commission),[4] suggest that some elements of the UK food supply chain, especially farming, have become detached from the rest of the economy, environment and society.

With regard to controversies in food and agricultural marketing, knowledge about consumer attitudes is important for anyone involved in the food supply chain, including the farmers who produce food, those involved in processing industries, wholesalers, retailers and those associated with policy generation and implementation.

This chapter reports the results of a longitudinal research project investigating various controversies in the food supply chain. The controversies covered include pricing of food and farm revenues; food farming practices and pollution; food safety (including food scares); organic food; genetic engineering and food production practices; food miles and regional production; and animal welfare and food production.

In addition to presenting the background of each controversy and identifying the key points arising from the survey data, we consider some implications for decision makers in the food supply chain. The practical limitations of producing a study based on such a large quantity of data mean that our discussion is confined to what we consider the critical points. It is important to note, however, that readers may be able to apply these data to the issues that they are facing and derive significant additional value from the results.

To begin, in the next section, we briefly outline the survey that generated the data and the way in which we will present these data.

Annual Consumer Survey

The data presented in this study were collected as part of an ongoing programme of research into consumer attitudes in the UK regarding food and the environment. An annual survey of the general public, covering the period 1996–2006, has been conducted each summer in four locations in the West Midlands region.

For this survey, a large number of attitude statements were generated from a search of the literature, then carefully reviewed by specialists in the field (including academics working in the agriculture and agri-food sectors) to ensure content (or face) validity.[5]

The many statements generated would lead to a questionnaire that takes an inordinate amount of time to complete, yet to omit any statements would seriously detract from the comprehensiveness of the study. As a result, half of the statements constitute one questionnaire (Questionnaire A), and the other half make up another questionnaire (Questionnaire B).

The questionnaires were subsequently completed in a series of street interviews conducted in Birmingham (city), Hanley (conurbation), Shrewsbury (large rural town) and Ludlow (small rural town). Although all the surveys were administered in the West Midlands region, the locations for the survey works may be considered representative of the UK. An interlocking quota sampling approach ensured half the respondents were male, and equal numbers of respondents fell into six predetermined age bands. Questionnaires A and B each were completed by 300 different respondents, which makes the results statistically significant at a 90 per cent level, given a ± 5 per cent margin of error.[6]

An explanation of the different potential approaches for analyzing the data and the approach actually employed to produce the data discussed in the remainder of this chapter appears in Figure 12.1.

Pricing of Food and Farm Revenues

According to the Policy Commission,[7] most people in the UK have higher disposable incomes, but the main factor determining what food they purchase is still price. This finding is supported by the Institute of Grocery Distribution,[8] which finds that price, sell-by-date and brand name are the three main factors influencing food purchases.

The price that consumers pay for food represents the revenue available to any entity that has played some role in the food supply chain. However, according to Kohls and Uhl, the allocation of consumers' food expenditures between farmers and the other members of the food supply chain, particularly retailers, is one of the most controversial aspects of food production.[9]

The Department for Environment, Food and Rural Affairs estimates that between 1988 and 2006, UK farmers' share of a basket of food staples fell by 23 per cent, and Hingley attributes this change to a power imbalance in the food supply chain.[10]

The impact of this reduction in farm incomes has been significant. Nevertheless, the Commission for Rural Communities states that there are still 162 000 farmers who manage approximately 75 per cent of the land in England and Wales, which means that farming is still the major land use in the UK and that food production remains the primary output of the land in rural areas.[11]

Farming incomes can be detrimentally affected by factors such as the weather or an outbreak of disease, but the pressure caused by reduced revenue share is an additional threat that has materialized in recent years and become an issue of considerable concern.

In practice, total income from farming in the UK in 2006 rose by an estimated 6.9 per cent in real terms compared with the previous year, but this rate is still 61 per cent below the high point of 1995.[12] The net result is that farm incomes and the rural economy as a whole remain under pressure.

According to the 1996 survey data (see Table 12.1) consumers believe that lamb, beef and free-range eggs are too expensive, but pork and (non-free-range) eggs are not. Despite some minor variations in the data dating back to 1996, these perceptions appear fairly consistent.

The five possible approaches to the analysis each has inherent advantages and disadvantages and is likely to provide a slightly different picture of the issue. Many studies that use attitude surveys employ a simple count of the frequency of responses to conduct the analysis (Treatment 1), but this approach can be difficult to implement when attempting to make judgements about the whole data set. As a consequence, it is often better to adopt an analytical technique that produces a summative evaluation of the data set, such as by calculating a mean score for each rating statement (Treatment 2). This approach is more systematic than simply trying to weigh the frequency data judgementally, but experience again suggests that that trade off ameliorates conflicting trends in the data, hiding them somewhat from the researcher. As a consequence, it would seem reasonable to adopt a balance of opinions approach that trades off agree and disagree data, such as responses (Treatment 3), but this method is somewhat simplistic because it does not take into account 'don't know' responses. Furthermore, even if 'don't know' responses were included (Treatment 4), the 'neither agree nor disagree' responses would still be omitted, which is again simplistic and not comprehensive. The most comprehensive and therefore useful balance of opinion approach includes both 'don't know' and 'neither agree nor disagree' data (Treatment 5). Thus, the output from Treatment 5 indicates that out of the 300 respondents, a reliable majority of 146 agree that British food farmers are good food producers. This approach is based on the principle of conservativism. The final measure depends not just on the proportion of the sample that agrees or disagrees with the statement but also is adjusted for the worst case scenario, in which respondents who claim to hold no opinion actually hold an opinion that is diametrically opposite that of the majority. This approach seems to offer the most prudent treatment of the data and therefore is used to analyze the data presented herein.

Treatment 1: Frequency analysis of responses

	Strongly agree	Slightly agree	Neither agree nor disagree	Slightly disagree	Strongly disagree	Don't know
	(1)	(2)	(3)	(4)	(5)	(6)
e.g. British farmers are	118	105	36	15	7	19
good food producers	(39.3%)	(35.0%)	(12.0%)	(5.0%)	(2.3%)	(6.3%)

Treatment 2: Mean

Calculation: $(118 \times 1) + (105 \times 2) + (36 \times 3) + (15 \times 4) + (7 \times 5) + (19 \times 6) / 118 + 105 + 36 + 15 + 7 + 19$

Output 2.15

Treatment 3: Balance of Opinion I

Calculation: (Strongly agree + Slightly agree) – (Slightly disagree + Strongly disagree)

$(118 + 105) - (15 + 7)$

Output: 2.01

Treatment 4: Balance of Opinion II

Calculation: (Strongly agree + Slightly agree) – (Slightly disagree + Strongly disagree) – Don't know

$(118 + 105) - (15 + 7) - 19$

Output: 1.82

Treatment 5: Balance of Opinion III

Calculation: (Strongly agree + Slightly agree) – (Slightly disagree + Strongly disagree)
– (Don't know + Neither agree nor disagree)

$(118 + 105) - (15 + 7) - (19 + 36)$

Output: .46

Figure 12.1 Possible treatments of the data

Table 12.1: Attitudes towards the pricing of food and farm revenues

	1996	1997	1998	1999	2000	2001	2002	2003	2004	2005	2006
Lamb is too expensive (b)	53	35	37	44	37	26	21	32	26	40	33
Pork is too expensive (b)	23	2	0	-3	-3	-2	-13	-19	-7	13	-20
Beef is too expensive (b)	43	34	34	34	37	27	25	9	11	12	23
Eggs are too expensive (b)	12	1	-6	4	-16	-23	-26	-24	-14	17	-37
Free range eggs are too expensive (b)	52	31	27	36	29	30	16	14	21	43	11
Farmers are only concerned with profit (a)	23	23	-5	13	-2	-10	7	-19	-2	13	-5
Many farms survive on bed-and-breakfast earnings (a)	-36	5	21	18	27	22	22	34	25	16	20
Farm incomes are declining (a)	-23	16	49	58	65	77	62	69	67	51	69
The farmer receives a fair price from the supermarket for his produce (b)	–	–	–	-51	-54	-53	-45	-46	-43	-23	-55

In terms of farm revenues, consumers' views on whether farmers are only concerned with profit vacillate, which may be a product of the issues in the media at the time of the annual survey. Consumers are more consistent in their views that many farmers survive on bed-and-breakfast earnings and that farm incomes are in decline. It is also possible that they relate this decline in income to a belief that farmers do not receive a fair price for their produce from supermarkets.

Several managerial implications derive from these survey results:

- Lamb, beef and free-range eggs are perceived as too expensive. If farmers, wholesalers or retailers wish to sell more of these products, they must work to overcome this perception. A reduction in price may appear to be the obvious strategy to follow, another option might be to enhance the perceived value of these products through a carefully conceived advertising campaign.
- Consumers are aware of the financial pressure on farmers and thus may be willing to accept higher prices, justified by the argument that the price premium supports UK agriculture. The entire market is unlikely to be willing to pay such a premium, but a certain market segment should. This segment may already exist in the form of consumers who only 'buy British'.

- Retailers may wish to address consumers' belief that farmers do not get a fair price from supermarkets for their products, perhaps through an advertising campaign. This perception may or may not be true, but it certainly is a pervasive belief that could create negative publicity and contribute to a poor image for retailers. A recent initiative by Tesco guarantees an increase in the returns to dairy farmers by raising the milk price, and the associated publicity connects to 'fair pricing'.

Food, Farming Practices and Pollution

Many consumers believe they have a reasonable to good understanding of food production methods, but this perception is not always the case. At the turn of the decade, the Institute of Grocery Distribution (IGD) reported that many consumers had a low, out-of-date, and often inaccurate understanding of the methods used to produce food.[13] Since then, globalization has expanded consumers' requirements and led to a greater increase in imports, but for many consumers, a vast gap still exists in terms of their knowledge of how food is produced.[14]

In reality, though the quality of the natural environment is improving, and this trend is likely to continue as agri-environment schemes such as LEAF Marque (Linking Environment and Farming)[15] become more widespread and agricultural practices more extensive, agricultural chemical usage remains high. Farmers struggle to maintain economically competitive production levels and extended growing seasons, but the use of chemicals to achieve these goals has had a significant impact on the environment. The effect of chemicals on habitat, soil and water quality means that population levels of certain species of bird and mammals remain low.[16]

Chemical usage and pollution in the countryside is something of an issue amongst the general public. For instance, though nearly 40 per cent of respondents to a DEFRA survey did not feel there was anything preventing them from enjoying the countryside, 23 per cent mentioned environmental problems or pollution as a concern.[17] The IGD also found that many members of the population wanted more information about the use of chemical sprays in agriculture.[18]

The survey data (see Table 12.2) indicate that consumers are strongly of the opinion that farmers still use too much pesticide and herbicide. They also believe farmers pollute the soil and water courses.

Although some views on pollution appear to be entrenched and consistent over the 10 years that the project has been running, the exception relates to farmers polluting the air. Consumer views about whether farmers pollute the air have changed over the period of the project. Whereas 10 years ago, consumers felt that farmers polluted the air, now they believe that they do not. This variable should be monitored in the future, especially since recent media coverage has highlighted the contribution that farm animals make to greenhouse gases and global warming.

The data suggest that consumers believe farmers contribute to pollution, but they also believe that farmers are making efforts to reduce this contribution. In this instance, they assert farmers should spend more money on the environment and that regulations regarding farm pollution should be tightened.

Table 12.2 Attitudes towards food, farming practices and pollution

	1996	1997	1998	1999	2000	2001	2002	2003	2004	2005	2006
Farmers use too much pesticide (b)	39	48	56	57	65	57	57	54	57	27	55
Farmers use too much herbicide (b)	33	38	48	46	0	40	35	40	43	26	32
Most farmers pollute the air (b)	22	13	2	0	0	8	-8	-2	1	-12	-10
Most farmers pollute the soil (b)	24	13	10	16	14	11	8	14	21	7	8
Most farmers pollute water courses (b)	30	14	14	22	17	17	11	14	13	13	9
Farmers are making efforts to reduce pollution (b)	18	20	42	32	31	27	26	32	41	36	38
Farmers should spend more money on the environment (b)	59	51	46	41	40	39	30	40	29	46	43
Regulations regarding farm pollution should be tightened (b)	74	72	59	72	71	67	63	60	63	38	61

The managerial implications of these findings include the following:

- The belief that farmers use too much pesticide and herbicide is probably a main reason for the growth in organic food purchasing. Given the strength and pervasiveness of public opinion on this matter, there may be further growth in the organic sector that farmers can take advantage of.
- The view that farmers contribute to pollution of the soil and water is pervasive. Farmers would be advised to continue with their efforts to reduce pollution at every opportunity.
- The government may be able to encourage the reduction of chemical usage and pollution on farms through the sympathetic development of appropriate legislation, which would appear to have public support.

Food Safety (Food Scares)

Recent food scares include Bovine Spongiform Encephalophy (BSE) via new variant Creutzfeldt-Jacob Disease (nvCJD) in beef cattle and salmonella in eggs. Combined with the escalating media attention that accompanies such events, public anxiety has spiralled, and food safety has become a major issue of public concern.

Knowles, Moody and McEachern postulate a typology of EU food scares and identify three categories: microbiological, contaminant and animal disease related (see Table 12.3).[19]

The Council for the Protection of Rural England (CPRE) identifies food scares as the main reason for increasing consumer interest in and awareness about food.[20] They also drive the increasing number of farmers' markets and the introduction of local and regional sourcing policies by major food retailers.[21]

Reduced levels of consumer confidence in food safety have led to the rapid evolution of legislation, practice and thinking about food safety in both the UK and Europe in general.[22] This evolution has shifted the focus of food policy from a production orientation to a risk-based orientation that, in turn, has developed into a tentative EU consumer-based policy, reinforcing the consumer's individual 'right to choose'.[23]

Yeung and Morris review the factors influencing consumer perceptions of food safety risks and their likely impact on purchasing behaviour.[24] The risk components include physical, performance, financial, time, social and psychological loss. When consumers perceive a possible food hazard, they reduce their purchase of the offending product. Concerns with respect to food safety also have led consumers to seek good quality food at affordable prices, high food hygiene standards in stores, and reliable and helpful information when food scares occur. Because the overall purpose of marketing is to build enduring, mutually beneficial relationships between suppliers and consumers, it is important that risk analysis and management adopt a whole supply chain perspective.

The survey data (see Table 12.4) reveal that consumers strongly believe that lamb, pork, beef and chicken originating from British sources are safe to eat but that consumer confidence in the same products coming from non-British sources is significantly worse. This point is interesting, because the data appear to indicate that these beliefs remain fairly constant, even when a food scare has been reported in the media.

Table 12.3 Food scares

Typology	Food scare	Year
Microbiological	*Salmonella* in eggs	1998
	Campylobacter in poultry meat	1995
	Salmonella Montevideo in chocolate	2006
Contaminant	*Benzene* in Perrier bottled water	1989
	Sudan1 used illegally as a food colorant	2004
	Para Red used illegally as a food colorant	2005
Animal disease related	*BSE* in beef cattle	1989
	CJD deaths from eating beef contaminated with	1996
	BSE	2006
	Avian Flu in the poultry flock	

Table 12.4 Food safety (food scares)

	1996	1997	1998	1999	2000	2001	2002	2003	2004	2005	2006
British lamb is safe to eat (b)	61	68	51	72	73	66	67	65	75	80	76
Non British lamb is safe to eat (a)	40	34	21	25	24	10	19	28	20	37	21
British pork is safe to eat (b)	66	66	69	68	67	67	66	62	69	75	69
Non British pork is safe to eat (a)	33	32	20	18	6	1	6	-2	2	9	3
British beef is safe to eat (b)	13	20	43	49	44	47	49	55	63	72	63
Non British beef is safe to eat (a)	10	1	4	9	2	-3	-6	-3	-6	14	10
British chicken is safe to eat (b)	78	65	63	66	71	54	64	49	56	61	67
Non British chicken is safe to eat (a)	32	20	16	10	10	-4	0	-6	-7	-28	-6

The survey data (see Table 12.5) also suggest that with respect to BSE, which caused a major food scare in the beef industry during the 1980s, consumers remain strongly of the opinion that it is not a naturally occurring disease and that it was a result of unnatural farming methods. They also strongly believe that intensive farming is not the only way to go about farming. Furthermore, though consumers strongly hold the view that too much food is imported, they believe British farmers are good food producers.

With regard to these findings, several managerial implications again emerge:

• The data show that consumer confidence is higher for meats produced in the UK than abroad, which should be particularly pleasing for domestic farmers. It provides further evidence of the importance of the country of origin in the consumer decision-making process, and in this case, it appears to be evidence of the positive impact of the domestic country of origin.

• Although the data suggest that country of origin is an important factor in the decision-making process, another factor that likely is at work is quality assurance. Discussion persists about the large number of quality assurance schemes to which farmers may subscribe and the potential confusion that this may cause in the minds of consumers, but it remains likely that the presence of a quality assurance label on a product serve to reassure consumer.[25] Farmers, retailers and intermediaries therefore should continue to promote such schemes.

Table 12.5 Food safety and farming practices

	1996	1997	1998	1999	2000	2001	2002	2003	2004	2005	2006
Intensive farming practices are the only way to meet demand (a)	0	-25	-26	-30	-24	-23	-8	-40	-23	1	-29
BSE is the kind of natural disease that happens every so often (b)	-36	-26	-26	-26	-34	-32	-37	-32	-39	-26	-6
BSE is the result of the unnatural way we farm animals today (a)	49	48	42	49	51	51	48	51	60	28	47
British farmers are good food producers (a)	66	62	69	68	65	67	65	72	71	60	70
Too much food is imported (a)	77	58	62	50	61	57	58	55	65	50	60

- The prejudice that consumers appear to show towards products sourced from overseas and the belief that British farmers are good food producers is likely to support the patronage of farmers' markets and continued demand for regional foods supplied by national retailers. Farmers and retailers should be aware of this trend and take steps to satisfy consumer demand.
- Despite a suspicion that some food scares have been aggravated by the media, it is important to remember that information is the basis of consumer confidence. Food retailers might establish a means of managing their suppliers, similar to the guidelines established by Tesco for working with meat suppliers[26] to be able to provide consumers with timely and helpful information.

Organic Food

Organic farming is a method of farming that requires farmers to operate based on ecological principles, with strict limitations on the inputs that may be used, which minimizes damage to the environment and wildlife. The emphasis is on natural methods of production and pest control. Approved UK organic certification bodies include the Soil Association Certification Ltd., Organic Farmers and Growers Ltd., Scottish Organic Producers Association and Organic Food Federation.[27]

According to Lampkin, Measures and Padel,[28] the UK organic retail market grew to £1.6 billion in 2005, with sales of organic food representing approximately 1.3 per cent of total food sales. Supermarkets have continued to be the dominant outlet for organic food and drink, accounting for 76 per cent of sales in 2005.

The Soil Association reported that in 2006, 65.4 per cent of consumers knowingly bought organic food, and only 29 per cent of respondents to a Mintel survey claimed never to buy organic produce.[29] The Mintel survey went on to report that among those purchasing organic food and drink, the most popular purchases were vegetables, fruit, meat, dairy and flour/breads/cereals. The propensity to purchase organic products was most prevalent among higher socio-economic and older consumers.

Several authors have attempted to explore the reasons behind organic food purchasing. In a study of regular and occasional purchasers of organic food, Padel and Foster find that most consumers initially associate organic food with vegetables and fruit and a healthy diet.[30] Fruit and vegetables are also the first, and in many instances the only, experience of consumers in buying organic products.

Rimal, Moon and Balasubramanian evaluate the role of consumers' perceived risks and the benefits of agro-biotechnology in shaping the purchase pattern for organic food among UK customers.[31] They find that only 4 per cent of the respondents to an online household survey purchased organic foods all the time, whereas 26 per cent never purchased. As the risk perception of agro-biotechnology increased, consumers became more likely to purchase organic food more often. Although premium prices of organic foods were of concern to many consumers, food safety was the most important consideration when making organic food purchase decisions.

Finally, in a survey reported by the Soil Association, 84 per cent of respondents thought organic food was too expensive, whereas 37 per cent of the public (and 65 per cent of regular organic consumers) agreed with the statement that 'organic food tends to be more expensive but I think it is a price worth paying'.[32]

The desire to eat more organic food has been a strong and consistent view offered by consumers throughout the period of the project (see Table 12.6), with a majority of consumers believing that organic food tastes as good as food produced by other farming methods.

The desire to eat more organic food is strong, but so too is the perception that organic food is too expensive. This point is therefore a factor that may constrain the development of the market for organically produced food. Consumers also believe chemicals are needed to produce food cheaply; this belief too may need to be overcome before the organic market sector can achieve its full potential.

Table 12.6 Organic food

	1996	1997	1998	1999	2000	2001	2002	2003	2004	2005	2006
I would like to eat more organic food (a)	30	54	44	61	48	50	45	53	51	48	59
Organic food tastes as good as food produced by other methods (b)	–	–	–	22	24	18	27	30	28	35	41
Organic food is too expensive (a)	83	69	59	75	70	64	71	78	78	65	66
Chemicals are needed to produce food cheaply (b)	-12	-2	3	5	8	14	22	9	18	5	17

In terms of the managerial implications, we note:

- Organic food appears to be popular amongst consumers, but they also believe that it is too expensive. If the sector is to grow to its full potential, all those involved in the production, distribution and sale of organic products need to work to overcome this price issue.
- The price of organic produce may fall as the sector grows and producers gain economies of scale, but it may also be useful for those involved in the supply chain to promote the benefits of organic produce as a means of justifying perceived product value.
- Many people, particularly the elderly, were subjected to prolonged exposure to the production orientation that developed during the war years and that has been challenged only recently. One characteristic of the production orientation was the ready availability of cheap food, which resulted from the judicious use of chemicals to enhance productivity and ward off disease. The data suggest that this view remains popular today and might be a specific target for advertising by organic suppliers.

Genetic Engineering and Food Production Practices

Even though many consumers know little or nothing about the scientific techniques in question, genetic engineering, or genetic modification (GM) as it is more widely known, has become a great cause for concern.[33] Davies, Richards, Spash and Carter argue that consumer perceptions of the advantages and disadvantages associated with GM and ultimately its acceptance depend on whether people see the technology as radically different science or simply the next step after 'traditional' breeding techniques.[34] These authors also believe that current legislation relating to GM and risk management is sufficient and that consumer disquiet is really related to wider issues, such as the economic and social impacts of new technologies, political decision making, the role of science in society and changes to institutional structures.

With regard to genetic engineering, consumers in the survey believe that its use is inevitable and increasing (see Table 12.7), which is somewhat ironic, because they think the genetic engineering of crops and, especially, of animals is unacceptable! This latter point supports research by the IGD that indicates a lower level of interest in the production methods associated with fruit and vegetables than those for animals.[35]

With regard to the managerial implications, we recognize the following:

- Although GM may offer significant economic gains for farmers and others involved in the food supply chain, there is strong resistance to it amongst consumers as a means of producing food. This resistance may or may not be ill founded, but farmers and others must be careful if they become involved with this form of food production.
- There is strong resistance to all forms of GM, but consumers appear particularly concerned about the GM of animals; anyone involved in this form of food production must be aware of the market's extreme sensitivity on this matter.
- Given the depth of feeling on this subject, any involved organization may need to take a long-term perspective and seek to change people's opinions about the wider issues identified by Davies and colleagues as a first step to achieving greater acceptance of GM.[36]

Table 12.7 Genetic engineering

	1996	1997	1998	1999	2000	2001	2002	2003	2004	2005	2006
Genetic engineering is inevitable (b)	26	19	18	16	25	28	24	8	39	21	14
Genetic engineering is increasing (b)	67	63	74	75	80	71	70	76	83	64	70
Genetic engineering of crops is acceptable (b)	-5	-6	-32	-29	36	-29	-41	-31	-36	-49	-27
Genetic engineering of animals is acceptable (b)	-42	-55	-61	-58	-65	-59	-55	-68	-62	-65	-57

Food Miles and Regional Production

The UK food supply chain has come to be dominated by a small number of large retail organizations that often compete on price; to secure price advantages, they often source products from great distances. However, for a growing number of consumers, the distance that food travels has become an issue because of its implications for carbon dioxide emissions, food quality and animal welfare.

One development in response to consumer concerns relating to the food miles issue, and especially their greater awareness of the origins and traceability of food,[37] has been the growth in the market for local and regional foods. This development has been seen as an opportunity for both the large multiple food retailers and farmers involved in direct marketing. As Padbury argues, local foods offer retailers and food service companies an opportunity to differentiate themselves from their competitors, as well as a valuable commercial proposition.[38]

Despite the market for local and regionally sourced food, it is not always possible for consumers to locate sources of such food.[39] Multiple food retailers such as Waitrose are attempting to promote regional and local products, but the structural problems within local food supply chains, such as fragmentation and small scale, mean that doing so is not always possible.[40] Furthermore, a report on UK grocery retailing concludes that increased local sourcing would probably reduce the average distance travelled by food but also would require smaller vehicles and less efficient loadings.[41] Jones, Comfort and Hillier thus conclude that though local and regional foods offer consumers an opportunity to support small, local producers and generally support the local economy and job market, they are unlikely to form more than a small part of the nation's future diet.[42]

Farmers have sought to take advantage of consumers' concerns by instituting alternative food supply chains and developing farm shops and farmers' markets.[43] Consumer interest in quality local food, food production techniques and traceability appears to have led to the establishment of an estimated 4000 farm shops and more than 500 farmers' markets with an annual turnover of £1.5 billion.[44] With the overall grocery market worth £128.2billion at the start of 2007, the contribution of farm shops and farmers' markets thus would be on the order of just more than 1 per cent.[45] This growth appears directly attributable to demand for local and regional food, which represent useful value-adding activity by which farmers can distribute local and regional foods directly to consumers.

Farm shops and farmers markets are not uniformly successful and tend to focus on horticulture, dairy, meat and poultry products, which can be sold directly to consumers with little or no processing or remain readily identifiable after processing, yet many continue to attract significant business from concerned consumers.[46] Prices are often higher than in the supermarkets, yet the market are perceived to sell fresh, quality, tasty, local produce in an attractive atmosphere while supporting local producers.[47] Holloway and Kneafsey suggest the success of farmers' markets is due to their ability to link localness with perceived quality, health and rurality.[48] They believe that in the minds of consumers, farmers' markets represent a diversifying rural economy arising in response to the difficulties being experienced by UK farmers and a more general perception of a countryside under threat.

The survey data (see Table 12.8) suggest that consumers are favourably disposed toward buying locally produced products. Existing literature suggests that social responsibility and the reduction in food miles is likely to be the main explanation, but the data suggest that support for farmers and the preservation of jobs in the locality are also important considerations.

When asked to respond to the statement, 'I would pay more for goods produced locally', many respondents indicated agreement. However, experience suggests that what people say they will do in response to a statement like this and what they actually do are often different. A market segment might be willing to pay a premium for local products, but some might not be able to afford the premium prices, which means that any indication of the size of the segment derived from the data is likely to be overstated.

In turn we consider several managerial implications:

- Interest in local and regional products is fairly high and likely to remain so in the future. Local and regional products therefore represent a viable market segment to target for both multiple retailers and farmers.

Table 12.8 Interest in local and regional products

	1996	1997	1998	1999	2000	2001	2002	2003	2004	2005	2006
I buy locally produced produce to support farmers (b)	–	–	–	38	46	43	46	52	32	5	55
I buy locally to preserve jobs in the area (b)	–	–	–	40	44	43	40	43	24	6	55
I would pay more for goods produced locally (b)	–	–	–	17	39	36	34	49	34	15	52

- If the segment is to be exploited to its full potential, the structural issues that currently serve as impediments to its development need to be addressed. One means of overcoming the issues surrounding fragmentation and scale might be some sort of cooperative mechanism.
- Another item that should receive further attention is the pricing of local and regional food. If the sector is to achieve its full potential, then the pricing of local and regional food must be correct; further research and a pricing sensitivity analysis appear required.

Animal Welfare and Food Production

As far back as 1995, observers recognized that animal welfare concerns were not a fad but a deep-seated issue that was not going to disappear. As UK livestock production, distribution and processing systems continued to evolve, it became clear that consumer concerns regarding welfare meant that farm animals had to enjoy a decent life. Livestock products that were animal welfare-friendly, produced with traditional or natural methods, that carried some form of certification offered important consumer benefits that could provide a competitive edge.[49] This view has not really changed in the intervening years, as public concern has prompted bans on intensive farming methods such as sow stalls, veal crates and (by 2012) battery cages.

Consumer concern for animal welfare appears to be supported in practice by their willingness to pay a premium for animal welfare-friendly products. Burgess and Hutchinson found that people are willing to pay extra on their weekly food bill to ensure that laying hens, broiler chickens, dairy cows and pigs have improved welfare conditions.[50] The willingness to pay a premium for animal welfare-friendly products probably derives from consumers' tendency to use it as an indicator of other, more important attributes, such as safety and impact on health.[51]

Producers, distributors and retailers involved in animal products have sought to exploit consumer willingness to pay a premium price for products derived from animals that have been farmed according to high standards of welfare. A number of high status welfare products have been developed, including the RSPCA-monitored Freedom Foods range, which sells at a price premium.

A review of the survey data (see Table 12.9) shows a clear belief amongst many consumers that a healthy diet contains red meat, that animals feel pain like humans, that farm animals need better living conditions and that battery egg production is problematic.

However, consumers are also of the opinion that British farmers have high standards of animal welfare, are not cruel to animals and that most farm animals have a happy and comfortable life.

It would appear therefore that though consumers have some specific concerns and wish animal welfare standards to be as high as possible, they are realistic and acknowledge the high standards that farmers already follow.

Another aspect of animal welfare relates to the impact that farming has on the natural wildlife that live on farms. From the data in Table 12.10, it appears that respondents believe that farming is detrimental to both birdlife and wildlife. They also find fox hunting (banned by legislation in 2005) and the shooting of game birds unacceptable.

Table 12.9   Animal welfare

	1996	1997	1998	1999	2000	2001	2002	2003	2004	2005	2006
A healthy diet does contain red meat (b)	46	31	39	36	32	39	28	46	52	53	43
Animals feel pain like humans (a)	89	79	84	83	76	80	79	84	67	53	87
Farm animals need better living conditions (a)	9	38	39	51	62	63	46	64	56	29	59
I do not like battery egg production (a)	55	66	66	64	52	5	53	74	72	70	72
British farmers have high standards of animal welfare (a)	20	28	43	29	37	31	40	28	34	45	37
Farmers are cruel to animals (a)	-62	-35	-44	-33	-31	-29	-30	-26	-34	-37	-30
Most farm animals have a happy and comfortable life (a)	-2	0	2	-8	-10	-4	6	-10	-12	23	–

Table 12.10 Wildlife on farms

	1996	1997	1998	1999	2000	2001	2002	2003	2004	2005	2006
Modern farming is detrimental to birdlife (b)	30	34	41	36	48	38	34	41	39	51	27
Modern farming is detrimental to wildlife (b)	39	36	36	42	42	40	40	40	41	36	33
Fox hunting is an acceptable activity (a)	-61	-49	-28	-37	-31	-33	-17	-25	-16	-41	-34
The shooting of game birds is an acceptable activity (a)	-44	-29	1	-16	-15	-11	3	-11	-5	-25	-13

These two tables summarize some managerial implications in both areas:

- Animal welfare is an important issue for many consumers, who are willing to pay premium prices for products that guarantee high standards of welfare. Products that subscribe to codes of conduct with regard to animal welfare are likely to remain popular in the foreseeable future.
- While many people are concerned to some extent about animal welfare, some are particularly sensitive. This sensitivity may result from 'anthropomorphism' amongst the UK population, yet the results suggest implications of these beliefs that range from increasing numbers of people becoming vegetarian to people becoming animal rights activists. These developments might be construed as threats to the industries associated with animal farming and meat production and retailing.
- Anthropomorphism may be responsible for respondents indicating a dislike for fox hunting and the shooting of game birds. This trend could be the result of consumers differentiating between activities that involve animals being reared and killed for food and activities that involve animals being reared and killed for sport. Alternatively, this view could be an indication of a gap opening between people who live in towns and cities and those who live in the rural areas. If so, the implications for the future of those involved in farming, food production, and the rural economy in general could be very significant indeed.

Conclusions

This chapter presents data that reflect current consumer views on a variety of food-related issues. The data have been subjected to a perceptive evaluation to derive specific managerial implications. However, because the data set is so very rich, it is impossible to discuss every aspect of the data in detail, which means that readers should reinterpret the data as they see fit. The chapter does however identify many important implications for farmers, food processors, retailers and policymakers.

References

1. Sloman, J. (2007), *Essentials of Economics*, 4th edn, Harlow: Pearson Education.
2. Malhotra, N.K. (2005), 'Attitude and affect: new frontiers of research in the 21st century', *Journal of Business Research*, Vol. 58, No. 4, pp. 477–482.
3. Duffy, R., Fearne, A. & Healing, V. (2005), 'Reconnection in the UK food chain: bridging the communication gap between food producers and consumers', *British Food Journal*, Vol. 107, No. 1, pp. 17–33.
4. Policy Commission (2002), Farming and Food – A Sustainable Future. Report of the Policy Commission on the Future of Farming and Food, Cabinet Office, London.
5. Churchill, Jr. G.A. (1992), *Basic Marketing Research*, 2nd international edn, London: The Dryden Press; Kinnear, T.C. & Taylor, J.R. (1996), *Marketing Research: An Applied Approach*, 5th International edn, New York: McGraw-Hill Inc.
6. West, C. (1999), *Marketing Research*. Basingstoke: MacMillan Press.
7. Policy Commission (2002), op. cit.

8. Institute of Grocery Distribution (2004), *The Consumer Tracker 2004*, [on-line] Available at: http//www.igd-interactive.com/index.aspx?ReportID=26&Lang=en8MainPage=renderConten t&StoryID=186261 (accessed 26/01/09).
9. Kohls, R & Uhl, J. (1998), *Marketing of Agricultural Products*, 8th edn, Upper Saddle River, NJ: Prentice Hall.
10. DEFRA (2006a), *Agriculture in the United Kingdom 2006*, Department for Environment, Food and Rural Affairs, The Stationery Office, London; Hingley, M.K. (2005), 'Power to all our friends? Living with imbalance in supplier-retailer relationships', *Industrial Marketing Management*, Vol. 34, pp. 848–858.
11. CRC (2007), *The State of the Countryside*, Commission for Rural Communities, Cheltenham.
12. DEFRA (2006a), op. cit.
13. Institute of Grocery Distribution (2001), *Consumer Watch*, Institute of Grocery Distribution, Watford.
14. Institute of Grocery Distribution (2005), *Food Consumption*, Institute of Grocery Distribution Business Publication, Watford.
15. Drummond, C. (2006), 'Integrated farm management systems – making sustainability accessible to all: the LEAF viewpoint', in Andrews, M., Turley, D., Cummins, S., Dale, M and Rowlinson, P. (eds), *Aspects of Applied Biology* 80, pp. 51–63.
16. CRC (2007), op. cit.
17. DEFRA (2002), *Survey of Public Attitudes to Quality of Life and to the Environment 2001*, Department for Environment, Food and Rural Affairs.
18. Institute of Grocery Distribution (2001), op cit.
19. Knowles, T., Moody, R., & McEachern, M.G. (2007), 'European food scares and their impact on EU food policy', *British Food Journal*, Vol. 109, No. 1, pp. 43–67.
20. CPRE (2001), *Sustainable Local Foods*, The Council for the Protection of Rural England, London.
21. Mintel. (2003), *Attitudes Towards Buying Local Produce*, Mintel International Group (Mintel Market Intelligence), London.
22. Shears, P., Zollers, F., & Hurd, S. (2001), 'Food for thought–what mad cows have wrought with respect to food safety regulation in the EU and UK', *British Food Journal*, Vol. 103, No. 1, pp. 63–87.
23. Knowles et al., op. cit.
24. Yeung, R. & Morris, J. (2001), 'Food safety risk: consumer perception and purchase behaviour', *British Food Journal*, Vol. 103, No. 3, pp. 170–186.
25. Walley, K., Parsons, S., & Bland, M. (1999), 'Quality assurance and the consumer – a conjoint study', *British Food Journal*, Vol. 101, No. 2, pp. 148–161.
26. Lindgreen, A. & Hingley, M. (2003), 'The impact of food safety and animal welfare policies on supply chain management: the case of the Tesco meat supply chain', *British Food Journal*, Vol. 105, No. 6, pp. 328–349.
27. DEFRA (2006b), *Economic Note on UK Grocery Retailing*, Department for Environment, Food and Rural Affairs, London; DEFRA (2007), *Approved UK Certification Bodies*, [on-line] Available at: http://www.defra.gov.uk/farm/organic/standards/certbodies/approved.htm (accessed 09/08/07).
28. Lampkin, N., Measures, M., & Padel, S. (eds) (2006), *2007 Organic Farm Management Handbook*. Aberystwyth: University of Wales and Organic Advisory Service.
29. Soil Association (2006), *Organic Market Report 2006*, Soil Association, Bristol; Mintel (2005), *Organics-UK*, Mintel International Group (Mintel Market Intelligence), London.

30. Padel, S. & Foster, C. (2005), 'Exploring the gap between attitudes and behaviour: understanding why consumers buy or do not buy organic food', *British Food Journal*, Vol. 107, No. 8, pp. 606–625.

31. Rimal, A.P., Moon, W., & Balasubramanian, S. (2005), 'Agro-biotechnology and organic food purchase in the United Kingdom', *British Food Journal*, Vol. 107, No. 2, pp. 84–97.

32. Soil Association, op. cit.

33. Morris, S.H. & Adley, C.C. (2001), 'Irish public perceptions and attitudes to modern biotechnology: an overview with a focus on GM foods', *Trends in Biotechnology*, Vol. 19, No. 2, pp. 43–48.

34. Davies, B., Richards, C., Spash, C.L., & Carter, C. (2004), *Genetically Modified Organisms in Agriculture: Social and Economic Implications*, Aberdeen Discussion Paper Series No 2004-1, University of Aberdeen, Aberdeen.

35. Institute of Grocery Distribution (2001), op. cit.

36. Davies et al., op. cit.

37. Groves, A. (2005), *The Local and Regional Food Opportunity*. Watford: Institute of Grocery Distribution.

38. Padbury, G. (2006), *Retail and Foodservice Opportunities for Local Food*, [online] Available at: http://www.igd.com/consumer (accessed 1/10/07).

39. Countryside Agency (2000), *Eat the View–Promoting Sustainable Local Produce*. The Countryside Agency, Cheltenham.

40. Waitrose (2009), *Regional and Local Sourcing* [online] Available at: http://www.waitrose.com/food/originofourfood/sourcingbritishfood/regionalandlocalsourcing.aspx (accessed 26/01/09).

41. DEFRA (2006a), op. cit.

42. Jones, P., Comfort, D., & Hillier, D. (2004), 'A case study of local food and its routes to market in the UK', *British Food Journal*, Vol. 106, No. 4, pp. 328–335.

43. Youngs, J. (2003), 'A study of farm outlets in North West England', *British Food Journal*, Vol. 105, No. 8, pp. 531–541.

44. Farmers Weekly (2005), *So You Want to Open a Farm Shop?* [online] Available at: http://www.fwi.co.uk/articles/2005/06/24/92680/so-you-want-to-open-a-farm-shop.htm (accessed 1/10/07); FARMA (2007), *Certified Farmers' Markets*, [online] Available at: http://www.farma.org.uk (accessed 28/9/07).

45. Institute of Grocery Distribution (2008), *UK Grocery Retailing*, [online] Available at: http://www.igd.com/index.asp?id=1&fid=1&sid=7&tid=26&cid=94 (accessed 26/01/09).

46. Ilbery, B., Watts, D., Simpson, S., Gilg, A., & Little, J. (2006), 'Mapping local foods: evidence from two English regions', *British Food Journal*, Vol. 108, No. 3, pp. 213–225.

47. Archer, G.P., Sánchez, J.G., Vignali, G., & Chaillot, A. (2003), 'Latent consumers' attitude to farmers' markets in North West England', *British Food Journal*, Vol. 105, No. 8, pp. 487–497.

48. Holloway, L. & Kneafsey, M (2000), 'Reading the space of the farmers' market: a case study from the United Kingdom', *Sociologia Ruralis*, Vol. 40, No. 3, pp. 285–299.

49. Hughes, D. (1995), 'Animal welfare: the consumer and the food industry', *British Food Journal*, Vol. 97, No. 10, pp. 3–7.

50. Burgess, D. & Hutchinson, G.W. (2005), 'Do people value the welfare of farm animals?' *EuroChoices*, Vol. 4, No. 3, pp. 36–43.

51. Harper, G.C. & Makatouni, A (2002), 'Consumer perception of organic food production and farm animal welfare', *British Food Journal*, Vol. 104, Nos. 3/5, pp. 287–299.

Appendix 1: Consumer Attitude Questionnaire (A)

Good morning/afternoon. Can you spare a few minutes to answer some questions concerning food, farming and environmental issues?

The work is being conducted by Harper Adams University College and the data will be used to improve farming practice. All the information which you provide will be treated in the strictest confidence.

Please say to what extent you agree or disagree with the following statements.

CIRCLE ONE NUMBER PER STATEMENT.

		Strongly agree	Slightly agree	Neither agree Nor disagee	Slightly disagee	Strongly disagree	Don't know
1.	Farmers are only concerned with profit	1	2	3	4	5	6
2.	Lamb is safe to eat	1	2	3	4	5	6
3.	Too much land is being built on	1	2	3	4	5	6
4.	Non-British pork is safe to eat	1	2	3	4	5	6
5.	Too much food is imported	1	2	3	4	5	6
6.	Fox hunting is an acceptable activity	1	2	3	4	5	6
7.	Non-British beef is safe to eat	1	2	3	4	5	6
8.	I would like to eat more organic food	1	2	3	4	5	6
9.	Animals feel pain like humans	1	2	3	4	5	6
10.	Organic food is too expensive	1	2	3	4	5	6
11.	Too much land is being turned over to golf courses	1	2	3	4	5	6
12.	Most farm animals have a happy and comfortable life	1	2	3	4	5	6
13.	Intensive farming practices are the only way to meet demand	1	2	3	4	5	6
14.	Pork is safe to eat	1	2	3	4	5	6
15.	BSE (mad cow disease) is the result of the unnatural way we farm animals today	1	2	3	4	5	6
16.	I do not like battery egg production	1	2	3	4	5	6
17.	Non-British chicken is safe to eat	1	2	3	4	5	6
18.	Farmers look after the countryside	1	2	3	4	5	6
19.	The shooting of game birds is an acceptable activity	1	2	3	4	5	6
20.	Chicken is safe to eat	1	2	3	4	5	6
21.	British farmers are wealthy landowners	1	2	3	4	5	6
22.	Non-British lamb is safe to eat	1	2	3	4	5	6
23.	British farmers have high standards of animal welfare	1	2	3	4	5	6
24.	British farmers are good food producers	1	2	3	4	5	6
25.	Farm animals need better living conditions	1	2	3	4	5	6
26.	Many farms survive on bed-and-breakfast earnings	1	2	3	4	5	6

27.	Beef is safe to eat	1	2	3	4	5	6
28.	Farmers are cruel to animals	1	2	3	4	5	6
29.	The Government does not care for the countryside	1	2	3	4	5	6
30.	Chemicals used in farming can still be on food when it is eaten	1	2	3	4	5	6
31.	Farm incomes are declining	1	2	3	4	5	6

FOOD PURCHASE FACTORS

How important do you consider the following factors when you buy food?
CIRCLE ONE NUMBER PER STATEMENT.

		Extremely important	Very important	Moderately important	Slightly important	Not important
32.	Price	1	2	3	4	5
33.	Taste	1	2	3	4	5
34.	Health	1	2	3	4	5
35.	Animal welfare	1	2	3	4	5
36.	Environmental issues	1	2	3	4	5
37.	Fashion	1	2	3	4	5
38.	Availability	1	2	3	4	5

DEMOGRAPHIC INFORMATION

39. Into which age group do you fall? (CIRCLE ONE)
    (1)   16–24   (3)   35–44   (5)   55–64
    (2)   25–34   (4)   45–54   (6)   65+
40. How would you describe the area in which you live? (CIRCLE ONE)
    (1)   City   (2)   Town   (3)   Village   (4)   Country
41. Are you the principal wage earner of the household? (CIRCLE ONE)
    (1)   Yes   (2)   No   (3)   Joint
42. What is your occupation? (CIRCLE ONE)
    (1)   Professional/managerial   (4)   Clerical   (7)   Retired
    (2)   Skilled manual   (5)   Self-employed   (8)   In education
    (3)   Manual   (6)   Housewife   (9)   Between jobs
43. What is the occupation of the principal wage earner? (CIRCLE ONE)
    (1)   Professional/managerial   (4)   Clerical   (7)   Retired
    (2)   Skilled manual   (5)   Self-employed   (8)   In education
    (3)   Manual   (6)   Housewife   (9)   Between jobs
44. Are you vegetarian ? (CIRCLE ONE)
    (1)   Yes   (2)   No

That is the end of the interview. Thank you for your time and cooperation.

45. Sex of interviewee? (CIRCLE ONE)
    (1)   Male   (2)   Female

Appendix 2: Consumer Attitude Questionnaire (B)

Good morning/afternoon. Can you spare a few minutes to answer some questions concerning food, farming and environmental issues?

The work is being conducted by Harper Adams University College and the data will be used to improve farming practice. All the information which you provide will be treated in the strictest confidence.

Please say to what extent you agree or disagree with the following statements.

CIRCLE ONE NUMBER PER STATEMENT.

		Strongly agree	Slightly agree	Neither agree Nor disagee	Slightly disagee	Strongly disagree	Don't know
1.	British chicken is safe to eat	1	2	3	4	5	6
2.	Regulations regarding farm pollution should be tightened	1	2	3	4	5	6
3.	A healthy diet does contain red meat	1	2	3	4	5	6
4.	Most farmers pollute the soil	1	2	3	4	5	6
5.	Eggs are too expensive	1	2	3	4	5	6
6.	Genetic engineering of crops is ok	1	2	3	4	5	6
7.	Genetic engineering is inevitable	1	2	3	4	5	6
8.	Lamb is too expensive	1	2	3	4	5	6
9.	BSE (mad cow disease) is the kind of natural disease that happens every so often	1	2	3	4	5	6
10.	Most farmers pollute water courses	1	2	3	4	5	6
11.	Modern farming is detrimental to birdlife	1	2	3	4	5	6
12.	The use of genetic engineering is increasing	1	2	3	4	5	6
13.	Free range eggs are too expensive	1	2	3	4	5	6
14.	Farmers should spend more money on the environment	1	2	3	4	5	6
15.	Most farmers pollute the air	1	2	3	4	5	6
16.	Young people have to leave rural areas for a decent job	1	2	3	4	5	6
17.	Pork is too expensive	1	2	3	4	5	6
18.	Organic food tastes better than food produced by other methods	1	2	3	4	5	6
19.	Chemicals are needed to produce food cheaply	1	2	3	4	5	6
20.	Agriculture is a struggling industry	1	2	3	4	5	6
21.	There is so much information concerning healthy eating that it is confusing	1	2	3	4	5	6
22.	People moving from urban areas have a beneficial effect on rural communities	1	2	3	4	5	6
23.	Genetic engineering of animals is ok	1	2	3	4	5	6
24.	The farmer receives a fair price from the supermarket for his produce	1	2	3	4	5	6
25.	Using radiation to preserve food is safe	1	2	3	4	5	6
26.	I buy locally produced products to support local farmers	1	2	3	4	5	6

27.	There is too much information concerning healthy eating	1	2	3	4	5	6
28.	British lamb is safe to eat	1	2	3	4	5	6
29.	Modern farming is detrimental to wildlife	1	2	3	4	5	6
30.	Farmers use too much pesticide	1	2	3	4	5	6
31.	British pork is safe to eat	1	2	3	4	5	6
32.	I am knowledgeable about farming and the environment	1	2	3	4	5	6
33.	Farmers remove too many hedges	1	2	3	4	5	6
34.	British beef is safe to eat	1	2	3	4	5	6
35.	Farmers are the financial backbone of the rural community	1	2	3	4	5	6
36.	There should be free access to the country side	1	2	3	4	5	6
37.	I buy locally produced products to preserve jobs in the area	1	2	3	4	5	6
38.	Farmers use too much herbicide	1	2	3	4	5	6
39.	I would pay more for food products grown locally	1	2	3	4	5	6
40.	Services in rural areas are not as good as in urban areas	1	2	3	4	5	6
41.	Beef is too expensive	1	2	3	4	5	6
42.	Farmers are making efforts to reduce pollution	1	2	3	4	5	6

DEMOGRAPHIC INFORMATION

43. Into which age group do you fall? (CIRCLE ONE)
   (1)   16–24   (3)   35–44   (5)   55–64
   (2)   25–34   (4)   45–54   (6)   65+

44. How would you describe the area in which you live? (CIRCLE ONE)
   (1)   City   (2)   Town   (3)   Village   (4)   Country

45. Are you the principal wage earner of the household? (CIRCLE ONE)
   (1)   Yes   (2)   No   (3)   Joint

46. What is your occupation? (CIRCLE ONE)
   (1)   Professional/managerial   (4)   Clerical   (7)   Retired
   (2)   Skilled manual   (5)   Self-employed   (8)   In education
   (3)   Manual   (6)   Housewife   (9)   Between jobs

47. What is the occupation of the principal wage earner? (CIRCLE ONE)
   (1)   Professional/managerial   (4)   Clerical   (7)   Retired
   (2)   Skilled manual   (5)   Self-employed   (8)   In education
   (3)   Manual   (6)   Housewife   (9)   Between jobs

48. Are you vegetarian ? (CIRCLE ONE)
   (1)   Yes   (2)   No

That is the end of the interview. Thank you for your time and cooperation.

49. Sex of interviewee? (CIRCLE ONE)
   (1)   Male   (2)   Female

13 Consumer Preferences for Food Quality: A Choice Experiment Regarding Animal Welfare and Food Safety in Chicken

BY MORTEN RAUN MØRKBAK,[*] TOVE CHRISTENSEN[†]
AND BERIT HASLER[‡]

Keywords

food safety, animal welfare, valuation, consumer preferences, chicken, choice experiment.

Abstract

The present study focuses on animal welfare and food safety as inherent attributes of consumer perceptions of animal products. Through a choice experiment, we find significant willingness to pay for animal welfare and food safety. As a novel contribution, consumer willingness to pay for a bundle of attributes exceeds the sum of the willingness to pay for the individual attributes.

[*]   Mr Morten Raun Mørkbak, Institute of Food and Resource Economics, Faculty of Life Sciences, University of Copenhagen, Rolighedsvej 25, 1958 Frederiksberg C, Denmark. E-mail: mm@foi.dk. Telephone + 45 3533 68 72.

[†]   Dr Tove Christensen, Institute of Food and Resource Economics, Faculty of Life Sciences, University of Copenhagen, Rolighedsvej 25, 1958 Frederiksberg C, Denmark. E-mail: tove@foi.dk. Telephone + 45 3533 6800.

[‡]   Dr Berit Hasler, Department of Policy Analysis, the National Environmental Research Institute, University of Aarhus, Frederiksborgvej 399, 4000 Roskilde, Denmark. E-mail: bh@dmu.dk. Telephone + 45 4630 1835.

Introduction

A wide variety of animal products appears in food markets, indicating that consumers are willing to pay for products that offer specific quality attributes, such as environmental friendliness, increased animal welfare or food safety. Markets for products with a particular focus on food safety may reflect the general increase in attention to quality characteristics. The existence of markets for safer food also might be seen as a reaction to the seemingly increasing numbers of food-borne zoonotic infections. In 2006 for example, almost 5000 human cases of the two most common zoonotic infections, *Campylobacter* and *Salmonella*, were registered in Denmark, which in turn implemented official action plans to control the risks.[1]

Products that promote attributes such as regard for the environment and animal welfare, as well as those that are free of bacteria, are often more costly than conventional products. Therefore, an essential concern is the extent to which consumers are willing to pay a price premium for goods offering these attributes. For example, the price premiums for organic meats averaged almost 50 per cent in 2006,[2] but we know little about the value that consumers place on the individual attributes of organic meat.

Danish markets for chicken meat include those for products with specific quality attributes, such as '*Salmonella* and/or *Campylobacter* free', free-range and organic. These markets are typically small, partly because informational problems create market failures, and partly because we lack knowledge about the potential interactions between food safety and other quality parameters, which makes it almost impossible for consumers, as well as producers and public authorities, to balance the benefits against the risks. Furthermore, practical problems of availability in the local shops, at least temporarily, cause additional failures. As a consequence, market prices do not reveal consumers' true valuation of animal products that offer increased safety or animal welfare, which suggests hypothetical valuation methods may be of great help to improve our knowledge about how consumers value such attributes.

STUDY OBJECTIVES

This study investigates consumer valuation of the inherent attributes of animal products. Specifically, we attempt to elicit the amount that consumers state they are willing to pay for individual attributes, identify the relative importance of those attributes and uncover any potential interactions among the different attributes.

To pursue these objectives, we perform a choice experiment to elicit estimates of consumers' willingness to pay (WTP) for animal welfare in broiler production and food safety in chicken meat. The case represents a very interesting trade-off, in that outdoor access is believed to increase chicken welfare but also increases the risks of *Campylobacter* infections. The food safety attribute is highly relevant, because chicken meat is believed to be the main source of *Campylobacter* infection. The welfare aspect of broiler production has been of greater concern, mainly because of increasingly large herds, restricted space and rapid growth of broilers, which cause leg problems, as well as the use of technical facilities to ensure climate and light for optimal growth conditions and the short production period.

The remainder of this chapter is organized as follows: After a short review of previous studies on food safety and animal welfare in animal products, we describe the case study and present the results. Next, we discuss the results, along with both marketing and policy implications.

Animal Welfare and Food Safety: A Literature Review

Prior stated preference studies regarding the economic valuation of meat products, animal welfare and food safety mainly use contingent valuation and choice experiments. We group these previous studies as follows: those estimating the value of both food safety and animal welfare, studies examining only microbiological food safety and those examining only animal welfare.

FOOD SAFETY AND ANIMAL WELFARE

Few studies attempt to determine the value of both attributes at the same time, and all use choice experiments. Meuwissen and van der Lans examine three characteristics (animal welfare, food safety and environmental concerns) of pork.[3] They find that respondents are willing to pay a price premium of 45 per cent for animal welfare, 42 per cent for food safety and 38 per cent for environmental concerns. Carlsson and colleagues include three different meat products (pork, beef and poultry) in their analysis, assessing animal welfare, place of origin and lack of genetic modifications (GM), and find that consumers on average are willing to pay price premiums ranging from -4–70 per cent for animal welfare, 10–20 per cent for knowing the place of origin and 30–60 per cent for being GM-free.[4] Finally, Kontoleon and Yabe estimate respondents' WTP for pesticide-free, GM-free, increased animal welfare eggs and obtain price premiums, of, respectively, 200 per cent, 50 per cent and 120 per cent.[5]

MICROBIAL FOOD SAFETY

Two stated preference studies focus on the WTP for microbial food safety. Using choice experiments and contingent valuation analyses, Goldberg and Roosen investigate 0 per cent, 40 per cent and 80 per cent reductions of *Salmonella* and *Campylobacter* risks in chicken breasts.[6] Consumers appear willing to pay to reduce these risks, ranging approximately 0–50 per cent price premiums for various risk reductions in the choice experiments and price premiums of 10–25 per cent according to the contingent valuation method. The WTP for reducing *Salmonella* risks is slightly higher than for reducing *Campylobacter* risks. Furthermore, Goldberg and Roosen show that the WTP for a joint reduction is smaller than the sum of individual WTP for *Salmonella* and *Campylobacter* risk reduction. They consider this result as a test of the additivity of WTP and test for cross-effects between attributes. Similarly, Hobbs and colleagues' experimental auction offers respondents a choice among four alternative sandwiches: one with increased food safety, one that could be traced back to the farm, one with farm information (animal welfare assurance) and one that combined all three attributes.[7] They find that the WTP for a sandwich offering all attributes – food safety, traceability and farm information – is less than the sum of the WTP for the sandwiches offering the individual attributes.

ANIMAL WELFARE

Both Bennett and Bennett and Blaney use a contingent valuation method to find the WTP for animal welfare related to the production of eggs.[8] They estimate the value of improved animal welfare, associated with battery cages for egg production, as equivalent to 0.5–1.2 Euro.

SUMMARY

In all the reviewed studies, consumers state that they are willing to pay for increased food safety and/or animal welfare in meat products. However, no clear pattern in the overall rankings of animal welfare and food safety emerges. Animal welfare and food safety might span a variety of underlying attributes, which requires caution when making comparisons across studies. We add to this sparse literature on joint valuation of two distinct quality attributes by focusing specifically on food safety and animal welfare.

Choice Experiment Method

The choice experiment method belongs to the family of stated preference methods. It provide a natural choice for analyzing consumers' WTP for specific attributes, because the method is based on Lancaster's attribute-based consumer theory[9] and random utility theory (RUT).[10] Similar to other choice modelling techniques, choice experiments originally were developed for market analyses,[11] specifically, in the car industry. Since the mid-1980s, choice modelling techniques increasingly have been used and developed for valuations of non-market goods, such as environmental issues and landscape restoratation.[12] During the past decade, interest in market analyses has returned – though now in relation to food quality.

In a choice experiment study, a representative sample of respondents chooses from a predefined set of alternatives (choice set). The alternatives consist of several attributes that can assume different levels. Usually respondents repeat the exercise several times with different choice sets. For our study, the attributes are food safety, animal welfare and price. Each choice provides information about the respondents' preferences and, because one of the attributes is presented in monetary terms, their WTP for the attributes. A choice experiment provides an indirect valuation method, because we must derive the WTP indirectly from the information inherent in respondents' choices. In contrast, respondents directly state their WTP in contingent valuation studies.[13] A choice experiment is recommended for valuing complex problems, because it reflects real market conditions better than other forms of valuation method, so the cognitive burden imposed on respondents by the valuation task diminishes. Another advantage of this method is that the focus remains on attributes of the goods rather than on the good *per se*, which facilitates more detailed analyses of changes in the marginal values of goods.[14]

METHODOLOGICAL FRAMEWORK

The random utility framework can be described as follows: Assume that individual i obtains utility U_{ij} from good j. Assuming that the utility consists of a deterministic part

(V) that depends on a vector S of product characteristics[§] and is partly stochastic (ε_i), it can be expressed as:

$$U_{ij} = V(S_{ij}) + \varepsilon_i \tag{1}$$

An important issue in any economic analysis of consumer choice behaviour is the need to obtain estimates of the WTP for certain attributes. Assuming that the random utility function is additively separable and linear in attributes, the estimated preference parameter for attribute k (β_k) represents the marginal utility of attribute k. Furthermore, marginal rates of substitution between any two attributes m and k can be computed as the ratio of the parameter estimates $\text{MRS}_{km} = \beta_k / \beta_m$. The parameter estimate of the price attribute β_p has a special status, because MRS_{kp} is the marginal value of attribute k, per DKK. This ratio between β coefficients is known as the implicit price, which also can be denoted as the WTP,[16] as can easily be shown when the utility is formulated as:

$$U_{ij} = \beta_1 S_{1j} = \ldots\ldots + \beta_k S_{kij} + \ldots\ldots + \beta_p S_{pij} + \ldots\ldots\ldots + \beta_K S_{Kji} + \varepsilon_j \tag{2}$$

and the implicit value or WTP is calculated as:

$$\frac{\partial U_{ij}}{\partial s_{kij}} \bigg/ \frac{\partial U_{ij}}{\partial s_{pij}} = \frac{\beta_k}{\beta_p} \tag{3}.$$

When interactions occur between the attributes k and m, the marginal utility of attribute k depends on the level of attribute m, such that:

$$\partial U_{ij} / \partial s_{kij} = \beta_k + \beta_{km} S_{mij} \tag{4}.$$

A random utility presentation of a discrete choice model typically is analyzed using logit or probit models, whether binary logit, multinomial logit (also denoted conditional logit), nested logit, mixed logit or (multinomial) probit.[17] A standard logit model requires independence between the ratios of probabilities of choosing any two alternatives of the availability of other alternatives, formulated as the model exhibiting independence from irrelevant alternatives (IIA).[18][¶] Mixed logit and probit models are more general models that allow for taste variations and correlation of unobserved factors over time and do not require the IIA property. In the present analysis, probit provides the best modelling fit and is used throughout.

[§] Characteristics that are not attribute-related can be included in the utility function, such as individual characteristics including gender, income and so forth.[15]

[¶] The restrictive nature of the IIA assumption is often illustrated by the red bus/blue bus problem.[19] Consider a choice between going to work by car or by a blue bus. For simplicity, assume that the utilities of the two means of transport are equal, implying equal choice probabilities (½). The probability ratio is 1 ($P_c/P_{bb} = 1$). Now introduce the choice of a red bus that, other than the colour, is identical to the blue bus. This introduction should not affect the probability of choosing the car, but the probability of choosing a bus should be shared between the two types of buses, $P_{rb} = P_{bb} = .25$. However, in a logit model, the probability ratio between the car and the blue bus must remain 1, and hence, $P_c = P_{rb} = P_{bb} = .33$. The logit representation only reflects a real-life situation if consumers care about bus colour. If consumers do not care about colour, a logit model overestimates demand for the two bus modes.

SURVEY DESIGN

A choice experiment was conducted using an Internet-based survey during February 2005. On the basis of a focus group meeting and two pilot studies, we determined to keep the choice experiment as simple as possible. Each chicken product (alternative) is described by only three attributes. Food safety consists of two levels of *Campylobacter* risk: a chicken without information about *Campylobacter* contents versus a chicken labelled *Campylobacter*-free. Animal welfare also consists of two levels, this time of production method: a chicken had had outdoor access versus a chicken raised indoors. The experiment also includes eight levels of the price attribute.

One whole chicken (average weight 1300g) is the product in the choice experiment, because pre-tests show it is the most homogeneous and well-known chicken product. In addition, the pre-tests indicate we need to specify whether the chicken is frozen or chilled, because they serve quite different purposes for consumers and different product substitutes. For example, when a consumer is shopping for a chilled chicken for that night's dinner, a frozen chicken is not a substitute, whereas chilled minced pork might be. The maximum price used in the choice experiment equals the price of an organic chicken in a Danish supermarket, because the organic chicken should be associated with a long range of attributes and thus charge a higher price than a chicken with fewer attributes. Respondents received no further information regarding the food safety or animal welfare attributes, because the study intention was to capture market behaviour as closely as possible. The levels of the attributes appear in Table 13.1.

To mimic actual shopping situations as closely as possible and obtain as much information about each choice as possible, we incorporate the respondents' usual purchases as their opt-out value, which we identified in a questionnaire previous to the choice experiment. Prior to each choice scenario, we reminded respondents that people often act differently in a hypothetical scenario or responding to a questionnaire than they would in a real situation when they faced a similar problem. According to Carlsson and colleagues, this reminder should reduce the risk that respondents state a higher WTP than they would in a real situation, due to some kind of moral satisfaction (that is, 'warm glow').[20] The reminder is denoted a 'cheap talk' script.

We use a full-factorial design, which allows us to estimate the two-way interaction effects. Each choice set contains two alternatives (three with the opt-out alternative). The 16 choice sets constitute four blocks, such that each respondent received four choice

Table 13.1 Attributes and their levels in the choice experiment

Attributes	Levels
Food safety	Not labelled *Campylobacter*-free, labelled *Campylobacter*-free
Animal welfare	Indoor produced, outdoor produced
Price (DKK)	40, 47, 55, 64, 74, 85, 97, 110

Note: DKK 10 correspond to €1.34.

sets. Three choice sets contain dominant alternatives[**] and serve as consistency tests. A total of 121 respondents chose a dominated alternative, which we interpret as a failure to maximize utility, so we exclude them from the estimation (inconsistent answers), as we note in Table 13.2. In addition, 246 respondents stated that they supported animal welfare or *Campylobacter*-free chickens but did not think that it was their duty as consumers or taxpayers to pay for these attributes. These statements indicate that the respondents are not willing to make monetary trade-offs to secure given characteristics, which this contrasts with the random utility framework that we apply. Therefore, we interpret these responses as protest answers and exclude them from the data analysis.[21] The final data set used for further estimation thus consists of 2301 respondents.

The sample comes from ACNielsen Denmark's online database. Estimates suggest that that 75 per cent of approximately 2.4 million private households in Denmark are online. Panel members are all at least 15 years of age, and the online panel is representative with respect to the 75 per cent of the Danish population who have Internet access in their homes.[††]

Results

The socio-demographic distribution of the sample of Danish consumers appears in Table 13.2, with respect to gender, education, household income, age and number of children. Compared with the Danish population in 2007, the socio-demographic characteristics are statistically different. The sample contains too many respondents under the age of 50 years and too many families with children. Finally, high-income groups are over-represented in the sample. However, these skewed distributions are not surprising, in that previous surveys also show that younger, highly educated respondents with high incomes tend to answer such questionnaires more frequently. Nonetheless, we keep this bias in the sample in mind when interpreting the results.

A model that allows for cross-effects between food safety and animal welfare provides better fit in terms of describing the respondents' choices than does a simple main effects model. Table 13.3 lists the model results in terms of the marginal utilities of the attributes, standard errors, *p*-values and WTP estimates,[‡‡] using a multinomial probit model.

We find a positive WTA for avoiding *Campylobacter* and allowing chickens outdoor access. An average consumer is willing to pay an additional 1.8 Euro for an outdoor-produced chicken compared with an indoor-produced chicken when none of the chickens are guaranteed *Campylobacter*-free. Similarly, an average consumer would pay an additional 2.3 Euro for a *Campylobacter*-free chicken compared with a chicken without

---

[**]   A dominant alternative is one where all attribute levels are better than the other alternatives in the choice set. In our analysis, consumers should consider animal welfare better when the chicken has had outdoor access. Similarly, *Campylobacter*-free should dominate no information about *Campylobacter* risks. Finally, respondents are assumed to prefer lower prices. Hence, a dominant chicken is cheaper than the alternative product, labelled *Campylobacter*-free, and outdoor bred. If a respondent chooses a dominated alternative, we characterize them as inconsistent. If a respondent does not agree with this ranking of attributes, an alternative that we consider dominant might not be dominant in their eyes, which suggests that removing these respondents might lead to an overestimated WTP. Nevertheless, we consider this potential bias small, as there are only 121 inconsistent answers (4.5 per cent), and so we do not run the model again including this group.

[††]   This limitation in representativeness is important to keep in mind when interpreting the results.

[‡‡]   Exchange rate: €1 corresponds to DKK 7.41.

Table 13.2 Socio-demographic distribution of the respondents in the sample

	Sample	STAT Denmark		Chi-test
	Frequency	Percent	Exp.	(chi-value)
Total	2668			
Protest answers	246			
Inconsistent answers	121			
Sample used in estimation	2301			
Gender				
Men	1074	0.495	1139	0.007
Women	1227	0.505	1162	
Education				
Lower	811	0.143	1420	
Higher	1316	0.240	329	0
Other	174	0.617	552	
Household income				
Lowest income group Income under 100 000	217	0.091	209	
Next lowest income group DKK 100 000–199 999	174	0.259	597	6.84E-136
Next highest income group DKK 200 000–399 999	553	0.283	652	
Highest income group DKK 400 000 or more	1357	0.366	843	
Age				
under 50	1420	0.545	1255	4.91E-12
Above 50	881	0.455	1046	
Children				
No children	1481	0.714	1642	1.13E-13
Children	820	0.286	659	

Notes: Low education = nine-year (compulsory) school. High education = upper secondary or university degree. Other education = vocational education.

Table 13.3 Consumer behaviour with respect to avoiding *Campylobacter* risks and increased chicken welfare using a cross-effects model

Choice	Coefficients	Std. err	*p*-value	WTP – € per chicken
Chicken produced with outdoor access and not *Campylobacter* free	0.4056	0.0583	0.000	1.8
Chicken produced indoors and *Campylobacter* free	0.5186	0.0592	0.000	2.3
Chicken produced with outdoor access and *Campylobacter*-free	0.4078	0.1073	0.000	5.8
Price	-0.0304	0.0008	0.000	-
ASC	-1.6602	0.1264	0.000	-
VAR(opt-out)	1.9192	0.1271	0.000	-
LRI	0.2694			
N	9204			
Log L	-7388			

Notes: The WTP estimates are marginal extra values in relation to a normally produced chicken at 5.3 Euro. The p-values indicate the significant levels of the parameters. ASC is an alternative specific constant for the status quo alternative, and VAR(opt-out) is the variance of the error component of alternative 3. LRI is the McFadden likelihood ratio index,[22] which equals 0.27, an acceptable value for a model.[23] N is the number of observations, and Log L is the log-likelihood of the final model.

information about *Campylobacter* when both chickens are produced indoors. The sum of the WTP for food safety and animal welfare is 4.1 Euro. However, the interaction effects imply that when food safety and animal welfare are offered jointly, the value of the product increases to 5.8 Euro. These results suggest the existence of an extra WTP for attributes when they are offered as a bundle.

Conclusions and Policy Perspectives

As we set out to do, we have estimated the amount that consumers state they are willing to pay for avoiding *Campylobacter* (1.8 Euro per chicken) and allowing chickens outdoor access (2.3 Euro per chicken). Consumers will pay more for allowing outdoor access than for avoiding *Campylobacter*; we also identify interactions between consumer valuations of avoiding *Campylobacter* and allowing chickens outdoor access.

These results indicate two main findings. First, the amount that consumers state they are willing to pay for avoiding *Campylobacter* and for outdoor-produced chickens exceeds market evidence. Second, the value of bundles of attributes cannot be inferred from knowledge about the values of individual attributes. We therefore address the market and policy implications of these findings.

WILLINGNESS TO PAY

The amount that consumers state they are willing to pay for avoiding *Campylobacter* and outdoor-produced chickens exceeds market evidence. There will always be a hypothetical bias in stated preference studies (for example, non-commitment, moral glow, yea-saying), but we also offer a range of likely explanations for this difference.

In particular, our results emerge from a choice situation in which *Campylobacter*-free chickens and chickens with outdoor access are readily available and accessible, because respondents are instructed to focus on food safety, animal welfare and price. In contrast, in a real market situation, these conditions may not be present. Consumers may not even recognize the trade-offs they are making in reality because of the myriad of trade-offs that they must consider and because they lack information. In such situations, their market behaviour will not reflect well-informed choices. We asked consumers to assess a single shopping situation involving the choice of chicken (a marginal valuation task); they did not need to assess all future choices of chicken. Hence, when consumers consider the budgetary effects of paying extra or paying more for not just their next chicken but all future chicken, they might be less willing to pay a premium for the additional attributes. Instead, they may wish to reduce their consumption of outdoor-produced, *Campylobacter*-free chickens and buy other chicken products or substitute other types of meat. The aggregate market implications thus should be expected to be lower than the marginal effects.

Moreover, in a market setting, consumers do not exclusively determine demand; supermarkets also have a great impact on product availability. In modern markets, the consumers and producers seldom meet. Rather, producers are represented by producer organizations that coordinate production. Consumer demand depends on what consumers can buy in the shops. Of course, retailers try to satisfy consumer demand, but they also have their own agenda, based on maximizing profits. Therefore, consumers' WTP moves through a filter (retailer preferences) before they reach the producers (organizations). In turn, the market implications of the results depend on how consumers' stated behaviour is perceived and incorporated into sales strategies by retailers and supermarkets.

Our results also suggest that in certain conditions (for example, consumers focused on just a few attributes, *Campylobacter*-free chickens readily available), the willingness to pay to avoid the risk of *Campylobacter* in chicken is considerable. Food risks might be reduced and animal welfare increased by providing labels that allow consumers to choose *Campylobacter*-free and/or animal welfare chicken. The present market shares of products offering increased food safety and animal welfare may persist below a critical level, which implies that practical factors such as availability, rather than underlying consumer preferences, dominate consumer choice.

BUNDLES OF ATTRIBUTES

As a separate issue, we find that the value of bundles of attributes cannot be inferred directly from the values of the individual attributes. This non-additivity effect has not received much attention in existing literature, though both Goldberg and Roosen and Hobbs and colleagues find a similar effect[24] – except they find the bundles are less valued than the sum of the individual attributes. The differences partly result from interpretations. Hobbs and colleagues find a diminishing marginal utility of income, whereas our result comes from explicitly taking the relationship between the levels of attributes into account. Our approach is similar to Goldberg and Roosen's, but differences may exist because of the close substitution between the attributes (*Salmonella* and *Campylobacter*) in their study, which gives rise to an embedding effect. In our study, the attributes (animal welfare and food safety) are not close substitutes.

Another related result appears in a previous survey of consumer perceptions of organic products, which indicates that consumers perceive the benefits of organic products to be higher than those guaranteed by organic regulations.[25] Public information about organic production previously has concentrated on the environmental and animal welfare benefits of organic farming, as reflected in the survey, because consumers indicated that organic production would provide environmental and animal welfare benefits. However, consumers also perceived organic products as fresher, tastier and healthier, even though those benefits has not been supported by public information campaigns. The results of the survey suggest that organic consumers perceive organic products to provide a bundle of attributes and that the value of the bundle exceeds the value of documented benefits. These findings deserve much more research attention.

Our findings also suggest that knowledge about how consumers value not only individual attributes but also bundles of attributes can provide valuable input into how to design future niche productions. Socially optimal provisions of public goods, such as environmentally friendly and animal welfare-improving production, depend heavily on how consumers value these attributes.

RESEARCH LIMITATIONS AND AVENUES FOR FURTHER RESEARCH

Our analyses point to several avenues for research that could improve the validity and policy relevance of stated preference studies. Hypothetical bias might be reduced but rarely can be completely eliminated. Great efforts to reduce bias already have been made, and we suggest that continued efforts should take high priority in future choice experiment studies. Astonishingly little attention centres on the significance of assessing marginal versus aggregate choice behaviour, just as we still know little about the actual impact of consumer preferences on how supermarkets determine their assortment of goods.

The results also indicate that the elicited stated preferences do not reflect observed market behaviour. At the same time, our results might imply the existence of a significant WTP for food safety and animal welfare when certain specific conditions exist. Whether the findings result from stated preferences that do not reflect actual behaviour or whether the differences are merely a result of different settings in choice experiments versus actual shopping situations has important policy and marketing implications. Further research into the differences between estimated WTP values and actual behaviour thus would be valuable.

References

1.   National Food Institute (2006), *Annual Report on Zoonoses in Denmark 2006*, The Danish Zoonosis Centre, Technical University of Denmark.

2. Denver, S.; Christensen, T. & Krarup, S. (2007), 'Organic consumption and nutritious diet', *Samfundsøkonomen*, forthcoming.

3. Meuwissen, M.P.M. & van der Lans, I.A. (2004), '*Trade-offs between consumer concerns: An application for pork production*', Paper presented at the 84th EAAE seminar, Food Safety in a Dynamic World, Zeist, The Netherlands, 8–11 February 2004.

4. Carlsson, F., Frykblom, P. & Lagerkvist, C.-J. (2005), 'Consumer preferences for food product quality attributes from Swedish agriculture', *Ambio*, Vol. 34, No. 4–5, pp. 366–370.

5. Kontoleon, A. & Yabe, M. (2003), 'Assessing the impacts of alternative 'opt-out' formats in choice experiment studies: Consumer preferences for genetically modified content and production information in food', *Journal of Agricultural Policy Research*, Vol. 5, pp. 1–43.

6. Goldberg, I. & Roosen, J. (2005), *Measuring consumer willingness to pay for a health risk reduction of Salmonellosis and Campylobacteriosis*. Paper presented at the 11th Congress of the EAAE.

7. Hobbs, J.E., Bailey, D., Dickinson, D.L. & Haghiri, M. (2005), 'Traceability in the Canadian red meat sector: Do consumers care?', *Canadian Journal of Agricultural Economics*, Vol. 53, No. 1, pp. 47–65.

8. Bennett, J. (1996), 'CVM, dichotomous choice. Willingness-to-pay measures of public support for farm animal legislation', *Veterinary Record*, Vol. 139, pp. 320–321; Bennett, J. & Blaney, R.J.P. (2003), 'Estimating the benefits of farm animal welfare legislation using the contingent valuation method', *Agricultural Economics*, Vol. 29, No. 1, pp. 85–98.

9. Lancaster, K.J. (1966), 'A new approach to consumer theory', *The Journal of Political Economy*, Vol. 74, No. 2, pp. 132–157.

10. Luce, R.D. (1959), *Individual Choice Behaviour*. New York: Wiley,; McFadden, D. (1974), 'Conditional logit Analysis of qualitative choice behavior', in P. Zarembka (ed.), *Frontiers in Econometrics*. New York: Academic Press.

11. Batsell, R.R. & Louviere, J. (1991), 'Experimental analysis of choice', *Marketing Letters*, Vol. 2, No. 3, pp. 199–214; Louviere, J., Hensher, D.A., & Swait, J. (2000), *Stated Choice Methods. Analysis and Applications*. Cambridge, UK: University Press.

12. Adamowicz, W.L. (1995), 'Alternative valuation techniques: A comparison and movement to a synthesis', in K.G. Willis & J.T. Corkindale (eds) *Environmental Valuation: New Perspectives*. Wallingford, UK: CAB International, pp. 144–159; Boxall, P., Adamowicz, W., Swait, J., Williams, M. & Louviere, J. (1996), 'A comparison of stated preference methods for environmental valuation', *Ecological Economics*, Vol. 18, pp. 243–253; Hanley, N., Wright, R.E. & Adamowicz, W.L. (1998), 'Using choice experiments to value the environment', *Environmental and Resource Economics*, Vol. 11, No. 3–4, pp. 413–428; Hanley, N., Mourato, S. & Wright, R.E. (2001), 'Choice modelling approaches: A superior alternative for environmental valuation?', *Journal of Economic Surveys*, Vol. 15, No. 3, pp. 435–462.

13. Adamowicz, W.L., Louviere, J., & Swait, J. (1998), *Introduction to Attribute-Based Stated Choice Methods*, National Oceanic and Atmospheric Administration, Washington, USA. [citeret d. 14-9-0004], http://www.darp.noaa.gov/library/pdf/pubscm.pdf.

14. Ibid.

15. Adamowicz, W.L., Louviere, J. & Williams, M. (1994), 'Combining revealed and stated preference methods for valuing environmental amenities', *Journal of Environmental Economics and Management*, Vol. 26, No. 3, pp. 271–292; Bateman, I., Carson, R.T., Day, B., Hanemann, M., Hanley, N., Hett, T., Jones-Lee, M., Loomes, G., Mourato, S., Özdemiroglu, E., Pearce, D. W., Sugden, R. & Swanson, J. (2002), *Economic Valuation with Stated Preference Techniques: A Manual*. Cheltenham, UK: Edward Elgar.

16. Hanley, N., Adamowicz, W.L. & Wright, R.E. (2002), *'Price vector effects in choice experiments: an empirical test'*, World Congress of Environmental and Resource Economists, Monterey, CA.

17. Train, K. (2003), *Discrete Choice Methods with Simulation.* Cambridge: Cambridge University Press.

18. Holmes, T. & Adamowicz, W.L. (2003), 'Attribute based methods', in *A Primer on the Economic Valuation of the Environment.* New York: Kluwer.

19. Ben-Akiva, M. & Lerman, S.R. (1985), *Discrete Choice Analysis. Theory and Application to Travel Demand.* Cambridge: MIT Press.

20. Carlsson, F., Frykblom, P. & Lagerkvist, C.-J. (2004), *Using cheap-talk as a test of validity in choice experiments,* Working Paper in Economics. No. 128, Department of Economics, Gothenburg University.

21. Freeman, A.M. (1993), *The Measurement of Environmental and Resource Values: Theory and Methods,* Resources for the Future, Washington DC.

22. McFadden, op. cit.

23. Louviere et al., op. cit.

24. Goldberg and Roosen, op. cit.; Hobbs et al., op. cit.

25. Wier, M., Andersen, L.M., & Millock, K. (2005), 'Information provision, consumer perception and values – the case of organic foods', in S. Krarup & C.S. Russell (eds). *Environment, Information and Consumer Behaviour.* Cheltenham: Edward Elgar Publishing, pp. 161–178.

14 *Consumer Demand for Ethically Improved Animal Production Systems*

BY ANA ISABEL COSTA[*] AND JOHN W. CONE[†]

Keywords

agri-business ethics, food consumption behaviour, willingness-to-pay estimation, experimental auctions.

Abstract

In this chapter, we review a considerable body of empirical research (both published and new), with the aim of producing a well-founded, up-to-date, in-depth analysis of the nature and extent of consumer demand for ethically improved animal production systems and derived foods.

In this chapter, we will first show that consumers do not differentiate well between animal production systems with different ethical standards. Second, we see that when positive differentiation takes place, it is because the system is perceived to impact individual consumption benefits positively. Third, we point out that consumers' preference and willingness to pay for foods of animal origin is driven by habit and hedonic preference, not ethical considerations. Fourth, we explain why consumers' stated preferences for ethically improved foods of animal origin almost always lead to an overestimation of true demand. Fifth, and finally, we reveal that consumers are willing to pay a premium for ethically improved foods only when consumption is perceived to lower personal health risks.

[*]   Dr Ana Isabel Costa, School of Economics and Management, Portuguese Catholic University, Palma de Cima, 1649-023 Lisboa, Portugal. E-mail: anacosta@fcee.lisboa.ucp.pt. Telephone: + 351 217 214 270.

[†]   Dr John W. Cone, Animal Nutrition Group, Wageningen University, P.O. Box 338, 6700 AH Wageningen, The Netherlands. E-mail: john.cone@wur.nl. Telephone: + 31 317 483 542.

Introduction and Background to the Research

By the late 1970s, Western agricultural production slowly started responding to a growing societal awareness of the potential influence of farming practices on environmental conservation and animal welfare. This response first took shape in the form of food cooperatives and small organic farms led by a handful of dedicated 'green' producers and consumers. In addition to concern for the environment, their actions were often rooted in deeply held convictions about the wider societal benefits of a more local, traditional and natural way of producing food.[1] By the mid-1990s, however, the proliferation of agricultural systems following different ethical standards (for example, free-range, grass-fed, fair-trade, organic, locally produced), together with the globalization of environmental and welfare concerns – undoubtedly fuelled by a series of food scandals bringing into question the safety of conventional farming practices – drove governments to initiate the laborious task of regulating the ethics of food production.[2]

As legislation and certification standards were being discussed, developed and enacted at (supra-) national levels, public interest in ethically produced food naturally grew, and so did the respective markets.[3] By the dawn of the twenty-first century, mainstream agri-business players had realized that there was a lot to be lost by not demonstrating sufficient concern for ethical principles in food production. With the advent of societal marketing, promoting the integration of social responsibility into commercial marketing strategies, many began to believe that the adoption of higher ethical standards could be a very profitable endeavour at any point of the food supply chain.[4] Nowadays, and judging by their current product development and marketing efforts, nearly all major international players in the food arena seem to believe that higher ethical standards of any kind (preferably associated with certification) will positively differentiate their products.[5] It appears thus that an appropriate answer to the long-standing call for more market-oriented food production,[6] that generates more consumer value and increases competitive advantage, has finally been found. But is this really the case?

Whatever the drive might be – that is, to promote the welfare of nature and society alike or to generate higher corporate profits – changes leading to ethically improved animal production systems and associated certification schemes undeniably come at a cost to food chain actors. It is therefore imperative to learn in advance whether these changes will induce sufficiently large consumer demand that is eventually willing to pay a higher price for the products and benefits they enable. In turn, the aim of this chapter is to provide a well-founded, up-to-date and in-depth analysis of the nature and extent of consumer demand for ethically improved animal production systems and derived food products in the European Union.

Based on our own research and published studies, we show that consumers in general do not differentiate well between animal production systems with different ethical standards (conventional rearing included). We also see that when positive differentiation of an ethically improved system takes place, it is because the system is perceived to impact consumption benefits positively by addressing basic, individual needs, not necessarily higher, societal ones. This finding is not entirely surprising, because most empirical evidence gathered so far recognizes that consumers' preferences and willingness to pay for foods of animal origin are driven by habit and hedonic preference, not ethical considerations. Furthermore, we try to explain why consumers' stated preferences and willingness to pay for foods of animal origin that are produced under improved ethical standards almost always lead to an overestimation of their true demand. The social desirability of their revealed attitudes, combined with a lack of incentive to reveal true preferences, and the implicit associations between higher ethical standards and higher personal benefits (or lower personal risks) play significant roles for estimating consumer demand for fresh meat and fish based on their stated preferences. As we will see, consumers reveal themselves willing to pay a premium for foods of animal origin produced under improved ethical standards mainly when they believe that such consumption will lower their personal health risk, not increase societal benefits.

In the final section of this chapter, we note the practical implications of our analysis of consumer demand for ethically improved animal production systems and derived food products, especially for food production and marketing. In addition, we briefly discuss the long-term implications of our findings for marketing strategies, the future of agri-food business, institutional regulations and society at large.

Literature Review and Main Findings

LOW AWARENESS OF ANIMAL PRODUCTION SYSTEMS AND THEIR ETHICAL STANDARDS

European consumers are, in general, poorly informed about the farming practices of contemporary animal production systems and the regulations that govern them.[7] For the most part, their daily lives unfold at a great spatial and psychological distance from the realities of today's agri-business sector – a fully industrialized, technologically sophisticated activity geared towards mass production. Unless they have some kind of personal link (directly or indirectly through relatives and acquaintances) with animal husbandry or animal health sectors, European consumers have very little actual contact or experience with modern animal farming activities. Therefore, they tend to have very vague, romantic and idealized notions of animal husbandry that are mostly based on historical knowledge or the rearing of pets.[8]

When their idyllic mental images about animal husbandry are confronted with real ones released by the media, usually in the context of some food safety scandal, European consumers understandably react with shock. This process leads them to form fairly negative and one-sided opinions about the ethical standards of conventional animal production systems,[9] as well as fairly high and wide-ranging expectations regarding the ethical features of alternative rearing systems. In this ideal, holistic view of ethically improved animal husbandry, environmental friendliness, animal welfare, regional small-scale production, food safety and quality assurance become virtually indissociable.[10]

Figure 14.1 represents a collage made by consumers depicting what a healthy method to produce meat would look like, according to their view. It was obtained during a combined collage and focus group study involving 45 participants from different cities in the Netherlands.

Figure 14.1 constitutes a compelling illustration of the mental images and beliefs European consumers in general hold regarding conventional and ethically improved animal production systems. The picture of a nuclear plant at the bottom right-hand corner symbolizes conventional rearing practices – extremely industrialized, highly pollutant and damaging to both nature and society – that must be abandoned to reach

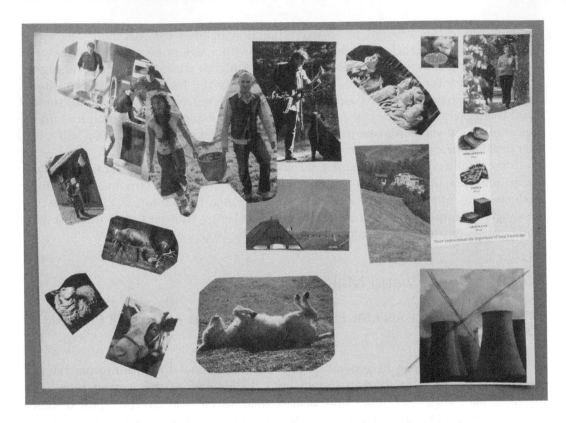

Figure 14.1 Dutch consumers' collage of natural meat production[11]

a healthier way of producing food. Directly above it, the picture with different types of bread from Asian countries symbolizes the drawbacks of globalization of food production, including the exploitation of natural and human resources in developing countries and the pollution caused by the transportation of foods over long distances, which also must come to an end. Meanwhile, the pictures on the left-hand side of the collage portray the way forward to a healthier way of rearing animals for food production. In the words of participants themselves:

Healthy meat comes from healthy and happy animals, growing free in their natural environment and eating only natural food. Only small-scale local production, which cares more for nature and animals than for profit, can produce meat which is safe and of good quality. If meat could always be produced like this we would all – farmers, butchers, retailers, consumers – benefit a lot from it.

Likewise, European consumers do not possess a lot of knowledge about non-conventional rearing practices and thus cannot distinguish very well between animal production methods with higher ethical standards.[12] For instance, they are largely unaware of the meaning and implications of certified organic farming, which they often confuse with free-range rearing.[13] They also seem to be poorly informed about the current possibilities for ethically improved fisheries and aquaculture, including organic or open-sea fish farming.[14] This lack of knowledge affects even their awareness of certification schemes that ensure the delivery of increasingly demanded ethical features, such as traceability, regional agricultural production and the preservation of local/national economies and cultural identities.

Survey studies pertaining to consumer perceptions of fresh beef, certified with a Protected Designation of Origin in Portugal and a Protected Geographical Identification in Spain, show that though individual brand names were easily recognized, only one-quarter of the respondents were aware of either type of certification scheme.[15] Focus group and survey research that we carried out in 2005–06 with 154 Portuguese consumers, however, indicates that this situation has recently improved (Figure 14.2). More than 50 per cent of these participants knew about the existence of certification, had tried certified beef at least once and were able to provide a reasonably accurate description of the features encompassed by this type of certification. Nevertheless, the same research showed that Portuguese consumers remain largely unable to distinguish between the ethical features of the animal production systems implied by certification and those implied by organic or free-range certification.

LINK BETWEEN POSITIVE DIFFERENTIATION OF ETHICALLY IMPROVED ANIMAL PRODUCTION SYSTEMS AND EXPECTATIONS OF SAFER AND TASTIER FOODS

When companies successfully promote higher levels of social responsibility, they not only improve their corporate image but also increase the value of their products, as perceived by customers.[16] A similar halo or spill-over effect is responsible for consumers' associations between improved ethical standards in animal production and the quality and safety of foods of animal origin. European consumers primarily associate higher levels of animal welfare with meat that is healthier and safer to eat and has better sensory quality relative to that produced under standard animal rearing practices.[17] A parallel

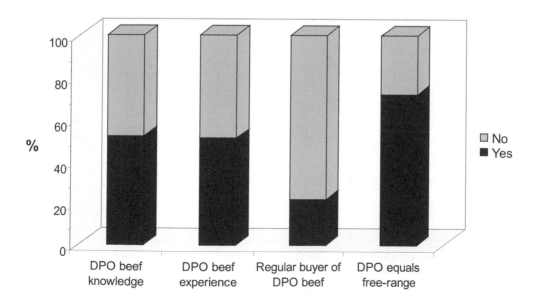

Figure 14.2 Indicators of Portuguese consumers' awareness of beef certified with a designation of protected origin (DPO)

Source: Focus groups and survey research (n = 154) carried out in 2005/06.

association occurs between organic farming practices and the perceived wholesomeness, authenticity, safety and quality of foods of animal origin.[18] Figure 14.3 illustrates this phenomenon by showing how positively Portuguese consumers, on average, judge the levels of safety and quality of certified beef compared with those of meat originating from conventional rearing. However, the data depicted in this figure also highlight that these European consumers automatically associate the positive individual benefits delivered by ethically improved animal production systems with high food prices and thus with high personal costs.[19]

Although deemed increasingly important, ethical benefits like improved sustainability and better living conditions for farm animals seem to play only a relatively minor role in the deliberations of most European food consumers.[20] For instance, when prompted about the features of organic farming during focus group discussions about animal production systems, Dutch consumers primarily mentioned aspects linked to improvements in the quality and safety of meat production and the consequent increase in fresh meat prices. Only afterwards did issues related to environmental protection, animal welfare or the promotion of small-scale regional farming surface in the discussion. These points nevertheless were extensively debated.[21] Likewise, Dutch consumers rated 'no genetic manipulation of fish species' and 'fish feed free from antibiotics and additives' – aspects clearly associated with perceived food safety – as the most desirable attributes of sustainable fish farming. Practices related to animal welfare, such as 'plenty of space to grow' or 'plenty of clean water,' were deemed relatively less important features of ethically improved fish farms.[22]

Nevertheless, when, for whatever reason, a positive differentiation of an ethically improved animal production system is established in consumers' minds, the risk of dissatisfaction and mistrust rises considerably. Disconfirmation of improved ethical

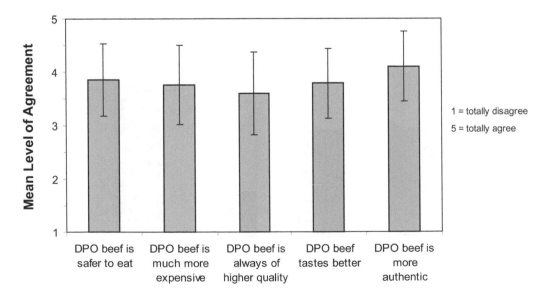

Figure 14.3 Portuguese consumers' evaluation of beef certified with a designation of protected origin (DPO) versus standard beef

Source: Survey research (n = 154) carried out in 2005/06.

standards (for example, by media exposure) or higher quality (for example, through a consumption experience) will reinforce lower perceived consumer value of foods produced with higher ethical standards.[23]

CONSUMER DEMAND FOR FOODS OF ANIMAL ORIGIN: DRIVEN BY HABIT AND HEDONIC PREFERENCE

Regardless of ethical improvements, people ordinarily do not consume animal production systems; they buy food and cook meals. Irrespective of the product category or production method, food or meal choice is determined mainly by perceived sensory quality, healthiness and convenience. The combined influence of these perceptions on food choice is mediated by powerful factors such as the consumers' hedonic preferences in a given context, eating habits, product price and availability.[24] Foods of animal origin, like meat or fish, are no exception to this rule, as repeatedly demonstrated by many studies on European consumers' food choice.[25]

Table 14.1 presents the results of a regression analysis conducted on beef consumption frequency at home, the importance of beef consumption attributes for purchase decisions and the degree of beef liking, as reported by 154 Portuguese meat consumers during a survey. From these findings, it is straightforward to conclude that individual hedonic preferences and eating habits remain the main drivers of beef consumption frequency, even when healthiness and ease of preparation are declared to be highly relevant factors in purchase decisions.

The collage in Figure 14.4 depicts the mental images Dutch consumers associate with the purchase and consumption of fresh meat. The core of this collage contains word clippings and pictures of cooked foods and shared meals, which together stress the vital

Table 14.1   Regression analysis results

| Independent variables | Beef consumption frequency at home | | | |
	Mean	STD	Beta	t
Beef Liking 1 = I don't like beef at all 5 = I like beef a lot	3.99	0.802	0.350	3.372*
Healthiness	3.97	0.969	-0.087	-0.994
Ease of Preparation	3.30	0.942	0.111	1.190
Habit	3.02	0.901	0.330	3.308*
Price 1 = Not at all important 5 = Extremely important	3.06	1.121	0.097	1.321
Constant			0.592	1.132
Adjusted R^2		0.304		
F-value		8.293**		

* $p < .01$ (two-tailed), ** $p < .001$ (two-tailed).

Source: Survey carried out with 154 Portuguese meat consumers in 2005–06.

Figure 14.4 Dutch consumers' collage of fresh meat consumption[26]

role of sensory quality, pleasure and enjoyment in fresh meat consumption. Also playing a central role in this consumer collage are beliefs that connect modest consumption of lean meat with health and well-being, as represented by the pictures of female figures. Finally, in a secondary position on the bottom of the collage appear visual depictions of beliefs that connect food safety concerns to the ethical standards of conventional animal production systems. A stack of meat cuts symbolizes mass production, the doll standing on a plate of fruit pieces signifies the power of mankind over nature, and the painting of a hunting party denotes concerns about animal welfare.

Food choice behaviour is often not aligned with attitudes and beliefs, even when they might be strongly held – a paradox both consumers and social scientists have grown painfully aware of.[27] The contents and structure of the mental images depicted in Figures 14.1 and 14.4 offer a striking illustration of how consumers' increasing concerns about the ethics of animal production systems spill over to their evaluation of foods of animal origin, but only to a limited extent. In spite of the negative images held regarding conventional animal rearing, most European consumers still view meat and fish consumption as having a central and positive role in their diets and daily lives alike. This perception does not mean, however, that these consumers are willing to forfeit a minimum level of quality and safety in their food choices. High-quality meat and fish are expected from ethically responsible farming practices, even if they do not constitute a sufficient enough guarantee of a high and stable demand for ethically improved foods.[28]

STATED PREFERENCE AND WILLINGNESS TO PAY, OVERESTIMATED DEMAND

An in-depth analysis of consumer demand and willingness to pay for ethically improved foods requires the existence of well-established and organized market institutions through which these products can be acquired. The European market for organic foods constitutes a good example of such an institution[29] and can therefore be used as benchmark in the current analysis. Judging by the most recent data available, the market share of organic foods of animal origin in Europe is around 2 per cent, and items are sold at a price that averages 60–65 per cent higher than conventional counterparts. It is important to notice that these estimates nevertheless can vary considerably across countries and product categories. In Denmark, for instance, the market share for organic dairy products is about 10 per cent at a premium price of 14 per cent, whereas in the Netherlands, organic minced beef is sold at a premium price of 94 per cent and commands a market share of 2 per cent.[30] These findings indicate that there is a small group of committed organic consumers in most European Union countries who are willing to pay a premium price for ethically improved foods of animal origin.

Although fairly accurate and realistic, analyses of shares and premium prices in actual markets only indicate the lower boundary of committed consumers' overall valuation of ethically improved foods at current levels of consumption. They do not tell us what the maximum willingness to pay for such products might be, how individual food attributes get valued against one another (for example, sensory quality versus environmental friendliness), or at what kind of premium price levels uncommitted consumers could be led to become ethical food users. Moreover, good estimates may be impossible to obtain if markets for the ethically improved foods are poorly developed or simply do not exist yet. One way to deal with this problem is to ask current and potential consumers of these goods to state how much they would be willing to pay for them, in a methodological approach generally known as contingent valuation analysis.[31]

Taking once more organic foods of animal origin as a benchmark, recent contingent valuation studies in Denmark and the Netherlands with representative samples show that consumers' stated willingness to pay for these products is, on average, 10–15 per cent and 20–25 per cent higher than for conventional ones, respectively. In both countries, the willingness to pay is higher for those already committed to organic food purchase than for those who are not. Willingness to pay is also higher among consumers who value the individual use benefits associated with these foods, irrespective of whether they also value their ethical features.[32] These findings confirm our earlier statements regarding the crucial role of perceived use benefits in shaping the demand for ethically improved foods. They also indicate that there is room for increased current market shares by lowering prices.

The stated willingness-to-pay approach allows insight into consumers' food demand, outside the scope of the products and prices featured by actual markets. However, many questions remain regarding the accurateness and reliability of consumers' own estimates of how they would behave in a real purchase situation, especially when they are elicited in hypothetical market circumstances. If no real products and no real money is being exchanged, and there is no way to hold individuals accountable for their stated valuation or buying behaviour, consumers have little incentive to reveal their true preferences and willingness-to-pay estimates.[33]

Contingent valuation studies give consumers the possibility to provide strategic answers when asked about their willingness to pay for ethically improved foods or animal production systems, because of the lack of accountability created by hypothetical markets.[34] If consumers underestimate the influence of their stated preferences on the course of market events and policy decisions, they might state willingness-to-pay estimates lower than their true valuations and attempt to free-ride on others who are willing to pay a higher premium for ethically improved foods. Conversely, if they overestimate their influence, they might feel tempted to provide willingness-to-pay estimates higher than their true valuation, in the hope that it eventually will lead others to benefit from the consequent rise in ethical standards. Examples of this discrepancy between private preferences and public choices recently have appeared for sustainable fish production systems and environmentally certified pork.[35]

The contingent valuation analysis methodology employed to elicit willingness-to-pay estimates also has become the subject of intense scrutiny and controversy. Different sources of bias have been identified that can lead to an overestimation of consumers' demand for ethically improved foods. Loureiro and Lotade, for instance, study interviewer effects and social desirability bias,[36] that is, the tendency of respondents to give answers they think the interviewer would like to hear or that are aligned with perceived social norms. In the same experimental settings, consumers provide significantly higher willingness-to-pay estimates for fair-trade coffee when questioned by a black African interviewer than when questioned by a white American one. Bennett and Blaney conduct a contingent valuation study of English consumers' willingness to pay for hen welfare legislation through a general increase of the price of eggs[37] and find that the warm glow bias (that is, purchase of moral satisfaction associated with contributing to a good cause) and part-whole bias (that is, perceiving willingness to pay as a contribution to the welfare of farm animals) results in an overstatement of willingness to pay of 50 per cent for the ethically improved eggs.

Overall, ample evidence reveals stated preferences and willingness to pay for ethically improved foods of animal origin, such as those obtained by contingent valuation studies, most likely lead to an overestimation of demand and market prices.[38] Such techniques normally do not encompass mechanisms that distinguish clearly between the share of stated willingness to pay that derives from the valuation of the product's perceived ethical attributes and that derived from the valuation of associated use benefits. As we discussed previously, this issue is highly relevant for food marketing based on higher ethical standards of production. Nevertheless, these difficulties can be largely overcome by the use of demand-revealing, experimentally induced markets, as we discuss in the next section.

PREFERENCE AND WILLINGNESS TO PAY, VALUATIONS OF SENSORY QUALITY, NATURALNESS AND SAFETY

As we highlighted previously, consumers' positive differentiation of ethically improved animal production systems, if it occurs, relates strongly to their expectations of superior food quality and safety compared with conventional products. It is therefore not surprising that their valuations of foods of animal origin produced under higher ethical standards also reflect these expectations.

Tables 14.2 and 14.3 present the results of a study aiming to uncover the determinants of Portuguese consumers' willingness to pay for certified beef. Respondents' answers regarding their level of agreement with statements about attitudes and beliefs potentially influencing their willingness to pay were factor analyzed to obtain determinant dimensions. Table 14.2 presents the outcome of this analysis, which uncovers four main factors: price sensitivity for beef, belief in certified beef's higher sensory quality, belief in certified beef's higher overall quality and attitudes towards the sustainability of animal production systems. For each of these factors, the associated statements' factor scores were correlated with respondents' willingness to pay for certified beef, obtained through the use of an iterative bidding method during face-to-face interviews.[39]

The correlations obtained and their statistical significance (see Table 14.3) indicate that, as theoretically expected, the income level of respondents is positively associated with their willingness to pay for certified beef, whereas their price sensitivity is negatively related. These results also confirm previous assumptions regarding the nature of the main determinants of consumers' willingness to pay for ethically improved meat. Respondents' beliefs regarding the sensory and overall quality of certified beef have significant positive associations with their willingness to pay for this meat. However, a significant relationship between the fairly positive attitudes of respondents toward the sustainability of animal production systems and a higher willingness to pay for certified beef cannot be demonstrated.

The iterative bidding method may reduce the uncertainty of respondents who provide their valuations for goods in a hypothetical setting. However, it does not eliminate other sources of biases that may seriously compromise contingent valuation approaches for determining consumers' willingness to pay.[40] Towards this end, alternative research methods have been proposed, namely, the use of demand-revealing laboratory auctions.[41]

Table 14.2 Factor analysis results

Extracted factors	Mean	SD	Factor loadings
Price Sensitivity for Beef			
Price is important in my beef purchase decision	3.06	1.12	.73
Price weighs heavily in my decision to buy beef	3.18	1.13	.82
I wait until beef is on special offer to buy it	2.67	1.17	.81
Variance Explained: 62%			
Belief in DPO Beef's Higher Sensory Quality			
DPO beef tastes better than standard beef	3.60	.77	.80
DPO beef is more tended than standard beef	3.32	.73	.93
DPO beef is juicier that standard beef	3.45	.74	.92
Variance Explained: 79%			
Belief in DPO Beef's Higher Overall Quality			
DPO beef is more authentic than standard beef	4.10	.65	.87
DPO beef's quality is more consistent	3.85	.68	.87
DPO beef is always safer than standard beef	3.74	.66	.83
Variance Explained: 74%			
Attitude towards the Sustainability of Animal Production Systems			
I do not mind paying more for animal welfare	3.75	.83	.79
Meat traceability is very important to me	3.97	.75	.75
DPO beef sales promote regional development	4.09	.72	.74
Variance Explained: 58%			

Note: Level of item agreement measured on a five-point Likert scale (1 = totally disagree, 5 = totally agree).

Source: Survey of 154 Portuguese meat consumers in 2005–06. DPO: Designation of Protected Origin.

Table 14.3 Correlation analysis results

Factors	Willingness to pay for DPO beef (Euro/kg)	
	Spearman's rho	Significance
Price Sensitivity for Beef	-.248*	.017
Belief in DPO Beef's Higher Sensory Quality	.222*	.033
Belief in DPO Beef's Higher Overall Quality	.318**	.002
Attitude towards the Sustainability of Animal Production Systems	.173	.097
Income Class (net household income/month)	.270**	.009

* p < .05 (two-tailed), **p < .01 (two-tailed).
Source: Survey of 154 Portuguese meat consumers in 2005/06. DPO: Designation of Protected Origin.

Laboratory or experimental auctions simulate active market environments and thus help estimate consumers' valuations of goods and uncover the determinants of these valuations. Relative to other methods for estimating consumers' willingness to pay, such as contingent valuation analysis and choice experiments, experimental auctions have the following advantages:

- They take place in a non-hypothetical context, with real products and real money being exchanged.
- They are incentive compatible when appropriately designed; that is, respondents' dominant strategy is to reveal their truthful valuation of the good in question.
- They take place in an active trading environment in which respondents can incorporate market feedback and become accountable for their revealed valuation through their buying behaviour.

They are also particularly suitable for exploring the effects of different attribute levels on willingness-to-pay estimates. Therefore, this type of experimental market institution has been employed often in the design of pricing and communication strategies for new and improved foods.[42]

Experimental auctions were conducted with Dutch consumers to elicit their preferences for ethically improved fishery and aquaculture systems.[43] Such preferences can be inferred from the differences in respondents' willingness to pay for sole and cod (in Euro/kg fresh fish), before and after they receive information about the ethical features of the associated production systems, as depicted in Figures 14.5 and Figure 14.6.[‡] These differences show that fish originating from conventional fisheries was always preferred by these consumers even when:

- The information provided highlights relatively low levels of the ethical attributes displayed by this alternative (including attributes respondents had classified as highly relevant immediately prior to the experiment).

---

‡   Further details on the experimental methodology employed can be viewed at http://www.fcee.lisboa.ucp.pt/docentes/url/anacosta/confer/AIAC_SAMM2005.pdf

- The information provided about fish from alternative, more sustainable production methods (ensured by the experimental market set up) stressed relatively higher levels of ethical attributes.

Regarding fish farming systems, fish from sea aquaculture was always preferred to fish from inland aquaculture, irrespective of the information provided on the ethical

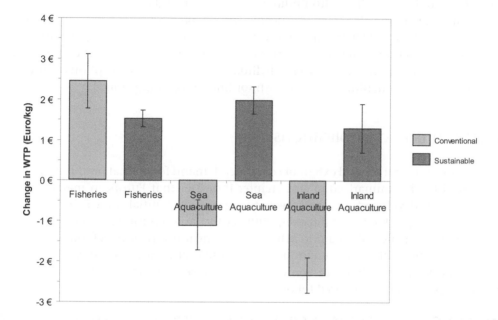

Figure 14.5 Effect of information about ethical standards of production on Dutch consumers' willingness to pay for sole

Source: Experimental auctions carried out in 2003/2004 (n = 90).

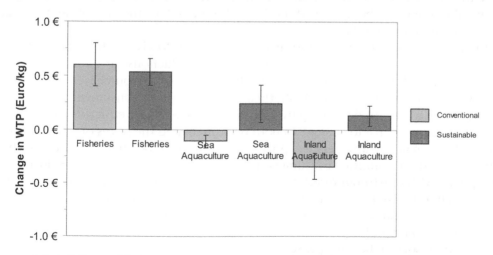

Figure 14.6 Effect of information about ethical standards of production on Dutch consumers' willingness to pay for cod

Source: Experimental auctions carried out in 2003/2004 (n = 90).

features of each available alternative. Because the information provided about fish from conventional and sustainable aquaculture (whether sea or inland) differs only in terms of whether the administration of feed additives and antibiotics were allowed, the respective differences in willingness to pay indicate that revealed preferences are strongly influenced by safety concerns. These findings agree with Dutch consumers' stated concerns about the safety of foods of animal origin immediately prior to the experimental auctions and in previous studies, as well as with findings from similar studies.[44]

Overall, the results illustrate the deeply-rooted preferences of consumers toward the provision of fresh fish through conventional fisheries and sea aquaculture, even when such production systems are overtly associated with low levels of environmental sustainability. These preferences are closely linked to consumers' perceptions of the sea as the most natural and habitual source of fish for human consumption.[45]

Conclusions and Implications

The analysis of the nature and extent of consumer demand for animal production systems and derived foods summarized in this chapter indicates that European markets are not yet sufficiently developed, on their own, to sustain substantial ethical improvements in farm animal production. Moreover, sufficiently large consumer demand for these improvements may be quite hard to achieve in the near future, unless food chain actors, together with the institutional environment, decide that a significant change in current policies and strategies is in order. Some guidelines regarding the main vectors on which such changes could be structured follow.

MARKETING STRATEGIES TO INCREASE CONSUMER DEMAND FOR ETHICALLY PRODUCED FOODS OF ANIMAL ORIGIN

Marketing strategies that might lead to increased demand for ethically produced foods of animal origin could be structured on the traditional 4Ps of the marketing mix: place, price, promotion and product.

Increased availability and consumer awareness of foods produced according to improved ethical standards on one hand and price reductions on another hand will increase demand.[46] More product exposure at the usual points of purchase will reduce the novelty of these products and facilitate their incorporation in consumers' evoked set of normal, habitual, day-to-day food purchase. Because demand for ethical foods of animal origin is highly sensitive to changes in both own prices and the prices of conventionally produced alternatives, any relative price reductions will be highly valued by both committed and uncommitted consumers. There seems to be room for such reductions in retail prices, with or without direct government intervention. Overall, food chain actors may do well to realize that, similarly to producers of any other innovation, it might pay off more to invest earlier on in foods with higher ethical standards at a cost, than catch up later at a smaller profit.

More than societal benefits per se, most European consumers expect to derive a high private value, namely, higher product quality and safety, from the consumption of more ethically produced foods. They are also willing to pay relatively more for products that they perceive bundle the highest number of individual and collective benefits.[47] Demand for the latter can be better supported by delivering products that consistently meet both expectations and communicate appropriately about them. Perceived societal benefits disconnected from experienced private value constitute a necessary but insufficient condition to raise consumer demand for ethically produced foods. Consequently, efforts should be made to design products and marketing communication strategies that align the private values experienced during consumption with previous expectations about both individual and societal benefits. Together with pricing decisions, the design of such strategies should be based on revealed (rather than stated) consumer preferences, which can be obtained through laboratory auctions and other market experiments.[48]

OTHER STRATEGIES TO PROMOTE CONSUMER DEMAND FOR ETHICALLY PRODUCED FOODS OF ANIMAL ORIGIN

The daily lives of modern European consumers unfold, for the most part, at a great spatial and psychological distance from the realities of the agri-business sector. Yet these same consumers increasingly are called upon to make consumption decisions about products that require a high level of knowledge about the characteristics of modern food production systems (ethically improved and not). Moreover, their food choice behaviour is supposed to reflect informed judgments about these characteristics and their broader societal implications. Taken together, these circumstances create a huge gap between what consumers are expected to know and what they actually know. Devising multiple effective strategies that can bridge this gap is essential to increase consumer demand for more ethically produced foods.

Food choice decisions might be empowered by reducing the lack of information and uncertainty regarding the characteristics of different animal production systems in general and their ethical standards in particular. The private sector can take the initiative by promoting greater transparency in the agri-business world through activities ranging from farm visits to self-enforced certification schemes that are clear, closely monitored and truly informative.[49] But the initiatives that can most effectively reduce consumer uncertainty and mistrust about the ethical standards of conventional and alternative animal production systems depend on national governments and other supra-national institutions.[50]

Governments should make clear to consumers that conventional food productions systems are highly effective in providing large amounts of good quality food at affordable prices, but that this method comes at the cost of current and future societal welfare. They should also make clear that all food chain actors, including consumers, are accountable for this state of affairs and call on them to share the burden of altering this situation. Many ways of sharing the burden could be devised, from raising the prices of conventional goods to reflect their true societal costs to reducing the prices of ethically improved foods by subsidizing their production or commercialization to taxing unethical food production and consumption to promoting private donations to support ethically improved animal production systems.

Irrespective of the formula chosen, both the institutional decision-making processes and the application of the resulting policies should be highly transparent and involve all relevant stakeholders. Governments should be able to offer enough credible guarantees to all those involved that their individual contributions matter and that they are proportional to their fair share of responsibilities and the public interest at stake.

Ultimately, relying on the market alone to correct the long-standing inefficiencies of an entire society will not work, especially if citizens perceive that those who demand rationality and coherence from their consumption behaviour do not behave according to these principles themselves.

Acknowledgements

The empirical data presented in this chapter about Dutch consumer samples were gathered in the context of a research project titled *Consumer-driven sustainable farming of fish*. This project was commissioned by and executed at the Animal Sciences Group of the Wageningen University and Research Centre in the Netherlands (ref. ASG2211842000). The empirical data presented in this chapter about Portuguese consumer samples were gathered in the context of a research project titled *A quality policy for beef consumption in Portugal*. This projected was carried out at the Faculty of Veterinary Medicine of the Technical University of Lisbon, Portugal (ref. AGRO 422). Ana Isabel Costa gratefully acknowledges the financial support of the Portuguese Foundation for Science and Technology in the preparation of this manuscript.

References

1. Ritson, C. & Oughton, E. (2007), 'Food consumers and organic agriculture', in Frewer, L. & van Trijp, H. (eds), *Understanding consumers of food products*. Cambridge, UK: Woodhead Publishing, pp. 254–272.

2. Heckman, J. (2006), 'A history of organic farming: transitions from Sir Albert Howard's *War in the Soil* to USDA National Organic Program', *Renewable Agriculture and Food Systems*, Vol. 21, No. 3, pp. 143–150(8).

3. Lohr, L. (1998), 'Implications of organic certification for market structure and trade', *American Journal of Agricultural Economics*, Vol. 80, No. 5, pp. 1125–1129.

4. Prothero, A. (1990), 'Green consumerism and the societal marketing concept: marketing strategies for the 1990s', *Journal of Marketing Management*, Vol. 6, No. 2, pp. 87–103.

5. Brady, D. (2006), 'The organic myth', *BusinessWeek*, October 16.

6. Costa, A.I.A. & Jongen, W.M.F. (2006), 'New insights into consumer-led food product development', *Trends in Food Science and Technology*, Vol. 17, No. 8, pp. 457–465.

7. Evans, A. & Miele, M. (2007), *Consumers' views about farm animal welfare. Part I: National Reports based on focus group research in seven EU countries*, Welfare Quality Reports No. 4, Cardiff University, Cardiff; Verbeke, W., Sioen, I., Brunsø, K., Henauw, S. & Camp, J. (2007), 'Consumer perception versus scientific evidence of farmed and wild fish: exploratory insights from Belgium', *Aquaculture International*, Vol. 15, No. 2, pp. 121–136.

8. Rathenau Instituut (2003), *Burgeroordelen over dierenwelzijn in de veehouderij*, Rathenauspecial, Rathenau Instituut, Den Haag.

9. Alvensleben, R. von (2001), 'Beliefs associated with food production methods', in Frewer, L., Risuik, E. & Schifferstein, H. (eds), *Food People and Society*. Berlin Heidelberg: Springer Verlag, pp. 381–400; Boogaard, B.K., Oosting, S.J. & Bock, B.B. (2006), 'Elements of societal perception of farm animal welfare: a quantitative study in the Netherlands', *Livestock Science*, Vol. 104, No. 1–2, pp. 13–22; Te Velde, H., Aarts, N. & Van Woerkum, C. (2002), 'Dealing with

ambivalence: farmers' and consumers' perceptions of animal welfare in livestock breeding', *Journal of Agricultural and Environmental Ethics*, Vol. 15, No. 2, pp. 203–19.

10. Scholderer, J., Nielsen, N.A., Bredahl, L., Magnussen, C.C. & Lindahl, G. (2004), *Organic in the head? Separating the effects of label information and actual meat type on consumer perceptions of pork quality*, Proceedings of the DTU Food Congress 2004, Lyngby; Rathenau Instituut, op. cit.

11. Costa, A.I.A. & Kole, A. (2004), *Ethical meat and fish production systems: consumers' mental images and beliefs*, Proceedings of the DTU Food Congress 2004, Lyngby.

12. Evans and Miele, op. cit.

13. Harper, G.C. & Makatouni, A. (2002), 'Consumer perception of organic food production and farm animal welfare', *British Food Journal*, Vol. 104., No. 3–5, pp. 287–299; Hoogland, C.T., De Boer, J. & Boersema, J.J. (2007), 'Food and sustainability: do consumers recognize, understand and value on-package information on production standards?', *Appetite*, Vol. 49, No. 1, pp. 47–57; Ritson and Oughton, op. cit.

14. Aarset, B., Beckmann, S., Bigne, E., Beveridge, M., Bjorndal, T., Bunting, J., McDonagh, P., Mariojouls, C., Muir, J., Prothero, A., Resich, L., Smith, A., Tveterasm, R. & Young, J. (2004), 'The European consumers' understanding and perception of the 'organic' food regime: the case of aquaculture', *British Food Journal*, Vol. 106, No. 2, pp. 93–105; Costa, A.I.A. (2005), *Measuring Dutch consumers' willingness-to-pay for ethically improved foods and supply chains through the performance of experimental auctions*, Proceedings of the 18th Scandinavian Academy of Management Meeting, Aarhus.

15. Marreiros, C. & Ness, M. (2002), *Perceptions of PDO beef: the Portuguese consumer*, Proceedings of the 10th EAAE Conference, Zaragoza; Loureiro, M. L. and McCluskey, J.J. (2000), 'Assessing consumer response to Protected Geographical Identification Labelling', *Agribusiness*, Vol. 16, No. 3, pp. 309–320.

16. Brown, T.J. & Dacin, P.A. (1997), 'The company and the product: corporate associations and consumer product responses', *Journal of Marketing*, Vol. 61, No. 1, pp. 68–84.

17. Phan-Huy, S.A. & Fawaz, R.B. (2003), 'Swiss market for meat of animal-friendly production – responses of public and private actors in Switzerland', *Journal of Agricultural and Environmental Ethics*, Vol. 16, No. 2, pp. 119–136; Verbeke, W. & Viaene, J. (2000), 'Ethical challenges for livestock production: meeting consumer concerns about meat safety and animal welfare', *Journal of Agricultural and Environmental* Ethics, Vol. 12, No. 2, pp. 141–151.

18. Harper and Makatouni, op.cit.; Ritson and Oughton, op.cit.; Scholderer et al., op. cit.

19. Hoogland et al., op. cit.

20. Harper and Makatouni, op. cit.; Scholderer et al., op.cit; Verbeke and Viaene, op. cit.

21. Costa and Kole, op. cit.

22. Costa, op.cit.; Kole, A., Haveman, D. & Costa, A.I.A. (2005), *Implicit attitudes, explicit attitudes, and consumers' willingness to pay for sustainably produced food* products, Proceedings of the 6th Pangborn Sensory Science Symposium, Yorkshire.

23. Costa, A.I.A., Teldeschi, E., Gerritzen, M.A., Reimert, H.G.M., Linssen, J.P. H. & Cone, J. W. (2007), 'Influence of flock treatment with antibiotic tylosin on poultry meat quality', *Netherlands Journal of Agricultural Science*, Vol. 54, No. 3, pp. 269–278; Ritson and Oughton, op.cit.; Scholderer et al., op. cit.

24. Costa, A.I.A., Schoolmeester, D., Dekker, M. & Jongen, W.M.F. (2007), 'To cook or not to cook: a means-end study of the motivations behind meal choice', *Food Quality and* Preference, Vol. 18, No. 1, pp. 77–88; Rozin, P. (2007), 'Food choice: an introduction', in Frewer, L. & van Trijp,

H. (eds), *Understanding Consumers of Food Products*. Cambridge, UK: Woodhead Publishing, pp. 3–29.

25. Grunert, K.G. (2005), 'Food quality and safety: consumer perception and demand', *European Review of Agricultural Economics*, Vol. 32, No. 3, pp. 369–391; Landbouw-Economisch Instituut (2006), *Een biologisch prijsexperiment; Grenzen in zicht?*, Landbouw-Economisch Instituut, Den Haag; Verbeke, W. & Vackier, I. (2005), 'Individual determinants of fish consumption: application of the theory of planned behaviour', *Appetite*, Vol. 44, No. 1, pp. 67–82; Wier, M., Anderson, L.M. & Millock, K. (2005), 'Information provision, consumer perception and values – the case of organic foods', in Krarup, S. & Russel, C. S. (eds), *Environment, Information and Consumer Behaviour*. Cheltenham: Edward Elgar, pp. 161–178.

26. Costa and Kole, op. cit.

27. Costa, A.I.A., Schoolmeester, D., Dekker, M. & Jongen, W.M.F. (2003), 'Exploring the use of consumer collages in product design', *Trends in Food Science and Technology*, Vol. 14, No. 1, pp. 17–31; Dagevos, H. & Sterrenberg, L. (2003), *Burgers en consumenten: tussen tweedeling en tweeëenheid*. Wageningen: Academic Publishers; Grunert, K.L. (2006), 'Future trends and consumer lifestyles in regards to meat consumption', *Meat Science*, Vol. 74, No. 3, pp. 149–160; Ngapo, T.M., Dransfield, E., Martin, J.F., Magnusson, M., Bredahl, L. & Nute, G.R. (2004), 'Consumer perceptions: pork and pig production. Insights from France, England, Sweden and Denmark', *Meat Science*, Vol. 66, No. 1, pp. 125–134.

28. Ritson and Oughton, op. cit.; Scholderer et al., op. cit.; Verbeke and Viaene, op. cit.

29. Ritson and Oughton, op. cit.

30. Hamm, U. & Gronefeld, F. (2004), *The European market for organic food: revised and updated analysis*, Organic Marketing Initiatives and Rural Development 5, University of Wales, Aberystwyth; Landbouw-Economisch Instituut, op.cit.; Wier et al., op. cit.

31. Carson, R.T., Flores, N.E. & Meade, N.F. (2001), 'Contingent valuation: controversies and evidence', *Environmental and Resource Economics*, Vol. 19, No. 2, pp. 173–210.

32. Landbouw-Economisch Instituut, op.cit.; Wier et al., op. cit.

33. Lusk, J.L. (2003), 'Using experimental auctions for marketing applications: a discussion', *Journal of Agricultural and Applied Economics*, Vol. 35, No. 2, pp. 349–360; Murhpy, J.J., Allen, G.P., Stevens, T.H. & Weatherhead, D. (2005), 'A meta-analysis of hypothetical bias in stated preference valuation', *Environmental and Resource Economics*, Vol. 30, No. 3, pp. 313–325.

34. Bateman, I.J., Langford, I.H., Turner, R.K., Willis, K.G. & Garrod, G.D. (1995), 'Elicitation and truncation effects in contingent valuation studies', *Ecological Economics*, Vol. 12, No. 2, pp. 161–179.

35. Costa and Kole, op.cit.; Costa, op.cit.; Lusk, J.L., Nilsson, T. & Foster, K. (2007), 'Public preferences and private choices: effect of altruism and free-riding on demand for environmentally certified pork', *Environmental and Resource Economics*, Vol. 36, No. 4, pp. 499–521.

36. Loureiro, M.L. & Lotade J. (2005), 'Interviewer effects on the valuation of goods with ethical and environmental attributes', *Environmental and Resource Economics*, Vol. 30, No. 1, pp. 49–72.

37. Bennett, R.M. & Blaney, R.J.P. (2003), 'Estimating the benefits of farm animal welfare legislation using the contingent valuation method', *Agricultural Economics*, Vol. 29, No. 1, pp. 85–98.

38. Carson et al., op. cit.; Murphy et al., op. cit.

39. Bateman et al., op. cit.

40. Bateman et al., op. cit.

41. Lusk, op. cit.
42. Costa, A.I.A. and Pires, C.P. (2007), *Economics for Marketing revisited*, CEFAGE-UE Working Paper 0207, University of Évora, Évora.
43. Costa, op. cit.
44. Hobbs, J.E., Bailey, D., Dickinson, D.L. & Haghiri, M. (2005), 'Traceability in the Canadian Red Meat Sector: Do Consumers Care?', *Canadian Journal of Agricultural Economics*, Vol. 53, No. 1, pp. 47–65; Lusk, J.L., Norwood, F.B. & Pruitt, J.R. (2007), 'Consumer demand for a ban on antibiotic drug use in pork production', *American Journal of Agricultural Economics*, Vol. 88, No. 4, pp. 1015–1033.
45. Costa & Kole, op. cit.
46. Landbouw-Economisch Instituut, op.cit.; Ritson and Oughton, op. cit.
47. Harper and Makatouni, op.cit.; Ritson and Oughton, op. cit.; Scholderer et al., op. cit.
48. Costa and Pires, op. cit.
49. Boorgaard et al., op. cit., Rathenau Instituut, op.cit.
50. Brom, F.W.A., Visak, T. & Meijboom, F. (2007), 'Food, citizens and the market: the quest for responsible consuming', in Frewer, L. & van Trijp, H. (eds), *Understanding Consumers of Food Products*. Cambridge, UK: Woodhead Publishing, pp. 610–623.

15 Beyond the Marketing Mix: Modern Food Marketing and the Future of Organic Food Consumption

BY HANS DAGEVOS*

Keywords

consumer behaviour, consumer typology, organic foods, citizen-consumers.

[I]t is quite clear that the future of organics will also be very much dependent on the motivations of end consumers. This is not to say that consumers will dictate that future, but that the success of [marketing] strategies (...) will be dependent on the ability to mobilise people as 'organic consumers' by providing foods that materially and symbolically satisfy and/or influence those peoples' needs, desires, pleasures and terrors' more successfully than other available foods.[1]

Abstract

The controversy we turn to in this chapter is about food marketing based on the idea of food products as real goods versus a focus on foods as feel goods. There are good reasons for modern food marketing to take the idea of foods as feel goods seriously. From this perspective, the well-known marketing mix appears rather poorly able to meet the modern desires and wishes of contemporary food consumers. In addition to the four supply-oriented Ps of the familiar marketing mix, four complementary and demand-centred Ps stemming from consumer wishes provide supplements. These eight Ps are introduced and specified in the context of organic food consumption.

* Dr Hans Dagevos, Agricultural Economics Research Institute (LEI), Wageningen University and Research Centre, The Hague, the Netherlands. E-mail: hans.dagevos@wur.nl.

Real Goods and Feel Goods

This chapter takes consumers' perspective as its starting point. We are interested in the motives underlying their behaviour – particularly with respect to their support of organic food consumption. This focus is evident enough, because consumers are the target group of marketing, and understanding consumers is vital to an effective marketing strategy. All this is true, without discussion or doubt. Nowadays, it is also almost a truism to suggest that the marketing mix is no longer suitable to approach modern food consumers. This chapter's position is that the well-known mix of product, price, place and promotion is becoming obsolete in the modern-day, affluent consumer society in which consumption is much more complex than that simply driven by price tags or tangible product characteristics.

New lines of thought in consumer-oriented studies proclaim, on the one hand, that contemporary (food) consumers are complicated, whimsical and elusive creatures.[2] Stereotypical images of food consumers as inveterate bargain hunters, greenies, gluttons or health freaks single out only one feature of their complex characters. As a result, such stereotypes are not very realistic. On the other hand, modern scholarly thinking concentrates on consumer goods in terms of identity and symbolic values. The appeal of consumer products derives not necessarily from tangible product features but from image, exclusivity or novelty. In other words, the emphasis goes from real goods to feel goods. A stream of studies maintains that today's consumption practices cannot be understood solely in terms of price, product qualities, appearance or availability but should also take into account that experiences, emotions, ethics, status or identity are of vital importance to understand people's motivations, sensibilities and doings in the present-day consumption age.[3] This avenue of research, within the disciplines of marketing and sociology in particular, focuses on intangible aspects of consumer goods. This so-called dematerialization of consumption is based on the belief that:

> The motivation of the modern consumer is predicted less upon real need than the emotional simulations the objects and experiences of consumer culture provide, the fantasies they engender, and the desires with which they are invested, whether 'realized' or not in actual patterns of consumption.[4]

Particularly with respect to food consumption, Alan Beardsworth and Teresa Keil argue:

> [W]hen we eat, we are not merely consuming nutrients, we are also consuming gustatory (i.e. taste-related) experiences and, in a very real sense, we are also 'consuming' meanings and symbols. (…) Thus, it is no exaggeration to say that when humans eat, they eat with the mind as much as with the mouth.[5]

In a similar vein, the main founder of the Slow Food movement, Carlo Petrini, claims passionately:

> But food (…) is far more than a simple product to be consumed: it is happiness, identity, culture, pleasure, conviviality, nutrition, local economy, survival. To think of stripping it of all these values, of all the connotations that a mouthful of food can immediately convey, to think

of mediating and reducing these connotations to the point where they disappear, is one of the greatest follies ever conceived by man.[6]

THE IMPORTANCE OF EXPERIENCE, EMOTION AND ETHICS

These kinds of quotations stress that it is not simply product quality or price–quality ratios that are important for consumers and their choices. Expanding the scope beyond the pragmatic or 'prosaic "expediency" factors such as price, accessibility and convenience,' to cite Charlotte Weatherell and her colleagues,[7] means that it becomes both possible and important to take into consideration such factors as the way consumer goods are produced, the image of the industry or business, the environment in which the goods or the service are consumed ('you are where you eat"), what the product can teach you, what it can tell you or what feelings it evokes. Stated succinctly, consumption situated in the realm of feel goods recognizes the importance of experiences, emotions and ethics, which go beyond mere tangible aspects that belong to the domain of real goods.

In bringing this section to a close, two necessary remarks relate to the remainder. Emotional and ethical bonds between consumers and a food product or a food production method potentially have particular importance when we think of organic food products and their production principles. More generally, the attention paid recently to the emerging experience economy, as well as the ethics of consumption, both invites and inspires the incorporation of non-materialistic aspects into our analysis. In our approach, we do so by including an underlying dimension that has materialistic and non-materialistic aspects as its poles.

Consumers as Citizens

The second dimension that underpins our approach to modern food consumption ranges from individualism to collectivism. This dimension is inspired by another line of reasoning in modern thinking about consumption. We refer to studies advocating that the distinction between the 'self-regarding' consumer and the 'public-minded' citizen is a problematic, not to say false, one. Recent discussions stress not the contrast but rather the combination and coalescence of consumption and citizenship.[8] They argue that it is over-simplistic to stipulate that consumer choices are, by definition, dictated by private and short-term interests, whereas public and long-term interests belong exclusively to the domain of individuals in their role as citizens (as it is to assume the other way around, that is, responsible consumption is defined as uniquely motivated by social or environmental concerns). Consequently, the consumer is not by definition selfish, and it is not only possible but also necessary to draw consumption into the moral and political domain. Introducing the notion of the citizen–consumer expresses how civic and ethical values can interfere with consumers' product preferences. The citizen–consumer notion also has received some scholarly attention, particularly in studies devoted to sustainable or green consumption and political and ethical consumerism.[9]

What such adjectives have in common is their suggestion of the importance of incorporating social, political, environmental and ethical concerns as critical components of consumer choice. Consumers are supposed to be interested in the collectivity, concerned about problems of the (global) environment, willing to base their choices on ethical values and virtues like justice or fairness, and inclined to undertake their consumption in line with the habits, mores and concerns of their social community. In brief, civic elements of consumption are stressed. This communitarian perspective of consumer behaviour also moves away from real goods towards feel goods, but at a food production process level rather than a food product level. Emphasizing such values as sustainability, integrity or authenticity implies immediately that food consumption is more than the mere gratification of private material needs. In the words of Michele Micheletti:

> For a growing number of people, particularly in the Western world, increased wealth implies the economic means to consider aspects other than the relationship between material quality and price in their marketplace transactions. Thus, their involvement with products concerns more than price and quality. These people politicize products by asking questions about their origin and impacts.[10]

FOOD CONSUMER CONCERNS

The approach of the consumer as a citizen is, first of all, in full accordance with the current discourse of the active consumer, instead of the consumer as a 'passive dupe,' in social scientific circles.[11] Moreover, and more specific to the world of food, consumer involvement and engagement that considers food as more than an object of material use and means of instant satisfaction has received academic attention for more than a decade, under the heading of consumer concerns. Food consumer concerns include issues such as food safety, animal welfare, fair trade and environmental issues.[12] Affluent food consumers tend, more and more often, to take the outside world into account in their food consumption choices. Feeling responsible for our choices and being aware that our choices influence the production practices and principles of the food system no longer are exclusively reserved for radical animal rights activists, fanatic environmentalists or hard-core world trade opponents. Food consumers' commitment and consciousness appear to have become more widespread with respect to environmental problems, such as deforestation or land degradation as an effect of expanding food, feedcrop or livestock production or the (un)friendly ways animals are treated, accommodated and transported. However modest the current fraction of involved consumers who regularly purchase food products on the basis of eco-friendly or ethical considerations, an emerging trend nevertheless reveals that pro-environmental consumer choices, fair trade consumption and buying local foods are gradually making their way into the mainstream food market in the affluent world. This tendency is meaningful to organic food consumption, which is all about (groups of concerned) consumers who really care about their food and its production conditions and who feel uncomfortable with a 'fuel approach' toward the things they eat. Citizen–consumers' uneasiness with treating food products and their consumption with indifference and thoughtlessness, rather than engagement and embeddedness, may have, in principle, a positive connection with their interest in organic food products.

Two Dimensions and Four Consumer Images

The previous sections summarize two strands of modern reflection on consumption that entertain the idea of widening the frame of analysis. As already indicated, both evolving patterns of thought are consistent with the development of an analytical framework that consists of two axes: a materialistic–non-materialistic dimension, rooted in the real good/feel good controversy, and an individualistic–collectivistic axis with roots in the consumer/citizen contrast (see Figure 15.1). Both dimensions form the basis of the so-called consumer images approach.[13] The materialistic pole refers to consumer behaviour that has price- and product-oriented factors as its key determinants. Non-materialism refers to emotional, ethical or ecological considerations with respect to food consumption. Individualism indicates the extent to which the behaviour of consumers is independent and self-centred, whereas collectivism represents consumption choices that take into account the socio-cultural or environmental consequences of consumption choices.

The horizontal materialistic–non-materialistic dimension of this framework gives us the opportunity to look at (organic) food consumption from the point of view of product qualities (freshness, user-friendliness or other functional product characteristics), price or availability on the one hand and emotional aspects attached to the consumption of the food product on the other. The vertical individualistic–collectivistic dimension ranges from autonomous consumer behaviour to embedded food consumption choices. The individualistic pole relates closely to notions of the consumer as a free agent, as well as to consumer sovereignty. The collectivistic pole highlights the process-related civic factors that are related and relevant to (organic) food consumption, such as concerns about environmental pollution, animal welfare, wasting natural resources, fair trade or the use of antibiotics, growth hormones or genetic engineering during the production process.

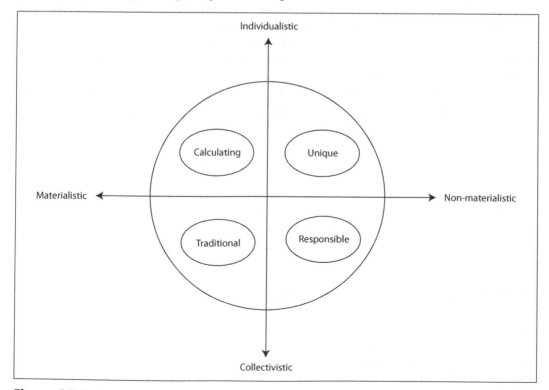

Figure 15.1 Consumer images continuum

CONSUMERS AS PLURALISTIC PERSONS

On the basis of this two-dimensional configuration, four consumer images emerge, representing four different food styles. Figure 15.1 shows that the four consumer images form a *continuum*, such that consumers can exhibit behaviour that fits into each of the four different consumer images. In other words, consumer images are distant from all kinds of one-dimensional portrayals (for example, the fabled *homo economicus*), which contradicts marketing techniques that attempt to enforce rigid consumer group segmentation. As mentioned in the beginning of this chapter, widespread acceptance suggests that consumers may engage in various types of behaviour, driven by multiple motives. Consumers thus should be considered pluralistic persons, whose varied identities require a multidimensional approach. The circular structure of the consumer images approach aims to meet such criteria.

The four consumer images can be characterized as calculating, traditional, unique and responsible. As depicted in Figure 15.1, calculating consumerism is situated in the quadrant formed by materialism and individualism: Self-interest is the main factor, and personal benefits prevail. Calculating consumption is aimed at effectiveness and efficiency; quantity is more important than quality, and these 'McDonaldized' consumers prefer uniformity and convenience. The traditional consumption style is comparable with calculation consumption in its materialistic dimension, with its understanding that monetary savings (price) weigh heavier than time savings (convenience). Traditional consumption attaches importance to collective traditions and eating habits. Modernization and change are only embarked upon with caution, because the comfort of the familiar is cherished. The unique consumption pattern is diametrically opposite to the food neophobia of traditional consumption. Unique consumers are neophilics, seeking change and variety, who are in the vanguard when it comes to new or adventurous products, shop formulae or consumption trends. Unique consumption represents modern consumerism, for which hedonism, conspicuousness, experience and status are keywords. Finally, the responsible consumption profile differs considerably from the unique consumer image, because individual pleasure or personal prestige does not gain the upper hand. Instead, consumption choices reflect environmental or communitarian interests and (future) consequences. The resemblance to what Paul-Marie Boulanger and Edwin Zaccaï call responsible consumption is too noticeable to be left unquoted:

> *The responsible consumer is thus the one who, having become aware of the public character of his or her consumption and of its impacts on others (directly or indirectly), subordinates his or her consumption choices to considerations other than the simple satisfaction of his or her needs and desires.*[14]

Eight Ps: The Marketing Mix and...

The four Ps (product, price, place, promotion) from marketing literature frequently are used as a point of departure for looking at products' market potential and growth. Although the customary four Ps of the marketing mix remain important for the success associated with bringing commodities to consumers, all four Ps are supply driven by definition. Because the modern food market is so eagerly typified as consumer-centric, the question arises whether a demand-based variation on the theme of the four Ps may be an appropriate adjustment to the increasingly consumer driven food system. A food market typified as consumer driven, however, does not imply that the supply side has become of only negligible interest. Likewise, the quartet of demand-oriented Ps does not intend to replace the quartet of supply-oriented Ps. We therefore present the eight Ps with the Ps of the marketing mix first and subsequently introduce those Ps based on consumer images.

P1–PRODUCT

Because all Ps are briefly discussed in the context of problems and possibilities regarding the consumption of organic food products, the product P gives a reason to note that food quality often is defined as equal to product quality. Therefore, the market chances for organics depend highly on the ability of organic food producers to compete with the standards of conventional food producers. Another important variable is the availability and continuity in the delivery of organic foodstuffs. Stock shortages irritate both retailers and consumers and diminish the market opportunities for organics considerably. Even for regular organic food consumers, non-availability is a main reason for not always buying organic. The extension of the organic product range is another issue. Ten years ago, 'the array of organic products has expanded well beyond fresh produce to include baby foods, dairy products, meats, and prepared convenience items',[15] and this development from the early 1990s, has continued. Nowadays, a wide range of organic food items can be found in food shops and supermarkets across Europe and the United States. In addition to the 'classic' supply of (unprocessed) organic fruits and vegetables, consumers may purchase organic wine, chocolate, eggs, bread, coffee, tea, rice, spices, sweets, ice cream, pizza, tinned vegetables and ready-to-eat meals.

P2–PRICE

The price factor frequently gets put into perspective in modern marketing. Price is neither automatically nor always the deciding factor in consumers' choices. In addition, many food consumers are not as price conscious as the stereotype of the cost-conscious consumer implies. In general, price awareness and accurate price knowledge of most food consumers should not be overestimated.[16] This point does not mean, however, that price is unimportant. What is more, price remains a prime obstacle to the breakthrough of organic products. The price difference between conventionally produced foods and organically cultivated counterparts is often unbridgeable for a large number of consumers. A price gap between organic and conventional product variants of 50 per cent or more is not unusual. Small wonder, then, that 'too expensive' is one of the most important reasons for not purchasing organic products or discouraging food consumers from buying larger quantities of organic food products.[17] The greater the price differences, the less organic foods represent a competitive alternative. The limiting effect of price differences for the consumption of organic foods increases even further, in that research suggests on the one hand that the demand for organic products is more sensitive to price than is demand for conventional food and on the other hand that the higher the price premiums, the lower the proportion of consumers who are willing to buy organic foods.[18]

P3–PLACE

The success of consumer goods reportedly depends on three factors: location, location and location. Although this exaggeration should not be taken too literally, the essence of the message is patently obvious. Availability and accessibility are essential to market opportunities. The more organic products are to be found at places visited by a lot of people, the better potential consumers can be reached. For this reason, the growth of organic market share depends very much on the availability of organic products in various food outlets. In this respect, a major step forward in recent decades has been that organics are no longer exclusively supplied by health food shops, natural foods supermarkets or organic speciality stores. At present, many shelves of conventional supermarkets are filled with organic food products. Increasingly, organic foods also have become available in canteens and in restaurants. In various European countries, organic food consumption is growing more due to the sale of organically grown produce in conventional supermarket chains than in specialist food shops.[19] Organic food products have even made their way into the discount supermarkets of such retailers as Aldi and Lidl. Other big supermarket chains, including those in Germany, the United Kingdom and the Netherlands, use organic as a quality mark for their own image by implementing their own brands of organic food products. Such private supermarket labels often cover a very broad group of different organic foods. These kinds of initiatives are undoubtedly essential to the expansion of organic food consumption.

P4–PROMOTION

The very same supermarkets are significant when it comes to promotion too, not only because supermarket chains are crucial to reaching a lot of consumers, but also because during the past 20 years, supermarkets have become increasingly powerful in agro-food chains. Their influence is a non-negligible factor when we try to understand changes in food production as well as consumption.[20] In this respect, it is meaningful that supermarket chains regularly promote sections of their organic range by providing information to their customers, allowing them to taste organic products, devoting more shelf space to organic products or displaying these products more prominently. Supermarket chains also may promote organic food through special offers, often accompanied by nationwide campaigns funded by national governments throughout the European Union. These campaigns use the mass media and modern communication to strengthen the recognizability and familiarity of organic products among consumers, usually based on long-term activity plans and a network of participating stakeholders. Thus, P4 embodies both government policies and private initiatives that attempt to encourage organic production and consumption.

...a Mix of Motives

The four people-centric Ps are inspired by the preceding consumer images. In this section, we introduce profit, pause, pleasure and principle. Profit is a particular feature of the calculating consumer style, whereas pause belongs to traditional consumption. Unique consumerism suggests the pleasure factor, and principle reflects the responsible consumer image. In comparison with the real goods focus of the four marketing mix Ps, this quartet of Ps shifts the emphasis to (organic) foods as feel goods. These additional four consumer-based Ps are also included to address the mix of motives found in recent research. Empirical evidence is mounting that organic food consumption is triggered by a wide variety of motives. This outcome therefore asks for a standpoint that is more explicitly oriented toward both consumers and feel goods than is the marketing mix perspective.

P5–PROFIT

The calculating consumption style represents mainstream consumer choice, driven by self-interest and personal gain. Calculating consumption is rational. Rational food consumers want the best for themselves and demand that the supplier provides them with good product quality and strong quality–price ratios. Easy-to-use products and easily accessible foods (for example, one-stop shopping), as well as nutritional value and guaranteed (standardized, predictable) quality are also favourite criteria that typify the pragmatic priorities and practical needs that dominate the calculating food consumption style. Although calculating consumption is frequently marked by indifference and carelessness toward food and the agri-food system behind it, more interest and emphasis centres on taste and personal health. Both are self-oriented aspects of food and surely belong to calculating consumption, yet we also should be aware that neither taste nor health is a distinguishing criterion in the consumer image approach. That is, these items cannot be reserved for just one of consumer image but instead apply to all of them to a certain extent. Taste is very important to the unique consumer image, and health receives less attention in the traditional consumption regime. For this reason, taste and health could be placed at the centre of the continuum. However, with respect to organic food consumption, research frequently shows that taste and healthiness play central roles in motivating organic food purchases, because consumers often perceive organic foods as more tasteful and healthy than conventional food products. A consistent finding across many consumer studies about organic food consumption notes that the health factor is all-important.[21] The stream of empirical research that supports the importance of health as a consumer motive for developing positive attitudes and buying intentions toward organic foodstuffs accords nicely with the established theoretical reflection of food marketing, which indicates health is one of the major motives of food demand.[22] This finding is also consistent with the opinion of Tim Lang and Michael Heasman[23] that health is the keyword in the food system of today and the future. To conclude, with respect to the relationship between calculating consumption and the demand for organic foods, the consumer's perception of healthiness is really an asset that has much appeal to calculating consumers who are keen on personal profit. The fact that health is frequently interpreted as an 'egoistic' motive[24] enables us to link of calculating consumption to organic food demand.

P6–PAUSE

The traditional consumption style points to an appreciation for organics that originates from conformist values and desires. The traditional consumer clings to the familiar. Well-known recipes, food products and food consumption habits are maintained to avoid change and create security and predictability. Organic food products are relatively new on the market, and therefore, conservative values suggest organic foods may be treated with suspicion (not to mention that, from the traditional consumers' point of view, feelings of mistrust grow stronger because of their cost sensitivity). This suspicion, however, can disappear quickly when these consumers take the organic production process into account. Organic production principles are very much in tune with traditional consumers' sympathies. Vegetable organics are produced without chemical pesticides or fertilizers, livestock farming is done without the routine use of drugs, and there are no artificial additives in organic food products. The symbolic value of such 'authentic' characteristics of organics matches the reactionary sensitivities and preferences of traditional consumption. As such, the organic food sector has a firm base on which to grow. Positioning organics on the food market as the protector of naturalness, integrity and (local) food and farming traditions serves the traditional consumption choice and is simultaneously a clear antipole to the global food system that thrives on accelerating the pace of change, high-quality technologies, mass production and economies of scale. In other words, organic foods appeal to the traditional consumption style because of their identity, which opposes the 'Fordist food' provided by the late modern agro-industrial food system.[25]

P7–PLEASURE

The unique consumer image, as noted previously, contrasts sharply with the traditional food style because fast changing needs, variety seeking and interest in product innovations and novelties are important components. The unique consumption pattern is conspicuous, competitive and cosmopolitan. Its profound desire to enjoy food makes the fun factor crucial. With regard to food, pleasure might derive from superior flavour, aesthetically attractive product features (freshness, colour), or putting something special on the table. The pleasure of enjoying foods increases easily in the domain of unique food consumption when food consumption is a means to the end of self-representation and social distinction (therefore, P7 is actually as much about prestige as about pleasure). The identity values of food and food consumption are particularly important for the self-interested unique food style. What, where and sometimes even with whom we eat are positional markers of status and lifestyle. Organic food will attract the attention of narcissistic unique consumers when it provides a status symbol or is superior to conventional food alternatives in terms of taste, nutritional quality or appearance. This perspective confirms the significance of marketing strategies that stress that organic are still relatively new and a niche market in the world of food, in which conventionally produced and fast foods are the mainstream. Despite the growth of organic demand in numerous Western food markets, from the viewpoint of unique food consumption, it remains important to keep the positive image – that buying and eating organics is 'different' – alive. Exclusivity is enhanced by price premiums, serving organic foods in gourmet restaurants or supplying them in gourmet stores, and labelling of certain organic foods with a special quality mark or brand. These forms of product marketing should be much more successful than focussing on process-oriented concerns, because unique consumption is hardly motivated by environmental friendliness or animal welfare.

P8–PRINCIPLE

The opposite holds true for the responsible consumer image, which particularly features an interest in foods that are bought and eaten and the production origins and processes used in their manufacture. The predominant motives for buying organic food are environmental and ethical aspects (caring animal husbandry, naturalness, no chemical applications). Responsible consumers care and are concerned about the production methods in modern agriculture and in the food industry. Organic foods thus are appetizing to the responsible food consumer because of the environmental rationale of organic production. Commitment and involvement are important in the province of responsible consumption. The consumers attracted to this food style take their consumer citizenship seriously, and for that reason, food consumption for them is not free of engagement. Consuming 'clean' organic foods is a matter of principle; eating is qualified as a moral act, an ecological act, a political act. The heavy users of organics are specifically motivated by awareness and concern for civic and ethical issues as well as ecological aspects. Responsible consumption is the home base, so to speak, of the outcomes of consumer studies that point to process-related motives as major determinants of buying organic food – among which environmental protection is often an important ethical issue for consumers. A study by Suzanne Grunert and Hans Jørn Juhl,[26] based on the value theory of Shalom Schwartz,[27] confirms that environmental concerns relate positively to the consumer motivation to purchase organically produced food. Collectivistic values such as universalism (responsible consumption) and tradition (traditional consumption) are favourable toward buying organic foods, whereas Schwartz's values, like hedonism and achievement, which serve individualistic interests (in the consumer image vocabulary: calculating and unique consumption, respectively), relate negatively to the purchase of organics. These results validate the hypothesis that suggests a more natural bond between organics and wider food-related aspects and food system awareness than between organic consumption and food-intrinsic, personal or pragmatic priorities.

Bringing the Discussion Full Circle: From Controversy to Complementarity

This chapter has shown that food, and by the same token food marketing, is not only about real goods but also increasingly about feel goods.

The domain of organics offers a clear example of the relevance of looking at foods from the perspective of feel goods. Two distinct dematerialization trends are of special interest to the world of food. Consumption as experience on the one hand and a politicization of consumption on the other have immediate significance with respect to organic food consumption, as the first two sections demonstrate. Modern theorizing stresses the emotional aspects of consumption as well as the relationships between consumption and citizenship, ethics and social solidarity. From the feel good point of view, organic food consumption can be explained by pointing to the enthusiasm of food consumers for buying organics for their (perceived) contribution to social and environmental sustainability. Roughly, engaged consumers link eating organics to a better world and a safe conscience. In addition, eating organic food is defined as a lifestyle experience. Organic food products catch consumers' attention primarily for emotional reasons.

It is possible to respond to these new perspectives with hostility and stick to the neatly arranged marketing mix. This kind of reaction sparks controversy as a consequence. In other words, controversy in organic food and agricultural marketing arises inescapably if we make a sharp distinction between organic food consumption in terms of real goods and feel goods.

Studies regarding organic food consumption do not support such a contrast. On the contrary, recent research makes it abundantly clear that consumers' decisions to buy organics depend on multiple motives. Consumer choices to purchase and ingest organic foods are influenced by a variety of egoistic and altruistic motivating factors, including health concerns, price sensitivity, consumers' attachment to nutritional and sensorial quality (or other food-intrinsic product qualities, like freshness or appearance), convenience, availability, food safety, sustainability, authenticity, community-building, environmental friendliness, respectful treatment of animals, fair trade, trust, food enjoyment and local production.

This multitude of (interwoven) motives requires a wide frame of analysis whose scope encompasses both the real goods and the feel goods perspective. In other words, the real challenge is not to compete but to complement. The combination of the marketing mix and the consumer images approach presented in this chapter attempts to make a contribution in this respect. As a consequence, the four consumer-based Ps have been introduced as *supplements* to the marketing mix, not *substitutes*. Taking both demand-oriented Ps and supply-oriented Ps seriously in the context of what we currently qualify as a consumer-driven food market is consistent with the idea that it is more nuanced and realistic to understand various movements and changes in the food market as consumer-*dependent* rather than consumer-*led*. We believe that this approach also accounts for organic food consumption and its growth in market size. That is, the increase of organic food consumption in the food market is substantially supply-driven and highly demand-dependent. As the theme of this chapter indicates, our main interest has been consumer-oriented, with a focus on (positive) motivations.

THE PROLIFERATION OF ORGANIC FOOD CONSUMPTION

With respect to the consumer images continuum, the broad range of motivations underlying food consumers' interest in buying and eating organics foods should remind us that the strong position of organics in the collectivistic and non-materialistic domains of the continuum is rather 'shaky' in the light of the proliferation of organic food consumption. That is, we should not lose sight of the importance of expediency factors – to borrow Weatherell's adjective again – such as price, value for money, lack of availability or guaranteed quality for the attitudes and actions of food consumers (note that engaged organic food consumers also are included here; price, convenience and availability are also motivating factors for their food choices). Whatever influential emotions and ethical considerations are to today's consumers and their appetite for organic foods, if the organic food sector is unable to make the connection with the other parts of the continuum, the expansion of organic food consumption will be limited and move only slowly away from the marginal share of total food sales that organics have today across Europe.[28] Put differently, the market's transformative potential for organic food consumption will be seriously obstructed if the emotional and/or ethical aspects are overstated. Consumers' passionate environmental consciousness and devotion to an organic philosophy are not enough to ensure substantial growth in the organic market. To assist the proliferation of organic food consumption, it is of vital importance that organics 'infiltrate' further into the (private, pragmatic and product-related) motivations of food consumers. Figure 15.2 illustrates this argument for bringing organic food consumption from the collectivistic–symbolic pole (eco and emo) toward the individualistic–materialistic quadrants (ego) of the continuum.

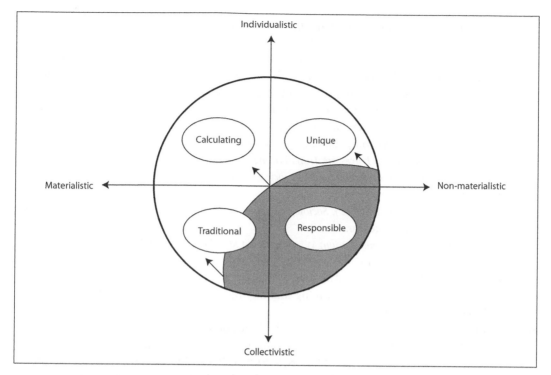

Figure 15.2 Proliferation of organic food consumption

Successful marketing strategies not only point at the civic virtues and public attributes of organic foods but also search for combinations that will appeal to self-interested benefits and private attributes. This argument is consistent with the conclusions drawn by various authors.[29] Moreover, this issue brings our discussion full circle: After emphasizing the feel good side and presenting four consumer-oriented Ps as key to modern food marketing and the future of organic food consumption, we return to stress the importance of real good factors for motivating organic food consumption, which are primarily the targets of the good old marketing mix. So, if one prefers, an alternative title for this chapter might be 'Beyond the Marketing Mix, and Back.'

References

1. Lockie, S., Lyons, K., Lawrence, G. & Mummery, K. (2002), 'Eating 'green': Motivations behind organic food consumption in Australia', *Sociologica Ruralis*, Vol. 42, pp. 23–40, see p. 24.
2. Gabriel, Y. & Lang, T. (1995), *The Unmanageable Consumer: Contemporary Consumption and its Fragmentations.* London: Sage; Lewis, D. & Bridger, D. (2000), *The Soul of the New Consumer: Authenticity – What We Buy and Why in the New Economy.* London: Nicholas Brealey Publishing; Ritzer, G. (2001), *Explorations in the Sociology of Consumption: Fast Food, Credit Cards and Casinos.* London: Sage; Szmigin, I. (2003), *Understanding the Consumer.* London: Sage.
3. Barnes, J.G. (2001), *Secrets of Customer Relationship Management: It's All About How You Make Them Feel.* New York: McGraw-Hill; Crocker, D.A. & Linden, T. (eds) (1998), *Ethics of Consumption: The Good Life, Justice and Global Stewardship.* Lanham: Rowman and Littlefield; Firat, A.F. & Dholakia, N. (1998), *Consuming People: From Political Economy to Theaters of Consumption.*

London: Routledge; Jensen, R. (1999), *The Dream Society: How the Coming Shift from Information to Imagination will Transform your Business*. New York: McGraw-Hill; Gilmore J.H. & Pine, B.J. (2007), *Authenticity: What Consumers Really Want*. Boston: Harvard Business School Press; Pine, B.J. & Gilmore, J.H. (1999), *The Experience Economy: Work is Theatre and Every Business a Stage*. Boston: Harvard Business School Press; Rifkin, J. (2001), *The Age of Access: How the Shift from Ownership to Access is Transforming Modern Life*. London: Penguin; Schulze, G. (1997), *Die Erlebnisgesellschaft: Kultursoziologie der Gegenwart*. Frankfurt: Campus Verlag; Slater, D. (1997), *Consumer Culture and Modernity*. Cambridge: Polity Press; Wolf, M.J. (1999), *The Entertainment Economy: How Mega-media Forces are Transforming our Lives*. Harmondsworth: Penguin.

4. Williams, S.J. (2001), *Emotion and Social Theory: Corporeal Reflections on the (Ir)Rational*. London: Sage, p. 112.

5. Beardsworth, A. & Keil, T. (1997), *Sociology on the Menu: An Invitation to the Study of Food and Society*. London: Routledge, pp. 51–52.

6. Petrini, C. (2007), *Slow Food Nation: Why Our Food Should be Good, Clean and Fair*. New York: Rizzoli Ex Libris, pp. 166–167.

7. Weatherell, C., Tregear, A., & Allinson, J. (2003), 'In search of the concerned consumer: UK public perceptions of food, farming and buying local', *Journal of Rural Studies*, Vol. 19, pp. 233–244, see p. 234.

8. Schudson, M. (2007), 'Citizens, consumers, and the good society', in D.V. Shah et al. (eds) *The Annals of the American Academy of Political and Social Science*. Los Angeles: Sage, pp. 236–249; Shah, D.V., McLeod, D.M., Friedland, L., & Nelson, M.R. (2007), 'The politics of consumption/ the consumption of politics', in D.V. Shah et al. (eds) *The Annals of the American Academy of Political and Social Science*. Los Angeles: Sage, pp. 6–13; Trentmann, F. (2007), 'Citizenship and consumption', *Journal of Consumer Culture*, Vol. 7, pp. 147–158.

9. Harrison, R., Newholm, T. & Shaw, D. (eds) (2005), *The Ethical Consumer*. London: Sage; Jackson, T. (ed.) (2006), *The Earthscan Reader in Sustainable Consumption*. London: Earthscan; Micheletti, M. (2003), *Political Virtue and Shopping: Individuals, Consumerism and Collective Action*. New York: Palgrave Macmillan; Nicholls, A. & Opal, C. (2006), *Fair Trade: Market-driven Ethical Consumption*. London: Sage.

10. Micheletti, op. cit., pp. 74–75.

11. Paterson, M. (2006), *Consumption and Everyday Life*. London: Routledge; Sassatelli, R. (2007), *Consumer Culture: History, Theory and Politics*. Los Angeles: Sage; Trentmann, F. (2006), 'Knowing consumers–histories, identities, practices: An introduction', in F. Trentmann (ed.) *The Making of the Consumer: Knowledge, Power and Identity in the Modern World*. Oxford: Berg, pp. 1–27.

12. Blandford, D. & Fulponi, L. (1999), 'Emerging public concerns in agriculture: Domestic policies and international trade commitments', *European Review of Agricultural Economics*, Vol. 26, pp. 409–424; Korthals, M. (2006), 'The ethics of food production and consumption', in L. Frewer & H. van Trijp (eds) *Understanding Consumers of Food Products*. Cambridge: Woodhead Publishing Ltd, pp. 624–642.

13. Dagevos, J.C. & Hansman, H.J.M. (2001), 'Towards a consumer images approach: Exploring the quirks of modern food consumer behaviour', in H. Tovey & M. Blanc (eds) *Food, Nature and Society: Rural Life in Late Modernity*. Aldershot: Ashgate, pp. 135–160; Dagevos, H. (2005), 'Consumers as four-faced creatures: Looking at food consumption from the perspective of contemporary consumers', *Appetite*, Vol. 45, pp. 32–39.

14. Boulanger, P-M. & E. Zaccaï (2007) 'Conclusions: The future of sustainable consumption', in E. Zaccaï (ed.) *Sustainable Consumption, Ecology, and Fair Trade*. London: Routledge, , pp. 231–238, see p. 233.

15. Thompson, G.D. (1998), 'Consumer demand for organic foods: What we know and what we need to know', *American Journal of Agricultural Economics*, Vol. 80, pp. 1113–1118, see p. 1113.

16. Vanheule, M. & Drèze, X. (2002), 'Measuring the price knowledge shoppers bring to the store', *Journal of Marketing*, Vol. 66, pp. 76–85.

17. McEachern, M.G. & McClean, P. (2002), 'Organic purchasing motivations and attitudes: Are they ethical?' *International Journal of Consumer Studies*, Vol. 26, pp. 85–92; Padel, S. & Foster, C. (2005), 'Exploring the gap between attitudes and behaviour: Understanding why consumers buy or do not buy organic food', *British Food Journal*, Vol. 107, pp. 606–625.

18. Ritson, C. & Oughton, E. (2006), 'Food consumers and organic agriculture', in L. Frewer & H. van Trijp (eds) *Understanding Consumers of Food Products*. Cambridge: Woodhead Publishing Ltd, pp. 254–272, see pp. 267–269.

19. Sylvander, B. & Kristensen, N.H. (eds) (2004), *Organic Marketing Initiatives in Europe*. Aberystwyth: University of Wales. For the same tendency in Australia, see Lyons, K. (2007), 'Supermarkets as organic retailers: Impacts for the Australian organic sector', in D. Burch & G. Lawrence (eds), *Supermarkets and Agri-food Supply Chains: Transformations in the Production and Consumption of Foods*. Cheltenham: Edward Elgar, pp. 154–172.

20. Lyons, op. cit.

21. Baker, S., Thompson, K.E., Engelken, J., & Huntley, K. (2002), 'Mapping the values driving organic food choice: Germany vs the UK', *European Journal of Marketing*, Vol. 38, pp. 995–1012; Chryssohoidis, G.M. & Krystallis, A. (2005), 'Organic consumers' personal values research: Testing and validating the list of values (LOV) scale and implementing a value-based segmentation task', *Food Quality and Preference*, Vol. 15, pp. 585–599; Hartman, H. & Wright, D. (1999), *Marketing to the New Natural Consumer: Understanding Trends in Wellness*. Washington: Hartman Group; Lockie et al., op. cit.; Magnusson, M.K., Arvola, A.A., Koivisto Hursti, U-K. Årberg, L., & Sjödén, P-O. (2003), 'Choice of organic foods is related to perceived consequences for human health and to environmentally friendly behaviour', *Appetite*, Vol. 40, pp. 109–117; Miele, M. (2001), *Creating Sustainability: The Social Construction of the Market for Organic Products*. Wageningen: Wageningen University; Padel & Foster, op. cit.; Torjusen, H., Sangstad, L., O'Doherty Jensen, K., & Kjaernes, U. (2004), *European Consumers' Conceptions of Organic Food: A Review of Available Research*. Oslo: National Institute for Consumer Research; Zanoli, R. (ed.) (2004), *The European Consumer and Organic Food*. Aberystwyth: University of Wales; Zanoli, R. & Naspetti, S. (2002), 'Consumer motivations in the purchase of organic food', *British Food Journal*, Vol. 104, pp. 643–653. Not all studies confirm that health is a predominant motive; see Michaelidou, N. & Hassan, L.M. (2008), 'The role of health-consciousness, food safety concern and ethical identity on attitudes and intentions towards organic food', *International Journal of Consumer Studies*, Vol. 32, pp. 163–170.

22. Von Alvensleben, R. (1997), 'Consumer behaviour', in D.I. Padberg, C. Ritson, & L.M. Albisu (eds) *Agro-food Marketing*, CAB International, Oxon, pp. 209–224.

23. Lang, T. & Heasman, M. (2004), *Food Wars: The Global Battle for Mouths, Minds and Markets*. London: Earthscan.

24. Magnusson et al., op. cit.; Michaelidou & Hassan, op. cit.

25. Fonte, M. (2002), 'Food systems, consumption models and risk perception in late modernity', *International Journal of Sociology of Agriculture and Food*, Vol. 10, pp. 13–21.

26. Grunert, S.C. & Juhl, H.J. (1995), 'Values, environmental attitudes, and the buying of organic foods', *Journal of Economic Psychology*, Vol. 16, pp. 39–62.

27. Schwartz, S.H. (1992), 'Universals in the content and structure of values: Theoretical advances and empirical tests in 20 countries', in M. Zanna (ed.) *Advances in Experimental Social Psychology*. San Diego: Academic Press, pp. 1–65.

28. Hamm, U. & Gronefeld, F. (2004), *The European Market for Organic Food: Revised and Updated Analysis*. Aberystwyth: University of Wales.

29. Magnusson et al., op. cit.; Ritson & Oughton, op. cit.; Wier, M., O'Doherty-Jensen, K., Andersen, L.M., & Millock, K. (2008), 'The character of demand in mature organic food markets: Great Britain and Denmark compared', *Food Policy*, in press, doi: 10.1016/j.foodpol.2008.01.002.

16 Green Consumerism: What Can We Learn From Environmental Valuation Surveys?

BY MEIKE HENSELEIT*

Keywords

externalities, willingness to pay, food supply chain, ethics, information.

Abstract

This chapter proposes some new ideas into the controversial discussion about whether consumers are interested in the environmental impacts of food production and consumption. Investigations of this issue usually concentrate on social and psychological issues, whereas this chapter gives priority to stated preferences for environmental goods.

In this chapter, the author will overview factors of environmentally friendly consumption behaviour; explain the externalities of food production; and provide the results of environmental valuation surveys. As well, the author will reconsider reasons for the gap between preferences for environmental goods and consumption behaviour, as well as discuss possible ways food businesses might apply more environmentally friendly production processes, combined with marketing activities.

Attitudes, Values and Food Choice

Green consumerism is a subset of responsible consumerism. It is based on the assumption that people have a moral responsibility in their capacities as consumers to avoid causing harm and bring about a just and sustainable world. The concept of green consumerism also applies to businesses and their survivability, depending on whether they respond

*   Dr Meike Henseleit, Institute for Agricultural Policy and Market Research, Senckenbergstraße 3, 35390 Giessen, Germany. E-mail: meike.henseleit@agrar.uni-giessen.de. Telephone: + 49 641 9937 7037.

quickly to the demands of consumers for products and services that are environmentally friendly.

Research into attitudes towards environmental concerns and investigations of the willingness to pay for environmentally friendly goods and services usually show a high level of awareness of environmental issues. In contrast, many investigations conclude that most consumers rarely take environmental issues into consideration when shopping for food, and a minority† consider ethical factors regularly.[1] Usually such research concentrates on social and psychological factors and, in particular, people's attitudes and concerns, because they are deemed important, if not the main, factors behind the choice of eco-friendly products.[2] In some studies, environmental concern provides a major determinant of buying sustainably produced food,[3] but more studies, and particularly more recent studies, are concluding that health concerns outweigh environmental motivations.[4] Altogether, no strong relationship between attitudes and knowledge about environmental issues on the one hand and consumption behaviour on the other hand can be confirmed, and a gap remains in the thorough understanding of the demand for environmentally friendly food.[5]

Accordingly, the critical question remains: Are consumers both willing and able to turn their expressed interest in environmental problems into actual purchasing habits? This issue is of interest for the food industry as it considers the adoption of environmentally friendly production and processing techniques, in combination with consumer communication strategies, such as labelling. This chapter cannot, of course, provide an ultimate answer to this question – this discussion has remained unresolved for many years – but it may provide some new perspectives on the issue.

PURCHASE DECISIONS AND UTILITY

The primacy of the social-psychological viewpoint in relation to the environmental consciousness of consumers while purchasing any product has been criticized by several authors.[6] They argue that this focus concentrates too much on attitudes and behaviour, while other factors receive little attention. Beyond question, psychological, socio-psychological and social investigations make important contributions to the understanding of the demand for 'green' products, especially when they take account of trends and circumstances, as explained subsequently. This chapter discusses the problem more from a utilitarian perspective. Utilitarianism is a form of consequentialism, meaning that the moral worth of an action depends on its outcome. Utility, the good to be maximized, can be defined as the satisfaction of preferences, such as the satisfaction that a consumer derives from the consumption of a good. According to this construct, the total outcome of a purchase decision is important, and this total also includes consequences for the environment. We can measure how strong someone's preference is for a good by ascertaining how much they would be willing to pay for its satisfaction at the margin.

---

† However, any categorization of consumers depending on stated or revealed environmental consciousness should be regarded carefully, because it is often based on meanings attached to some variables. Thus, it may be the same as the basis for the segmentation, and the results become tautological. The categorization is itself an interpretation and may be a useful basis for strategic marketing decisions, but it is not suitable for explaining or understanding the changing patterns of consumption in the longer term.

This chapter concentrates on preferences for environmental goods that are affected by the food supply system. More precisely, it examines people's willingness to pay (WTP) for the preservation of environmental goods, which could be influenced by any stage of the food supply chain. Furthermore, this investigation surveys food labels noting both animal welfare and environmental issues to determine whether people are principally enabled to consider environmental issues in their preferences during shopping decisions.

Preferences for Environmental Goods

A wide variety of conceptual frameworks assess the environmental impacts of food production and consumption, usually based on an enormous variety of indicators.[7] This chapter concentrates on consumers' preferences rather than on the degree of externalities of single food products. Thus, a simple model of environmental externalities meets its requirements. When considering the ecological footprint of food products, every input and output along the whole supply chain must be taken into account, including used resources, required energy, the environmental impact of harmful emissions associated with food production, processing, transportation and consumption.[8] This chapter also considers influences on farm animal welfare, because several recent surveys show that this issue is an important criterion for consumers, and it does not appear clear whether animal welfare should be assigned to social or environmental concerns.[9]

Figure 16.1 illustrates the food supply chain, on the left-hand side, as well as the likely environmental impacts of the food chain in the middle. These impacts may affect various environmental goods and ethical values, shown on the right-hand side. The arrows to and from the box in the middle are not bound to single effects; for example, energy is used at each stage of the food supply chain.

The environmental impact of food production, processing and consumption can be described as the sum of the influences on environmental goods and values. About one-third of households' total environmental impact can be related to food and drink

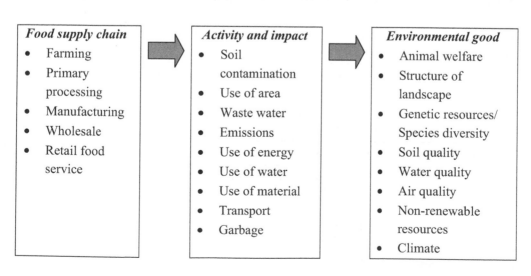

Figure 16.1 Potential environmental implications of food production

consumption.[10] The magnitude of environmental externalities is largely determined by producing methods but also by processing, packaging, retailing, preparation and waste management. Thus, the summarized impact on the environment is different for each food product and can hardly be determined exactly.

It is not the purpose of this chapter to evaluate the magnitude of externalities of single food products but rather to discuss the consideration of environmental impacts according to preferences in the food supply system. The following sections present selected surveys from the field of environmental valuation. The data come from Europe, Canada, the United States and Australia, which means that the findings are limited to wealthy countries; preferences for environmental goods are likely to differ in developing countries. To ease the comparison, monetary values are converted into US$ by applying the consumer price index (CPI) of the year 2000.

WILLINGNESS TO PAY FOR FARM ANIMAL WELFARE

It is quite difficult to estimate the WTP for animal welfare-oriented husbandry, especially because knowledge about farming conditions is not homogenous, and preferences differ for the various livestock animals.[11] Hence, various surveys of the WTP for improved farm animal welfare standards are difficult to compare. Nevertheless, in this section, some exemplary results (which are listed in Table 16.1) provide an impression of the dimension of stated preferences.

Bennett and Blaney estimate an average weekly WTP for a tax that goes to raising the welfare standards of veal and hens as US$13.42, whereas the estimated WTP for weekly changes in food expenditures ranged from a minimum of $1.60 per week for a ban on egg cages to a maximum of $4.68 for slaughtering pigs in a more humane way.[12] In another

Table 16.1   Willingness to pay for farm animal welfare[13]

Study	Payment vehicle and feature	Mean WTP per household/ week (US$)
Bennett and Blaney, 2002, UK	Tax for raising welfare standards of veal and hen	13.42
	Increase in food expenditures for a ban on egg cages	1.60
	Increase in food expenditures for slaughtering pigs more humanely	4.68
Burgess et al. 2003, UK	Improved welfare for laying hens	4.24
	Improved welfare for diary cows	4.15
	Improved welfare for chickens	3.78
	Improved welfare for pigs	3.02
Nocella et al. 2007, UK, Italy, Germany, Spain, France	Increase in food expenditures for improved animal welfare standards	12.73

UK study, Burgess et al.[14] estimated weekly individual net benefits from schemes aimed at improving the welfare of laying hens, dairy cows, broiler chickens and pigs. In this study, improvements for laying hens were most preferred ($4.24), followed by enhanced living conditions for dairy cows ($4.15), chickens ($3.78) and pigs ($3.02). However, tests of the statistical significance of this ranking indicate equal preference for laying hens, chickens and cows, whereas the improvement scheme for pigs appears to be less preferred.[15] This range may be caused at least partly by the image of the living conditions of farm animals. For instance, in a 2005 European survey, people supposed welfare to be worse for laying hens than for pigs and diary cows.[16] In the same survey, most citizens (57 per cent) declared their willingness to pay an additional price premium for eggs sourced from an animal welfare-friendly production system: One-quarter of respondents agreed to a price premium of 5 per cent, 21 per cent to an increase of 10 per cent, and 11 per cent of respondents stated that they would be prepared to accept an increase of 25 per cent or more, which seems a quite remarkable proportion.[17]

In a study conducted in 2006 in five European countries (UK, Italy, Germany, Spain, France), the estimated average weekly WTP for animal-friendly products compared with conventional ones was about $12.73.[18] Also, a recent European survey showed that the welfare of farmed animals is a matter of concern for the vast majority of Europeans: 19 per cent declared themselves 'very worried' about the welfare of farmed animals, and 41 per cent said that they were 'fairly worried'.[19] To sum it up, these surveys indicate that people in rich countries are concerned about farm animal welfare, and a high proportion seem willing to pay a premium for the enhanced living conditions of farm animals. However, attributes such as healthiness, leanness and safety are estimated more important than animal welfare by the vast majority.[20]

WILLINGNESS TO PAY FOR LANDSCAPE QUALITY

Land cultivation is the basis for the production of food and energy. Agricultural land, comprising arable and grassland, covers about 40 per cent of the terrestrial land surface. Many questions concerning biodiversity and the functioning of water and biogeochemical cycles have their foundation in the human practice of cultivating the land, particularly in the intensity of agricultural management by humans. Thus, the character of an area is widely affected by land cultivation systems, which represent the first stage of the food supply chain. Buildings for processing and retailing, as well as transportation routes, which are necessary at every stage of the food supply system, also are part of the landscape. Particularly in industrialized countries, the structure of scenery depends heavily on the food production system.

The valuation of landscape thus relates to several problems. First, landscape can be defined as a composition of many elements, whose value depends on not only their own estimation but also their proportion in any considered scenery. Second, preferences for sceneries depend on many factors, such as the kind of landscape of the regional provenance and cultural backgrounds. Third, aesthetic preferences are rather individual, which makes it difficult to estimate general preferences for various types of landscape. Many people express a preference for specific landscapes but for very different reasons; some may want to protect a landscape because it serves human utilitarian needs, whereas others may emphasize ecocentric values.[21]

To understand WTP for the protection or amelioration of a landscape and its elements, the results of recent surveys offer some guidance, as listed in Table 16.2. The selection is limited to investigations of cultivated land, because in most cases, sensitive areas are only marginally influenced by food production in the short term.

Table 16.2 Willingness to pay for landscape quality[22]

Study	Payment vehicle and type of landscape	Mean WTP per household/year
Drake 1992, Sweden	Income tax for the preservation of agricultural landscape	US$340
Tronstad 1993, Norway	Municipal tax for the preservation of agricultural landscape	100
Bonnieux and Le Goffe 1997, France	Income tax for the restoration of local scenery	46
Roschewitz 1999, Switzerland	Income tax for the preservation of agricultural landscape	256
Brink et al. 2000, Austria, Sweden, Netherlands, UK	Payments for the preservation of agricultural landscape	68
Israelsson 2001, Sweden	Single payment for enhanced landscape appearance	21 (one-off)
Moran et al. 2004, UK	Payments for the improvement of agricultural landscape	46
Sirex 2004, France	Income tax for the preservation of open landscape	15
Ollikainen and Lankoski 2004, Finland	Income tax for the preservation of open landscape	112

Not surprisingly, the stated amounts vary significantly across the surveys, because they differ in many respects, such as the characteristics of the landscape and the socio-economic characteristics of the survey respondents. Nevertheless, traditionally cultivated land appears to be highly estimated. Usually traditional farming is connected to structurally diverse sceneries with different kinds of land use, like grassland, woodland and arable farmland close together. According to the preceding results, the average WTP for the preservation of agricultural landscape is approximately US$80–90 per capita per year.[23] Benefits arising from visitors to the countryside, who usually travel to a certain area mainly because of its recreational values, are remarkable. Many of the sceneries that people highly prefer as a destination for holidays, such as the Alps or Tuscany, are characterized by agricultural activities. Pretty calculates that in the UK, day and overnight visitors account for some 433 million visit-days to the countryside, with an average amount spent per day of about $28 by day visitors and $97 by overnight visitors from overseas.[24]

In addition to the cultivated landscape as a whole, certain features are also particularly influenced by food production systems, such as hedgerows and orchards. Table 16.3 lists the WTP for different landscape features.

Table 16.3 Willingness to pay for landscape elements[25]

Study	Payment Vehicle	Feature	Mean WTP per Household/Year
Santos 1997, UK	Income tax to protect landscape elements	Farm terraces	US$46
		Meadows	24
Fischer et al. 2003, Germany	Donation to a fund for local environmental amenities	Hedgerows	36–58
Hanley et al. 2004, UK	Increase in general taxation to avoid losses of 10–50%	Hedgerows	20–25
Dachary-Bernard 2004, France	Local tax to avoid losses	Hedgerows	22
Henseleit 2006, Germany	Income tax to avoid losses	Grassland biotopes	30

One of the most comprehensive studies in this respect, which is not included in the table, is the final report of the Environmental Landscapes Features (ELF) project in the United Kingdom.[26] In this project, researchers used stated WTP to price the benefits of seven environmental features, such as hay meadows, woodlands and the like. The estimated national WTP was $198 for hay meadows, $127 for neutral, acid and calcareous grassland and $31 for hedgerows.[27]

In summary, there appears to be a high preference for traditionally cultivated land, as well as for landscape elements like hedgerows and meadows. The highest estimation may belong to well-balanced sceneries with woodland, flowery meadows, fields, orchards and single trees. Therefore, small and extensively used acreages appear to have special value.

WILLINGNESS TO PAY FOR GENETIC RESOURCES AND SPECIES DIVERSITY

The impact of food production, mainly agriculture, on biodiversity is rather complex. It can be both positive and negative, because some species rely on certain types of agricultural activities, whereas the status of others is declining due to farm practices.[28] Furthermore, the impact of the processing and retailing sector on biodiversity is essentially negative, not so much with regard to species diversity itself, but rather to the habitats and basic resources that are affected. Table 16.4 contains selected results of surveys considering preferences for species diversity.

From the multitude of investigations available in this field, some authors have combined the results of several studies to estimate the WTP for the preservation of species. For example, Brink et al.[29] estimate an average WTP for the protection of endangered species of some $130 per capita per year, and Nunes and van den Bergh[30] use a

Table 16.4 Willingness to pay for genetic resources and species diversity[31]

Study	Feature	Mean WTP per household/year
Holm-Müller 1991, Germany	Prevention of the extinction of species	US$144
Hampicke 1991, Germany	Preservation of endangered species	106–209
Spash and Hanley 1995, UK	Animal rights	16
	Biotic rights	12
	Ecosystem rights	13
Jakobssen and Dragun 1999, Australia	Preservation of all endangered species in Victoria	96
Lundhede et al. 2004, Danmark	Small biodiversity	9
	Large biodiversity	84

meta-analysis to calculate an average WTP of $5–126 for single species, $18–194 for multiple species, and $27–101 for ecosystems and natural terrestrial habitat diversity. Also using a meta-analysis, Travisi and Nijkamp[32] calculate an average WTP for a reduction in pesticide risk exposure that enhances biodiversity of $14 per annum.

It does not seem useful for the purpose of this chapter to regard evaluations of individual species, because they often suffer from the so-called embedding problem, which leads to biased values, mostly due to the part–whole effect, sequencing and nesting.[33] However, species diversity and some charismatic species, including bears, butterflies and orchids, typically are highly estimated. In addition, people tend to value endangered species and bigger species like mammals more than they do others.[34] In all, there are clear preferences for the preservation of species, but the WTP varies strongly depending on the volume and characteristics of the evaluated population(s), circumstances, applied methods and the study sample.

WILLINGNESS TO PAY FOR SOIL, WATER, AND AIR QUALITY

Most of the surveys that consider values for the quality of air, water and soil refer to certain regions, rivers or lakes, in which there is a distinct level of pollution of these resources. This tendency makes it difficult to conclude an average WTP for either the preservation or the improvement to a certain level of quality, particularly in terms of air and soil. However, surveys about preferences for water quality are more comparable because, on the one hand, there are many of them, and on the other hand, the value of water quality is more likely to be comparable according to its purpose, which features direct use values such as drinking or recreation. Furthermore, the results of studies about the perceived risks of pesticides indicate that concerns about the contamination of drinking water seem most significant, followed by concerns about possible adverse effects on ecosystems.[35] Anxieties about risks for human health from pesticide residues in food and exposure to residues in water, soil and air rank third. Thus, it is important to note that the order of magnitude of a WTP estimate relates not only to the specific type of risk and the nature of the risk scenario considered, but also to laypeople's subjective perception of risks. Table 16.5 contains some exemplary results from soil, water and air quality valuation surveys.

Table 16.5 Willingness to pay for soil, water and air quality[36]

Study	Feature	Mean WTP per household/year
Boyle et al. 1994, USA	Groundwater quality	US$65–1,341
Crutchfield et al. 1997, USA	50% reduction of nitrate in the drinking water	58–77
Brouwer et al. 1997, various countries	Groundwater quality	209
	Fresh water quality	97
	Riverine quality	113
Carlsson et al. 2000, Sweden	Income-related charge for 50% reduction of harmful aerial substances	312
Bateman et al. 2002, UK	Scheme to reduce toxicity of vehicle emissions	112
Bateman et al. 2006, UK	River Tame water quality improvements	Small: 13 Medium: 31 Large: 48
Colombo et al. 2006, Spain	Soil conservation programme	18.5–34.5

Colombo et al.[37] estimate social benefits of a soil conservation programme of approximately $18.5–34.5 per capita per year, using conjoint analysis and contingent valuation. It may be somewhat challenging for laypeople to quantify a WTP for distinct improvements of air quality though, because the average citizen is not aware of the health risks of certain degrees of air pollution. Hence, it is nearly impossible for them to calculate their own individual benefits from improvement activities. Carlsson and Johansson-Stenman[38] estimate a mean WTP for a 50 per cent reduction of harmful substances where the respondents live and work of about $26 per person per month as a charge related to income, whereas Diener[39] ascertains that $59 is the mean WTP per month in a similar investigation. In contrast, Bateman et al.[40] estimate a mean WTP of approximately $112 by asking the following question: 'The toxicity of vehicle emissions may be reduced in a number of ways, but all of these cost money. Please consider the maximum amount you would be willing to pay per year to fund a scheme which reduced the toxicity of vehicle emissions to the level where these respiratory and health effects no longer occurred.'

Altogether, the results indicate the difficulties people have in valuing unfamiliar goods, such as air and soil quality, so the stated amounts must be interpreted with care. However, people appear highly concerned about the quality of air and water, whose direct use cannot be avoided. This perception is confirmed by the higher WTP for the quality of drinking water and groundwater compared with that for the purity of lakes and rivers.

WILLINGNESS TO PAY FOR NON-RENEWABLE RESOURCES

Non-renewable resources are mainly used for the production of energy, which is required at each stage of the supply chain. Depending on the technology, the generation of energy relates to several impacts on the environment, as well as diverse levels of health risks. Likely effects resulting from the substitution of fossil fuels by renewable resources include changes to the landscape, wildlife and pollution levels, particularly air pollution. Furthermore, the change of technologies might be compared according to the creation of long-term employment opportunities and potential increases in electricity prices to pay for renewable sources.[41] Hence, this context seems to provide an interesting means to learn more about preferences for different types of energy used in the food supply chain.

A Scottish mail survey asked consumers about their acceptance of an increase in electricity charges per annum per household because of a change in the attributes involved in substituting fossil fuels with renewable energy resources.[42] According to their choices from various hypothetical energy programmes, people were willing to pay $13.27 more for the prevention of high impacts on the landscape and a premium of $6.96 to turn a slight increase in the harm to wildlife into no harm at all. Furthermore, Scottish households would be willing to pay $19.87 per annum to change a slight increase in harm to wildlife from renewable projects to an improved level of wildlife, compared with the current status. Most important seems to be the impact on air pollution: In this survey, participants stated they were willing to pay $23.45 to have renewable energy projects that did not increase in air pollution compared with a programme that resulted in a slight increase in pollution. However, the respondents did not seem to care about employment effects, up to a certain point. The estimated values of this survey for both unaffected landscape and wildlife improvement generally comply with results of other surveys.[43]

In a representative survey from 2006, German citizens were asked about important issues of environmental policy. They gave first priority to the aim of independence from fossil fuels and the use of renewable resources of energy instead. Atomic power seemed to be disapproved of by the vast majority of Germans.[44] Similarly, in a recent European survey, more than half of the respondents thought that the risks posed by nuclear energy were greater than the advantages it offers.[45] Furthermore, they expressed highly positive attitudes about the implementation of renewable energy technologies: Regarding energy use, 80 per cent supported solar energy, 71 per cent wind energy, 65 per cent hydroelectric energy, 60 per cent ocean energy and 55 per cent biomass energy. Only a marginal number of respondents opposed these energy sources. When asked to identify from a list the two issues that should be given top priority in the national government's energy policy, 45 per cent of EU citizens mentioned guaranteeing low prices, and 35 per cent chose continuous energy supply, and 'protecting the environment' earned only the third highest share, with about 29 per cent. To summarize, people seem to appreciate the use of renewable resources for energy, but it is difficult to determine a reliable WTP for the substitution of fossil fuels by renewable energy technologies.

WILLINGNESS TO PAY FOR CLIMATE PROTECTION

The debate over climate change is relatively new compared with other environmental problems. A Canadian survey of 2004 showed that citizens at that time had only little knowledge about climate change and its causes.[46] Information about the impact of climate change on daily life is still diverse and uncertain, which makes it difficult to state an order of preferences for avoiding climatic change. Therefore, many researchers refuse the application of contingent valuation or related methods to estimate preferences for activities to avoid global warming.[47] According to recent polls, global warming has gained importance though, especially in the last 4 years, which have seen a substantial change in perceptions of the severity of this problem.[48] Thus, it is not expedient to refer to surveys before at least 2003. Table 16.6 contains the results of selected surveys.

In a 2006 US study of public attitudes about energy and the environment, the mean WTP a premium on an electricity bill to address problems of climate change was estimated to be approximately $14 per month, which is 50 per cent more than it was 3 years ago in the same survey.[49] Similarly, the European Flash Survey of February 2007 shows that half of European citizens are 'very much concerned about the effects of global climate change' and a further 37 per cent are 'not indifferent about the issue'.[50] Polls pertaining to this subject also show that government is not doing enough to address global warming, according to the general opinion.[51] The prevention of climate change seems very important in the public view, and awareness about global warming is still growing.

Interpretation of the Valuation Surveys

The vast majority of these surveys indicate high preferences for the protection of environmental resources, landscapes, species and habitats. Admittedly, the interpretation of the results of environmental valuation surveys must be cautious though. Especially in the early surveys, the study design may have caused severe biases in answering behaviour. Stated WTP often diverges highly within the same environmental amenities. The amounts stated for the prohibition of negative environmental externalities and

Table 16.6 Willingness to pay for climate protection[52]

Study	Payment vehicle and feature	Mean WTP per household/month
Berk and Fovell 1999, USA	Payments to prevent changes in climate at the location	US$13.70
Berrens et al. 2004, USA	WTP for ratification of the Kyoto Protocol to reduce global warming	17
Curry 2004, USA	Premium on the electricity bill to address problems of climate change	13.10
Curry et al. 2005, UK	Premium on the electricity bill to address problems of climate change	18.50
Johnsson and Reiner 2007, Sweden	Premium on the electricity bill to address problems of climate change	5.60
Curry et al. 2007, USA	Premium on the electricity bill to address problems of climate change	18

the supply of positive ones might be biased by several factors, such as embedding or prior information. Differences in WTP also might reflect survey methods and, in particular, the payment vehicle, as well as the question format or the sample selection method.

The validity of the numbers generated therefore depends on the rigour and logic of the methodology. Furthermore, stated preferences depend not only on the characteristics of the considered object, such as beauty and uniqueness, but also on the socio-demographic characteristics of the questioned people, especially their income, age and environmental concerns.[53] More or less significant differences in people's preferences also emerge between countries.[54] Moreover, there are various motivations behind people's preferences for environmental goods and services.[55] These factors all make it difficult to interpret the survey results correctly. Accordingly, WTP answers cannot be treated as absolute values in economic calculations, through they can be applied to some extent for comparison purposes, as well as an indication that people hold significant preferences for such environmental goods.[56]

Regarding the variation in stated preferences for environmental goods and services and the potential biases such as the embedding effect, it seems inappropriate to conclude the total value on the basis of a quoted WTP for the impact of the environmental footprint of a single product. However, the stated amounts offer evidence to suggest that people appreciate a reduction of the various negative environmental impacts of food production and processing.

Not surprisingly, people seem willing to pay initially for the prevention of threats to vital resources such as the climate, water and air. Farm animal welfare and the omission of chemicals in general are valued highly, perhaps because they may affect the quality and safety of food directly. This result is consistent with common findings of investigations regarding the demand for organic food; that is, the choice of a food product made by environmentally friendly farming and processing methods tends to relate to some kind of health concerns.[‡57] The surveys also indicate a clear desire to prevent rare goods from extinction.[58] This motivation may be driven by moral values (for example, every creature's right to life) or by risk aversion, which means that people are afraid of losing something forever.[59]

Physical product characteristics will not necessarily be altered by eco-friendly methods, so the environmental compatibility of food products is a credence attribute. Thus, extrinsic cues like well-designed product labels should be used to communicate the impact on nature and animals. Recent surveys considering the question of what people want to know about the food they eat show that the majority of consumers seem interested in the treatment of animals and environmental impacts.[60] According to Howard,[61] more than 80 per cent of respondents prefer product labels as a source of information, in accordance with the findings of most other surveys in this field.[62]

The following section provides an overview of some of the most common eco-labels. The aim of this section therefore is to address environmental preferences with available information and discuss how consumers interpret that information. In this way, this chapter raises some new aspects of labelling.

---

‡   At the time of this writing, there is still no evidence that organic food is healthier or more nutritious than conventional food.

Eco-labels and Consumer Perceptions

The most common eco-labels are organic food labels, such as the European 'Organic Farming–EC Control System'. One of the main issues in the labelling of organic products is an effort to generate market transparency, but the extent to which it enables consumers to make informed choices at the point of sale is debatable, because organic labels usually provide information about neither health gains nor the environmental footprint of the labelled products. For example, environmentally conscious consumers would have difficulty deciding between organic apples from overseas and conventionally produced ones from their local region, due to the lack of clear information about their environmental externalities. Various labels refer to different kinds of organic farming, such as those from organic producer organizations, organic brands, retailers' own organic labels and national and federal organic labels. It is questionable if consumers can differentiate among these diverse organic labels, though the labels often indicate diverse farming systems with different impacts on the environment.

Some labels indicate farm animal-friendly production systems. The most popular labelling applies to different husbandry systems for laying hens. Other farm animals have just a few labels that indicate living conditions, which are of limited availability and fairly unknown. Currently the most common labels are the Freedom Food Label in the UK, the Free Farmed and the Certified Humane Raised and Handled Scheme in the US and the labelling and certification programme of the British Columbian Society for the Prevention of Cruelty to Animals.

However, confusion remains about the difference between farm animal-friendly products and organic food. Several European surveys reveal that the lack of labelling on production methods regarding animal welfare prevents consumers from possibly shifting toward products with higher animal husbandry standards.[63] Furthermore, 51 per cent of European citizens declared that they could very rarely or never identify husbandry conditions from the label, and almost one-third of the citizens of the European Union claimed it was impossible to acknowledge the rearing conditions of animal food products. Surveys investigating the reasons for the demand for food produced in animal-friendly ways usually reveal similar factors to those that emerge from investigations of the demand for organic products. Accordingly, people seem to assume a high correlation between the welfare of farm animals and the quality of food products.

With regard to climate change and greenhouse gas emissions, at the time of this writing, no food labels exist, but considerations in a number of countries propose to launch labels that will list the carbon footprint of food products. According to a survey undertaken by the Carbon Trust[§] of the UK, 66 per cent of consumers would like to know the carbon footprint of the products they buy.[64]

Various labels refer to biodiversity in some way, such as those that claim 'bird friendly', 'dolphin safe' and 'GMO free', but most of them remain rather unknown. Furthermore, most labels are difficult to associate with definite benefits for a single species or biodiversity.

Very few food makers emphasize the correlation between food production and the preservation of regional scenery (for example, Breisgaumilch company in the Black Forest, Germany; Meat Dartmoor Group in Dartmoor, UK). The basic requirements for doing so

---

§　The Carbon Trust is a private company, is set up by the UK government in response to the threat of climate change with the goal of accelerating the transition to a low-carbon economy.

include outstanding scenery, evidently characterized by agricultural activities that are widely associated with the product range. However, at this time, no official labels or legislative restrictions on this type of claim exist. Thus, it is the companies' responsibilities to convey the likely benefits of the characteristic landscape through purchasing their products.

Altogether, eco-labels do not necessarily provide the information consumers need to make environmentally friendly product choices. This lack of evidence might contribute to the gap between stated preferences and consumption behaviour. Furthermore, frequently cited barriers to the demand for green products can be commodity related, such as price, value, convenience, availability in shops, promotion activities and product labelling; they also may reflect the characteristics of consumers, like mistrust about the credence attributes of products and processes, lack of information, habits and social contexts.

However, rarely mentioned in the literature are trends in economy and society. In addition to the already mentioned reasons, an increasing remoteness of consumers from the food production process, due to the declining number of farms and people working on them, as well as the growing industrialization of farming, may be responsible for differences between values and attitudes on the one side and purchase behaviour on the other. Accordingly, the concentration process in food processing and retailing may diminish the possibility that consumers can gain insights into processing methods and obtain information about the conditions of food production. These trends make it easy for food companies to exploit a romantic image of agricultural production and traditional processing techniques for advertising issues. However, such changes may create problems for companies that sell more sustainably produced goods, because it is getting more and more difficult for them to demonstrate the differences of their own products from customary goods. Thus, together with an increasing variety of products on the shelves, it has become complicated for consumers to identify the advantages of environmentally friendly goods compared with those of conventional products and then to choose the product that suits them best.

Conclusions and Recommendations

Regarding the WTP for environmental goods and the basis of information on which consumers choose food products, there seems to be a potential to affect the food market by applying more sustainable production techniques, combined with reliable product information. An alternative would be to introduce more informative eco-labels to gain consumers' trust and assure demand in the long run. A non-negligible proportion of ethical consumers care about this, and it is possible that this proportion will grow.[65] At least this fraction of consumers likely would appreciate more information about the environmental consequences of their product choices. Considering the increasing competition and diversification of the food market, reduced environmental impacts as a reliable characteristic could create a crucial advantage over competing products. Labelling the environmental impacts also could provide an opportunity for suppliers to differentiate themselves from competitors by applying technologies that are less harmful to the climate, water and other environmental goods or that provide improved farm animal welfare.

However, several important questions need to be considered in terms of eco-labelling. First, to spread moral environmental reasoning to buying decisions, the characteristics that connect the purchase to environmental problems must become salient during the buying situation. Therefore, other characteristics of the purchase should not be too highly involving, to avoid 'monopolizing' the consumer's attention, as price often does. Second, the consumer should feel a high degree of concern for an environmental issue that is associated with the particular buying decision.[66] Third, it is also important to consider the amount of information people can take into account when purchasing food. Usually consumers do not spend much time on their daily shopping, because food is a low-involvement good, and a limited number of product characteristics are crucial for the buying decision. Fourth, the level of knowledge about environmental issues is very different across people. Therefore, the kind of information and its presentation requires more careful elaboration to convey the benefits of sustainable food production and processing methods.

It is also important to anticipate the abuse of green claims and misleading advertising, because consumers' confidence in environmental certificates still must be consolidated.[67] Consumers are often distrustful whether products labelled 'organic' are produced according to the rules of organic farming. The use of pictures and images on products and for promotion also is often misleading. For example, many diary products have lucky cows, green meadows and flowers on the packages, even though their milk comes from industrialized farming systems without free range. Because these symbols stand for animal welfare, healthy nature and nice scenery, consumers do not always get the right impression about the conditions of production.

As can be derived from both environmental valuation surveys and investigations of demand factors for sustainably produced food, labelling about ecologically sound production and processing methods is probably most effective when these characteristics indicate product quality. On the one hand, it can stimulate demand, but on the other hand, intangible characteristics such as a reduced application of pesticides and fertilizers can become experience attributes, which means that expectations can be confirmed after purchase. Such an association might raise barriers to increasing demand, because consumers could reject perhaps unrealistic expectations about better flavour or the positive health effects of eco-friendly products after consumption. Thus, marketing experts should communicate eco-friendly characteristics with a maximum of transparency but without creating unrealistic expectations.

Finally, the extent to which environmental and ethical issues grow in importance for consumers remains to be seen. The aspects may become more important as the incidence of perceptible implications of environmental problems for daily life increase. Labelling about environmental impacts may then influence not only on product choices of ethical consumers but also consumption behaviour in the mass market. Further research is necessary to understand consumers' conceptions of environmental sustainability, quality and healthiness. The effects of more transparency regarding the externalities of food production, as well as labelling strategies, have not been studied very well so far, so more investigations again are required.

References

1.   Birner, R., Bräuer, I., Grethe, H., Hirschfeld, J., Lüth, M., Wälzholz, A., Wenk R. & Wittmer, H. (2002), 'Ich kaufe, also will ich? – Eine interdisziplinäre Analyse der Entscheidung für oder gegen den Kauf besonders tier- und umweltfreundlich erzeugter Lebensmittel', *Berichte über Landwirtschaft*, Vol. 80, No. 4, pp. 590–613; Codron, J., Sirieix, L. & Reardon, T. (2006), 'Social and environmental attributes of food products in an emerging mass market: Challenges of

signalling and consumer perception, with European illustrations', *Agriculture and Human Values,* Vol. 23, No. 2, pp. 283–297; Halkier, B. (2001), 'Risk and food: Environmental concerns and consumer practices', *International Journal of Food Science and Technology,* Vol. 36, No. 8, pp. 801–812; Thøgersen, J. (1999), 'The ethical consumer: Moral norms and packaging choice', *Journal of Consumer Policy,* Vol. 22, No. 4, pp. 439–460; Verbeke, W. & Vermeir, I. (2006), 'Sustainable food consumption: Exploring the consumer attitude – behavioural intention gap', *Journal of Agricultural and Environmental Ethics,* Vol. 19, No. 2, pp. 169–194.

2.   Belz, F. (2001), *Integratives Öko-Marketing: Erfolgreiche Vermarktung von ökologischen Produkten und Leistungen.* Wiesbaden: Deutscher Universitäts-Verlag; Honkanen, P., Verplanken, B. & Olsen, S.O. (2006), 'Ethical values and motives driving organic food choice', *Journal of Consumer Behaviour,* Vol. 5, No. 5, pp. 420–430; Lintott, J. (1998), 'Beyond the economics of more: The place of consumption in ecological economics', *Ecological Economics,* Vol. 25, No. 3, pp. 239–248; Rubik, F. & Frankl, P. (2005), *The Future of Eco-Labelling. Making Environmental Product Information Systems Effective.* Sheffield: Greenleaf Publishing Ltd; Weber, C. (1999), 'Economic and socio-psychological models of consumer behaviour: Can "limited rationality" bridge the gap?', *Energy Efficiency and CO2 Reduction: The Dimensions of the Social Challenge,* 1999 Summer Study Proceedings, part 2, panel III, Paris.

3.   Brombacher J., & Hamm U. (1990), 'Ausgaben für eine Ernährung mit "Bio-Lebensmitteln"', *Agra-Europe 07/90, Markt und Meinung,* pp. 1–11; Grunert, S.C. (1993), 'Everybody seems concerned about the environment but is this concern reflected in (Danish) consumers' food choice?', *European Advances in Consumer Research,* Vol. 1, No. 4, pp. 428–33; Honkanen et al., op. cit.; Van Dam, Y.K. (1991), '*A conceptual model of environmentally conscious consumer behaviour*', Marketing Thought around the World, Proceedings of the 20th European Marketing Academy Conference, Vol. 2, No. 4, pp. 463–83.

4.   Bruhn, M. (2001), *Verbrauchereinstellungen zu Bioprodukten – der Einfluss der BSE-Krise 2000/2001,* Kiel: Arbeitsbericht Nr. 20 des Lehrstuhls für Agrarmarketing der Universität Kiel; Codron et al., op. cit.; European Commission (2007a), *Attitudes of EU Citizens towards Animal Welfare,* Special Eurobarometer 270; Halkier, op. cit.; Nocella, G., Hubbard, L. & Scarpa, R. (2007), *Consumer Trust and Willingness to Pay for Certified Animal-Friendly Products,* Working Paper, Department of Economics, University of Waikato, Waikato; Sirieix, L. & Schaer, B. (1999), *A Cross-Cultural Research on Consumers' Attitudes and Behaviors towards Organic and Local Foods,* Paper presented at the Seventh Symposium on Cross-Cultural Consumer and Business Studies, Cancun, Mexico, December 10–13.

5.   Martin, B. & Simintiras, A.C. (1995), 'The impact of green product lines on the environment: Does what they know affect how they feel?', *Journal of Marketing Intelligence and Planning,* Vol. 13, No. 4, pp. 16–23; Torjusen, H., Sangstad, L., O'Doherty Jensen, K., & Kjærnes, U. (2004), *European Consumers' Conception of Organic Food: A Review of Available Research,* National Institute for Consumer Research, Oslo; Verbeke & Vermeir, op. cit.; Weber, op. cit.

6.   Belz, op. cit.; Birner et al., op. cit.; Rubik & Frankl, op. cit.

7.   Gerbens-Leenes, P.W., Moll, H.C. & Schoot Uiterkamp, A.J.M. (2003), 'Design and development of a measuring method for environmental sustainability in food production systems', *Ecological Economics,* Vol. 46, No. 2, pp. 231–248.

8.   Tanner, C., Kaiser, F.G. & Wölfing Kast, S. (2004), 'Contextual conditions of ecological consumerism. A food purchasing survey', *Environment and Behaviour,* Vol. 36, No. 1, pp. 94–111.

9.   Bennett, R. (1995), 'The value of farm animal welfare', *Journal of Agricultural Economics,* Vol. 46, No. 1, pp. 46–60; European Commission, 2007a, op. cit.; Harper, G. C. & Henson, S.J. (2001),

Consumer Concerns about Animal Welfare and the Impact on Food Choice – The Final Report, Centre for Food Economics Research, The University of Reading, EU FAIR CT98-3678; Kjærnes, U., Miele, M. & Roex, J. (2007), *Attitudes of Consumers, Retailers and Producers to Farm Animal Welfare*, Welfare Quality Reports 2, Cardiff University, Cardiff; Nocella et al., op. cit.

10. Danish Environmental Protection Agency (EPA) (2002), 'Danske husholdningers miljøbelastning', Arbejdsrapport fra Miljøstyrelsen nr. 13, 2002. (updated 16 Jan. 2008), <http://www.mst.dk/default.asp?Sub=http://www.mst.dk/udgiv/publikationer/2002/87-7972-094-3/html/>; Grunert, op. cit.

11. Bennett, R. & Blaney, R. (2002), 'Social consensus, moral intensity and willingness to pay to address a farm animal welfare', *Journal of Economic Psychology*, Vol. 23, No. 4, pp. 501–520; European Commission, 2007a, op. cit.; Nocella et al., op. cit.

12. Bennett & Blaney, op. cit.

13. Bennett & Blaney, op. cit.; Burgess et al., op. cit.; Nocella et al., op. cit.

14. Burgess, D., Hutchinson, W.G., McCallion, T. & Scarpa, R. (2003), *Investigating Choice Rationality in Stated Preference Methods for Enhanced Farm Animal Welfare*, Working Paper ECM 03-02, CSERGE, University of East Anglia.

15. Ibid.

16. European Commission (2005), *Attitudes of Consumers towards the Welfare of Farmed Animals*, Special Eurobarometer 229.

17. Ibid.

18. Nocella et al., op. cit.

19. European Commission (2006), *Risk Issues*. Special Eurobarometer 238.

20. European Commission, 2007a, op. cit.; Nocella et al., op. cit.

21. Kaltenborn, B.P. & Bjerke, T. (2001), 'Associations between environmental value orientations and landscape preferences', *Landscape and Urban Planning*, Vol. 59, No. 1, pp. 1–11.

22. Brink, B. ten, Vliet, A. van, Heunks, C., Haan, B. de, Pearce, D. & Howarth, A. (2000), *Technical Report on Biodiversity in Europe: an Integrated Economic and Environmental Assessment*, RIVM report 481505019; Bonnieux, F. & Le Goffe, P. (1997), 'Valuing the benefits of landscape restoration: a case study of the Cotentin in Lower-Normandy, France', *Journal of Environmental Management*, Vol. 50, No. 3, pp. 321–333; Drake, L (1992), 'The non-market value of agricultural landscape', *European Review of Agricultural Economics*, Vol. 19, No. 3, pp. 351–64; Israelsson, T. (2001), *Valuing Natural Heritage – A Pilot Application of a Choice Experiment*, Licentiate thesis, Report 12, Department of Forest Economics, Swedish University of Agricultural Sciences, Umeå; Moran, D., McVittie, A., Allcroft, D. & Elston, D. (2004), *Beauty, Beast and Biodiversity: What Does the Public Want from Agriculture?*, Report to The Scottish Executive Environment and Rural Affairs Department 2004; Ollikainen, M. & Lankoski, J. (2004), *Multifunctional Agriculture: The Effect of Non-Public Goods on Socially Optimal Policies*, MTT Discussion Papers 1/2005; Roschewitz, A. (1999): *Der monetäre Wert der Kulturlandschaft. Eine Contingent Valuation Studie*, Vauk, Kiel; Sirex, A. (2004), *Le Paysage Agricole: Un Essai d'Évaluation*, These de Doctorat. Université de Limoges, Limoges; Tronstad, A. (1993), *Valuation of the Cultural Landscape – Holdninger og betalingsvillighet for jordbrukslandskapet*, Norwegian University of Life Sciences UMB.

23. Brink et al., op. cit.; Pretty, J. (2004), 'A perspective on environmental externalities and side-effects in the food chain', in Waldron et al. (eds), *Total Food*. University of Norwich, UK: Institute of Food Research, pp. 8–15.

24. Pretty, op. cit.

25. Dachary-Bernard, J. (2004), *Approche Multi-Attibut pour une Evaluation Economique du Paysage*, These de Doctorat, Cemagref; Fischer, A., Hespelt, S. & Marggraf, R. (2003), 'Ermittlung

der Nachfrage nach ökologischen Gütern der Landwirtschaft – Das Northeim-Projekt', *Agrarwirtschaft,* Vol. 52, No. 8, pp. 390–399; Hanley, N., Bergmann, A., & Wright, R. (2004), *Valuing the Environmental and Employment Impacts of Renewable Energy Investments in Scotland,* Research Report, Scottish Economic Policy Network, Stirling/Glasgow; Henseleit, M. (2006), *Möglichkeiten der Berücksichtigung der Nachfrage der Bevölkerung nach Biodiversität.* Göttingen: Cuvillier Verlag; Santos, J. (1997), *Valuation and Cost-Benefit Analysis of Multi-Attribute Environmental Changes,* Ph Thesis, University of Newcastle upon Tyne, Newcastle.

26.   Department for Environment, Food and Rural Affairs (2005), 'Environmental Landscape Features Model', https://statistics.defra.gov.uk/esg/reports/elf/default.asp.

27.   Oglethorpe D., Hanley, N. & McVittie, A. (2001), *Estimating the Value of Environmental Features: Stage Two,* Report to MAFF, London.

28.   EFTEC/IEEP (2004), *Framework for Environmental Accounts for Agriculture,* final report submitted to Department for Environment, Food and Rural Affairs, London.

29.   Brink et al., op. cit.

30.   Nunes, P.A.L.D. & van den Bergh, J.C.J.M. (2001), 'Economic valuation of biodiversity: sense or nonsense?', *Ecological Economics,* Vol. 39, No. 2, pp. 203–222.

31.   Hampicke, U. (1991), *Naturschutz-Ökonome.* Stuttgart: Eugen Ulmer; Holm-Müller, K. (1991), *Die Nachfrage nach Umweltqualität in der Bundesrepublik Deutschland,* Berlin; Jakobsson, K.M. & Dragun, A.K. (1996), *Contingent Valuation and Endangered Species. Methodological Issues and Applications.* Cheltenham: Edward Elgar Publishing Company; Lundhede, T., Hasler, B. & Bille, T. (2005), *Værdisætning af naturgenopretning og bevarelse af fortidsminder i Store Åmose i Vestsjælland,* Rapport udgivet af Skovog Naturstyrelsen, København; Spash, C. L. & Hanley, N. (1995), 'Preferences, information and biodiversity preservation', *Ecological Economics,* Vol. 12, No. 3, pp. 191–208.

32.   Travisi, C.M. & Nijkamp, P. (2004), *Willingness to Pay for Agricultural Environmental Safety,* Discussion Paper, Tinbergen Institute TI 2004-070/3, Amsterdam.

33.   Carson, R. T. & Cameron Mitchell, R. (1995), 'Sequencing and nesting in contingent valuation surveys', *Journal of Environmental Economics and Management,* Vol. 28, No. 2, pp. 155–173; McFadden, D. (1994), 'Contingent valuation and social choice', *American Journal of Agricultural Economics,* Vol. 76, No. 4, pp. 689–708.

34.   Loomis, J.B. & White, D.S. (1996), 'Economic benefits of rare and endangered species: Summary and meta-analysis', *Ecological Economics,* Vol. 18, No. 3, pp. 197–206.

35.   Wronka, T. (2004), *Ökonomische Umweltbewertung: vergleichende Analyse und neuere Entwicklungen der Kontingenten Bewertung am Beispiel der Artenvielfalt und Trinkwasserqualitä.,* Kiel:Verlag Vauk.

36.   Bateman, I.J., Cole, M.A., Georgiou, S. & Hadley, D.J. (2006), 'Comparing contingent valuation and contingent ranking: A case study considering the benefits of urban river water quality improvements', *Journal of Environmental Management,* Vol. 79, No. 3, pp. 221–231; Bateman, I.J., Georgiou, S., Langford, I.H., Poe, G.L. & Tsoumas, A. (2002), *Investigating the Characteristics of Expressed Preferences for Reducing the Impacts of Air Pollution: a Contingent Valuation Experiment,* CSERGE Working Paper EDM 02-02; Boyle, K.J., Poe, G.L. & Bergstrom, J.C. (1994), 'What do we know about groundwater values? Preliminary implications from a meta analysis of contingent valuation studies', *American Journal of Agricultural Economics,* Vol. 76, No. 5, pp. 1055–061; Brouwer, R., Langford, I.H., Bateman, I.J., Crowards, T.C & Turner, R.K. (1997), *A Meta-Analysis of Wetland Contingent Valuation Studies,* Centre for Social and Economic Research on the Global Environment, Working Paper GEC 97-20; Colombo, S., Calatrava-Requena, J. & Hanley, N. (2006), 'Analysing the social benefits of soil conservation measures using stated preference

methods', *Ecological Economics*, Vol. 58, No. 4, pp. 850–861; Crutchfield, S.R., Cooper, J.C. & Hellerstein, D. (1997), *Benefits of Safer Drinking Water: The Value of Nitrate Reduction*, U.S. Department of Agriculture: Food and Consumer Economics Division, Economics Research Service; Carlsson, F. & Johansson-Stenman, O. (2000), 'Willingness to pay for improved air quality in Sweden', *Applied Economics*, Vol. 32, No. 6, pp. 661–669.

37. Colombo et al., op. cit.

38. Carlsson & Johansson-Stenman, op. cit.

39. Diener, A. (1999), *Valuing Health and Air Quality Using Stated Preference Methods*, Dissertation, McMaster University, Canada.

40. Bateman et al., 2002, op. cit.

41. Bergmann, A., Hanley, N., & Wright, R. (2004), 'Valuing the attributes of renewable energy investments' *Energy Policy*, Vol. 34, No. 9, pp. 1004–1014.

42. Ibid.

43. Israelsson, op. cit.; Lundhede et al., op. cit.; Sirex, op. cit.; Travisi & Nijkamp, op. cit.

44. Kuckartz, U., Raediker, S. & Rheingans-Heintze, A. (2006), *Umweltbewusstsein in Deutschland*, Bundesministerium für Umwelt, Naturschutz und Reaktorsicherheit (BMU) Referat Öffentlichkeitsarbeit, Berlin.

45. European Commission (2007b), *Europeans and Nuclear Safety*, Special Eurobarometer 271.

46. Consumers Council of Canada (2004), *Consumers' Willingness to Pay for Climate Change*, Research Brief of April 2004.

47. Clarkson, R. & Deyes, K. (2002), *Estimating the Social Costs of Carbon Emissions*, Government Economic Service Working Paper 140, Environment Protection Economics Division, Department of Environment, Food and Rural Affairs, London.

48. Bowman, K. (2007), *Polls on the Environment and Global Warming*, AEI Public Opinion Study, (updated 20 Apr. 2007) http://www.aei.org/publication14888/; Curry, T.E., Ansolabehere, S. & Herzog, H.J. (2007), *A Survey of Public Attitudes towards Climate Change and Climate Change Mitigation Technologies in the United States: Analyses of 2006 Results*, MIT LFEE 2007-01 WP; European Commission (2007c), *Attitudes on Issues Related to EU Energy Policy*, Flash Eurobarometer 206a.

49. Curry et al., op. cit.

50. European Commission, 2007c, op. cit.

51. Bowman, op. cit.; Curry et al., op. cit.; European Commission, 2007c, op. cit.

52. Berk, R.A. & Fovell, R.G. (1999), 'Public perceptions of climate change: A "willingness to pay" assessment', *Climatic Change*, Vol. 41, No. 3–4, pp. 413–446; Berrens, R.P., A.K. Bohara, H.C. Jenkins-Smith, C.L. Silva &. Weimer, D.L (2004), 'Information and effort in contingent valuation surveys: Application to global climate change using national Internet samples', *Journal of Environmental Economics and Management*, Vol. 47, No. 2, pp. 331–363; Curry et al., op. cit.; Curry, T.E. (2004), *Public Awareness of Carbon Capture and Storage: A Survey of Attitudes toward Climate Change Mitigation*, Master Thesis, Laboratory for Energy and the Environment, Massachusetts Institute of Technology, MA, USA; Curry, T.E., Reiner, D.M., de Figueiredo, M.A. & Herzog, H.J. (2005), *A Survey of Public Attitudes towards Energy & Environment in Great Britain*, Publication No. LFEE 2005-001 WP, MIT Laboratory for Energy and the Environment, Cambridge, MA, USA; Johnsson, F. & Reiner, D.M. (2007), *Public and Stakeholder Attitudes towards Energy, Environment and CCS*, AGS Pathways report 2007: E2. AGS Office at Chalmers GMV, Chalmers SE – 412 96 Göteborg, Sweden.

53. Elsasser, P. & Meyerhoff, J. (2001), 'Ökonomische Bewertung von Umweltgütern. Methodenfragen zur Kontingenten Bewertung und praktische Erfahrungen im deutschsprachigen Raum', *Metropolis Ökologie und Wirtschaftsforschung*, Vol. 40, pp. 351–381.

54. European Commission, 2007a, op. cit.; Nocella et al., op. cit.
55. EFTEC/IEEP, op. cit.
56. Bateman et al., 2002, op. cit.
57. Birner et al., op. cit.; Torjusen et al., op. cit.; Verbeke & Vermeir, op. cit.
58. Henseleit, op. cit.
59. Ibid.; Karkow, K. (2003), *Wertschätzung von Besuchern der Erholungslandschaft Groß-Zicker auf Rügen für naturschutzgerecht genutzte Ackerstandorte in Deutschland.* Greifswald: Diplomarbeit; Carlsson, F., Frykblom, P., & Lagerkvist, C.J. (2007), 'Farm animal welfare-testing for market failure', *Journal of Agricultural and Applied Economics*, Vol. 30, No. 2, pp. 61–73.
60. European Commission, 2005, op. cit.; European Commission, 2007a, op. cit.; Burgess et al., op. cit.; Howard, P. (2005), *What Do People Want to Know About Their Food?*, Research Brief 5, Winter 2005, The Center for Agroecology & Sustainable Food Systems, University of California, Santa Cruz.
61. Howard, op. cit.
62. McCluskey, J.J. & Loureiro, M.L. (2003), 'Consumer preferences and eillingness to pay for food labeling: A discussion of empirical studies' *Journal of Food Distribution Research*, Vol. 34, No. 3, pp. 95–102.
63. European Commission, 2005, op. cit.; Harper & Henson, op. cit.
64. Carbon Trust (2007), '*Carbon Trust Launches Carbon Reduction Label*', Press Release (updated 16 Jan. 2008), http://www.carbon-label.co.uk/pdf/release.pdf.
65. Honkanen et al., op. cit.; Johansson-Stenman, O. (2006), *Should Animal Welfare Count?*, Working Paper, Department of Economics, Göteborg University; Tallontire, A., Rentsendorj, E. & Blowfield, M. (2001), *Ethical Consumers and Ethical Trade: A Review of Current Literature*, Policy Series 12, Natural Resource Institute, University of Greenwich.
66. Thøgersen, op. cit.
67. Henseleit, M. & Herrmann, R. (2007), 'Qualität als Einflussfaktor der Nachfrage nach Nahrungsmitteln' *Ernährung im Fokus*, Vol 11, No. 7, pp. 340–345

IV *Fair Engagement?*

17 The Elusive Written Contract: Dependence, Power, Conflict and Opportunism Within the Australian Food Industry

BY MELINA PARKER* AND JOHN BYROM†

Keywords

relationship marketing, relationship complexities, retailing, Australia.

Abstract

In this chapter, we explore buyer–supplier relationships within the Australian food industry. From a buyer–supplier perspective, this industry can be considered increasingly adversarial. Fifty in-depth dyadic interviews were conducted with suppliers and their trade partners. Using this research, we will investigate the role of relationship complexities, and specifically dependence, power, conflict and opportunism. Furthermore, we provide real-life examples of how buyers and suppliers cope with and react to relationship issues; and, finally, illustrate the implications of this research, namely, that managers and policymakers should pay attention to the less savoury aspects of relationship development in an attempt to build sustainable partnerships.

* Dr Melina Parker, Milton Farm Pty Ltd, Don, Tasmania 7310, Australia. E-mail: melina.parker@miltonfarm.com.au. Telephone: + 61 4 1751 9093.

† Dr John Byrom, School of Management, University of Tasmania, Locked Bag 1316, Launceston, Tasmania 7250, Australia. Email: john.byrom@utas.edu.au. Telephone: + 61 3 6324 3797.

Introduction and Background

Despite difficulties in defining relationship marketing and the various schools of thought that have contributed to current thinking, practitioners and academics generally agree that trade partnerships offer a great advantage to both buyers and suppliers.[1] Indeed, relationships are one of the most durable aspects of competitive advantage, because relationship dynamics are practically impossible for competitors to emulate.[2] Kasabov notes that the relationship marketing literature is dominated by 'positiveness',[3] calling for more research to consider relationship complexities such as power, conflict and disputation, to facilitate a more holistic understanding of relationship intricacies. Although consideration of negative aspects may not be as well researched as more favourable relationship traits, some academics have considered their importance. Hunt, for example, refers to dependence, power, conflict and opportunistic behaviour as attributes of relationship failure.[4] This chapter focuses on these types of complexities in the context of buyer–supplier relationships. Adopting a qualitative approach, we consider the impact of the complexities on the development of relationships between primary producers and their direct trade partners and take a dyadic approach. The context of the research is the Australian food industry, to which we now turn.

CONTEXT

The Australian food industry has changed significantly in recent decades, affecting producers, consumers and trade: 'The new food economy is bigger, broader, faster and more demanding.'[5] The industry has moved away from independent buyers and suppliers toward a highly integrated supply chain and highly integrated networks of buyers and suppliers.[6] The historic emphasis on production, adopted across the broader food industry, also is shifting in favour of relationship development.

The Australian retail environment

Although the Australian supermarket industry previously may have been considered uninteresting and unsophisticated,[7] the very high level of concentration within the marketplace (dominated by two main players, Woolworths‡ and Coles) makes for an interesting discussion. In 2004–05, the two major players within the 'Supermarket and Other Grocery (not Convenience) Stores' industry accounted for more than 70 per cent of total market share.[8] This high level of concentration has increased steadily during the past 30 years, such that the market share of the major chains increased from 40 per cent in 1975 to 60 per cent in 1985.[9] There is some concern that these major supermarkets have too much power. For example, Harvey notes that the two main consequences of buying power centralization are the introduction of higher levels of private branding and the emergence of dedicated, long-term supply relationships.[10]

---

‡   Whilst the Woolworths Group owned the variety stores which were until recently a feature of British High Streets, Woolworths Limited is a completely separate grocery business which continues to trade in Australia and New Zealand. Given that there are also several other retail fascias which go by the name 'Woolworths', we therefore feel that an explanatory footnote is not necessary.

Much industry discussion focuses on the two major supermarkets within the Australian retail sector, yet a variety of buyers operate within the marketplace, including supermarket retailers, independent retailers, greengrocers, wholesalers and specialist importers. These varied buyers have different strategies to complement their core business,[11] related not only to the type of retail organization but also to their size.[12] Traditional retail outlets are now smaller in number and perhaps declining in importance, largely due to supermarkets' ability to remain open longer and provide more stockkeeping units (SKUs) at cheaper prices.

Suppliers

Farmers continue to face both short- and long-term challenges.[13] Supply and demand issues, exacerbated by droughts and the increasing power of manufacturers and retailers, provide a difficult environment in which to succeed. The supply side is highly fragmented; suppliers range from large corporations to small family-owned and -operated farming enterprises. The desire of processors and retailers to source their product from fewer but larger suppliers has also been acknowledged.[14]

Suppliers recognize that vertical coordination provides them year round access to large retailers, greater security, additional information, feedback on variety acceptability and new product development and programming advice,[15] yet the power balance between the suppliers and the two major retailers remains important. Although there are positive outcomes from the duopolistic nature of the retail industry, such as demand stability, price stability, increased flow of information, and supply chain efficiencies, anecdotal evidence indicates that Coles and Woolworths are asserting their power to get what they want. As some suppliers divert their produce to export markets, it becomes evident that the major supermarkets can affect trade and domestic agricultural production significantly.[16] Although the presence of smaller buyers operating within the market allows suppliers to spread their risk, their continued survival remains threatened as the major supermarkets continue to drive down product and transaction costs. Thus, producers must retain their focus on price competitiveness and maintaining ever increasing quality and food safety standards, as specified by the supermarkets.[17]

Literature Review

Previous research shows the importance of dependence, power, conflict and opportunism within horticultural buyer–supplier relationships.[18] Evidence of asymmetrical dependence, coercive power, dysfunctional conflict and/or opportunistic behaviour should result in the demise of successful relationships between suppliers and buyers. We also might expect that relationship complexities would not have the same detrimental effect in circumstances in which power is non-coercive and conflict is considered functional. Relationship development further implies that trade partners accept a certain level of interdependency,[19] which can lead to potential power inequalities that result in conflict, opportunism and, in the worst cases, relationship dissolution.

DEPENDENCE

Dependence is a central defining construct of the buyer–supplier relationship,[20] related to trust, commitment and subsequent investment.[21] Both trade partners may perceive their or their partner's dependence differently than their partner does.[22] Transactions and goals should be mutually beneficial;[23] otherwise, there may be some indication that the asymmetrical dependence of partners renders them vulnerable to the misuse of power and the likelihood that the relationship will fail.[24] Dependence also should be greater if there are no apparent alternatives.[25] If dependence is not apparent, there will be little incentive for either party to invest in the relationship. When interdependency exists, cooperation is inherent.[26] However, if dependence seems overtly asymmetrical, distrust will result.[27]

Dependence also can be considered in terms of the cost of replaceability,[28] such that the trade partner that is most dependent more readily complies with the trade partner's requests. Although some elements of interdependency can contribute positively to overall relationship development, for this research, we consider interdependency a form of relationship complexity, that is, a factor that creates complications in terms of developing a trade relationship. For example, within the context of business-to-business relationships, asymmetrical dependence is more likely to exist than is symmetrical dependence. Therefore, dependence issues may affect the ability of trade partners to develop trust, commitment, cooperation or a long-term orientation.

POWER

Core definitions of power infer control, influence or direction of one party's behaviour by another.[29] Hingley addresses the impact of power on asymmetrical relationships, noting that its presence is not necessarily negative.[30] However, the modern approach to power is perhaps not as positive, and most research considers power a negative force within relationships.

Power as a topic of concern within the global retail milieu is not a new concept. In their qualitative study of buyers and suppliers in the UK and Australian food retailing sectors, Dapiran and Hogarth-Scott find Australian retailers are more ready to rely on coercive power to influence supplier behaviour.[31] Retailers also often hold a higher level of power within trade relationships because they control access to the consumer.[32]

Respect and fair treatment between trade partners are important if relationships are to succeed.[33] In contrast, the continual 'exercise of power to gain acquiescence ... destroys trust and commitment which decreases cooperation and inhibits long-term success.'[34] Even if it remains in control of policies and procedures, the more powerful partner should accept some responsibility for the other's profitability[35] and sense of worth. Even when there is an imbalance of power, 'each firm is gaining control of at least one part of its environment while giving away some of its internal control.'[36]

Procedural fairness binds relationship partners together,[37] and for this reason, partners may be wary of exploiting their power.[38] However, as evident from the preceding consideration of power, when one organization has a power advantage over the other, relationship problems and, in some cases, relationship dissolution can result. Therefore, we consider the negative impact that power can have on relationship development. This effect may be particularly important for our research context, given the undifferentiated nature of the industry, which typically provides a buyer with greater power over the supplier.

CONFLICT

There will always be conflict within relationships.[39] Conflict can be either dysfunctional, and therefore destructive to the relationship, or functional and facilitative of amicable dispute resolution.[40] Although conflict may be amicably resolved (functional conflict) and ultimately assist relationship development,[41] the possibility also exists that it will result in relationship dissolution.

Some scholars propose relationships between the optimum levels of conflict and performance, including Rosenbloom's U-curve,[42] according to which moderate levels of conflict result in channel efficiency, and Duarte and Davies's empirically supported linear model.[43] Acknowledging the vast literature pertaining to conflict, Zhou and colleagues address the need for academia to consider conflict by defining buyer–supplier relationship constructs such as dependence.[44] Although their exploratory empirical investigation relies on a small sample, Zhou and colleagues find support for their proposition of a difference between buyers' and suppliers' perceived dependence and that this perceived difference relates positively to channel conflict.[45] The only relationship variable they consider is dependence, yet they suggest that its fundamental role may mean that other variables similarly relate to channel conflict. The likelihood of dyadic relationships experiencing both functional and dysfunctional conflict necessitates their consideration within our proposed conceptual framework.

OPPORTUNISM

Opportunism is a fundamental principle underlying the trend toward buyer–supplier relationship development. When relationships develop, organizations increase their vulnerability to opportunistic behaviour.[46] Opportunism also has been defined as 'self-interest seeking with guile',[47] and the conditions that allow such behaviour stem from the development of relationships that create routinized transactions, subsequent mutual dependencies, asset specificity and small numbers conditions. Provan argues that relationships within the larger network structure minimize the ability of organizations to act opportunistically.[48]

Which partner is most open to opportunistic behaviour depends on the context. For example, network-suppressed opportunism is less likely to have an impact in the Australian food industry because of the lack of competition from the buyer's perspective. As Gulati and colleagues suggest,[49] when relationship-specific investment occurs, transaction cost economics dictate that the cost of the transaction increases, because the relationship-specific investments increase the likelihood of a hold-up situation, such that the party that has undertaken the investment becomes subject to the risk of opportunistic behaviour by their trade partner. The likelihood that buyers or suppliers exhibit some opportunistic behaviour within dyadic relationships prompts us to include opportunism as a form of relationship complexity. Because of the negative impact opportunism can have on relationships, we assume that when opportunistic behaviour is apparent within a dyad, relationship development may suffer.

Methodology

We consider a qualitative methodological approach most appropriate for this research. Lindgreen calls relationship marketing a 'contemporary, pre-paradigmatic and on-going phenomenon' that requires consideration within its real-life context.[50] Qualitative research facilitates this consideration, by allowing an exploration of the relationship with all its rich and meaningful characteristics intact.[51] This chapter draws on findings from 50 in-depth, dyadic, semi-structured interviews conducted across food-based organizations in five states of Australia. We provide the characteristics of the interviewees in the Appendix. The dyadic approach involves both buyers and suppliers, who provide their own views of the relationship between them (unit of analysis). A snowball sampling method was used. The interviews include both buyer and supplier perspectives of the relationships between direct trade partners and took place in 2005 and 2006. With the in-depth interviews, we can follow up and explore emergent ideas, though we are guided by a semi-structured interview guide. Probe questions were used throughout the in-depth interviews. To illustrate the key themes, we include selected excerpts from the interviews in the Findings section. We also employ pseudonyms to maintain the confidentiality of participants.

Findings

In this section, we consider how complexities may affect the development of relationships within trade dyads in the Australian food sector.

DEPENDENCE: LARGE RETAILER AVOIDANCE

Many partnerships are characterized by asymmetrical dependence, which left the more dependent partner in a relatively weaker position. Although some buyers claim they are dependent on their supply partners, this situation is relatively rare. For the buyer, dependence usually is manifested by having only one supplier from which it could source a particular product. As stated by Aileen (gourmet retailer), 'There's one I deal with only because I can't get the product anywhere else. If I could get it somewhere else, I wouldn't deal with them.' Such dependence on a particular product or brand is a situation that larger supermarkets are reluctant to develop.

Although respondents discuss their dependence in terms of interdependent trade relationships, we also find evidence that some actively seek to avoid such situations, as illustrated by the use of multiple suppliers of particular products and a reluctance to allow supplier brands into retail stores. Despite its dependence, the buyer's power to pick and choose suppliers is thus maintained. Therefore, our research reiterates the recognized tendency for suppliers to represent the dependent party in trade relationships.

POWER: DOES SIZE MATTER?

As expected, buyers are the more powerful trading partners, yet the buyers' perceived level of power within a relationship does not increase as the organizations' size increases. Larger retailers often are assumed to have greater levels of power within a relationship, yet small retailers also indicate that they regard themselves as more powerful than their supply partners. As Paulo (small retailer) notes, if a supplier starts to cause problems, he has no qualms about letting it go. Such an attitude reinforces the belief that he is the more powerful party within the relationship:

A couple of them [suppliers] can get very demanding. ... Suppliers that take it out on our staff, we don't deal with them anymore; it happens with customers too, if they have a complaint and they yell and scream at a staff member. But ... you have to be a little bit diplomatic with the customers ... you don't have to with the suppliers.

(Paulo, small retailer)

It is unlikely that Paulo would have developed a relationship with any suppliers that directly exerted such aspects of power.

Inherent amongst most buyers is the assumption that they have more choice than their trade partners and could exercise their power in the relationship by switching suppliers. However, there are some exceptions to this behaviour. In some circumstances, the switching costs are too high, such as when suppliers own exclusive access to a product, which allows them to establish a higher level of power within the relationship. This power relates to the dependence of the buyer on that product: 'They're [supplier] just hopeless, they're totally disorganized but ... you can't get it any other way' (Aileen, gourmet retailer). Aileen recognizes that this particular product is integral to her retail mix and that she must therefore deal with a trade partner with which she otherwise might not. However, this manifestation of power, as a result of dependence, does not mean that a close relationship has developed. Instead, the relationship may exist in a more transactional form, with no strategic relationship development. Relationships are more likely to develop in circumstances in which the trade partners recognize the strategic importance of the other and attempt to nurture the relationship to protect this asset. In our research, this finding emerges from examples of exclusive arrangements. Requests for exclusive arrangements come from both buyers and suppliers. Few retailers ask for exclusivity in their relationships with suppliers, recognizing the difficulty of committing to only one buyer if the buyer cannot purchase all of the supplier's product. However, some retailers state that they give preference to products for which they have exclusive arrangements, as Emily (gourmet retailer) notes:

Trying to be competitive and stuff, I always say, 'Look, am I the only person in here who is gonna sell this?' If I am, then yep I would take it on. You know if every other shop is gonna sell it then I'm more inclined to say no because what's the point?

(Emily, gourmet retailer)

The majority of retailers, however, accept that to survive, small suppliers must sell to multiple buyers.

Large buyers invariably get implicated when power becomes a more important issue. Suppliers experience the impact of both real and perceived power when dealing with major supermarkets and independent supermarket chains. Supermarkets exert power over trade partners in various forms, including pressure on suppliers to abandon their own brand strategy and pursue the supermarket's retail brands and encouragement of suppliers to commit significant financial resources to servicing the relationship, often without written contracts. Brian (seafood supplier), however, fears that supermarkets would attempt to trap him in a supply contract and then change the terms, leaving him with no alternative but to continue dealing for limited return. This fear represents a real risk, given the dependence that many supplier organizations had developed by building their capacity to deal with supermarkets. After making such an investment, suppliers would find it difficult to find alternative markets for excess supply, should they lose their major supermarket account:

> We don't like supermarkets … they'll get you into a contract, lock you in and then lower the price and cut the floor from underneath you and you'll be stuffed. Then you're in a contract with them and you can't go anywhere. We don't want to go there.
>
> (Brian, seafood supplier)

In an attempt to regain some of the power in the relationship, Ada (fruit supplier) had established a coordinated group of growers to supply a major retailer. However, this option met with some challenges. The major supermarket that the group dealt with was doing everything in its power to separate the growers and attempt to force them to negotiate on an individual basis. Maurice (vegetable supplier) tells a similar story, whereby a tight-knit group of growers shared information and ultimately developed greater strength in their negotiations with large buyers:

> The relationship we've had with them hasn't been too bad. … They try the old tricks. … I mean there's only two growers in Tasmania and they say, 'The grower up the road said he'll give them to me for $20. Can you give it to me for $19?' Well, I know for a fact the grower up the road won't give it to him for $19 because our industry is too close-knit.
>
> (Maurice, vegetable supplier)

Faith's (fruit supplier) large organization can exert more power over its major supermarket buyer because of its ability to provide large volumes of fruit. In addition, its exclusive access to particular varieties affords this supplier more control within the relationship. She discusses her organization's ability to let the retailer think that it was making strategic decisions and exerting power within the relationship, whilst Faith actually made the decisions for them:

> We let them think that that's what they do but … we've never had a product deleted. We've taken products off the shelf when they're not performing but we've never had a retailer come to us and say, 'Hey, listen, that baby spinach, it ain't working for us. We want more margins or we'll get it off the shelf.' We go there and we say, 'Hey, listen, the baby spinach, it's not giving you enough margin so what we're gonna do, we're gonna tart it up, we're gonna do this, do that or we're gonna delete it; and if not we're gonna replace it.' So we work really, really hard on understanding the stats; understanding the consumer trends to make sure that we're actually in the driver's seat.
>
> (Faith, fruit supplier)

Faith could create a more symmetrical power base by being more strategic and staying ahead of the buyer in terms of the direction of the category. Her organization acts in a way her trade partner should perceive as in the long-term best interests of the dyad. In addition, she appears committed to the relationship through her new product development and the tailoring of the product mix to suit the consumer statistics provided to her organization. In contrast, organizations that approach power differences in a more reactive fashion appear less likely to develop stronger relationships. Instead, the nature of power to destroy trust and commitment, decrease cooperation and inhibit long-term success reduces any such opportunity.[52]

CONFLICT: ALWAYS PRESENT, MOSTLY FUNCTIONAL

Conflict emerges as a natural component of negotiations. Respondents cite various reasons for conflict, including price and quality, exclusive arrangements and a perception of unreasonable requirements. The majority of the respondents perceive conflict as inevitable and suggest that strong relationships allow partnerships to recover from incidents that potentially strain the relationship. Thus, most conflict could be considered functional with amicable dispute resolution, not a detrimental impact.

George (fruit supplier) anticipates conflict in his relationship because of exclusive supply arrangements. In considering such conflict, he acknowledges that supplying multiple retailers within one area is not in the best interests of the relationship; rather, he would prefer to support one or two retailers. In addition to limiting the number of outlets that sell his product within each retail precinct, George provides added incentives, such as greater margins, which he clearly communicates to his buyers:

> It is inevitable to have conflict. If we are going to have conflict I suspect it is going to come in the area of, 'Well, I don't want you to sell it to my competitor,' because they might be based in the same shopping mall or something like that. But I go to the markets, the Central Market, and we are only in one stall there and we want one more and then we'll limit it to that. We want people to push our product, if we are in 10 stalls there well they'll say, 'Why should I push it?' But if there is one, they'll push it. I also try to make it so that they can make a good gross margin so that if they sell ours they make more money.
>
> (George, fruit supplier)

James (meat supplier) experienced dysfunctional conflict with a major retail buyer that led to the dissolution of the relationship. He believes that the organization made unrealistic demands on his organization. In this specific instance, he refers to the unreasonable quality assurance requirements of his buyer:

> We've got a licenced abattoir … Australian standard, inspected by the DPI [Department of Primary Industries] twice a year or, if anything is wrong, it would be more often which costs us $1300. [Major supermarket] wanted to fly an auditor in from New South Wales at our expense to look at it and we weren't prepared to do that. They either accept the Australian standard or they don't.
>
> (James, meat supplier)

Demands perceived as unreasonable, especially those incurring additional financial commitments at the supplier's expense, were the cause of many conflicts that sometimes result in relationship dissolution. Yet Spencer (major supermarket) claims his organization has developed a means by which suppliers would become so entrenched in its quality assurance programme that the possibility of confusion regarding quality requirements is minimal. Spencer also considers communication vital in avoiding conflict:

> By the time they get to the stage where they're supplying a lot of animals they would have become very skilled at grading the animals. So the conflict doesn't happen. As I said, the communication is there constantly so there's never any hiccups with them.
>
> (Spencer, major supermarket)

Cash flow and a perceived lack of honesty also can cause conflict. Cash flow remains a very important issue for suppliers, and many relationships have been dissolved because of non-payment or late payment:

> *I had a buyer in Sydney that I wasn't getting on with and I basically got rid of him, well not got rid of him, we are currently in the process of finding someone else and that was primarily because of paying ... That is a really big part of the problem that we've got, cash flow. You can't wait 4, 5 months to be paid.*
>
> (Sam, nut supplier).

Brand strategy support also causes conflict within some relationships when gourmet retailers request exclusive rights to the suppliers' brands. This request creates tension in the relationship, a risk that suppliers need to manage actively. If conflict is functional, it can prompt trade partners to open the lines of communication, share information and develop greater understanding and respect for the partner's position. The trade partners we interview primarily note functional conflict, which does not result in the break-up of the relationship.

OPPORTUNISM: SQUEEZING THE SUPPLIER?

Opportunism is inherent to most discussions of buyer–supplier relationships and even more apparent when major buyers are involved. Unsurprisingly, buyers do not acknowledge explicitly that they engage in opportunistic behaviour, whereas suppliers clearly indicate the presence of such behaviours. Opportunism predominantly occurs when buyers place pressures on suppliers in terms of quality requirements, price reductions and relationship-specific investments. These demands are expected to come at the supplier's expense, without any contracts to bind the relationship legally.

When suppliers deal with major supermarkets in Australia, relationship-specific adaptation generally is required to meet their stringent quality assurance specifications and expand production facilities to meet the large volumes required. Because there are only two major supermarkets in Australia, if suppliers expand their production capacity to supply large supermarkets, they usually need to do everything in their power to ensure they can sustain that relationship, because of the capital investment required. In such a situation, opportunism clearly occurs; major supermarkets put supply contracts out to tender on a regular basis, which pressures the supplier to maintain low prices to retain the contract and thus be able to pay off their capital investment.

Carlos (independent supermarket) discusses the ability of his organization to assist suppliers in obtaining volume sales and increase their production efficiency. However, Carlos's organization also put each contract out to tender regularly, ensuring that it continued to receive the product at a price that may have been too low for the supplier to maintain:

> *We have a relationship with the suppliers. I mean it's in their interest to drive the volume through our stores. It's in our interest to drive the loyalty through the customers. We will send out and push promotional programmes with the suppliers but they will be very quick to respond and knock on our door to promote [their product]. Where they can get volume and*

sales, they will take advantage of that ... because we have the vehicle which they can use to drive volumes.

(Carlos, independent supermarket)

Thus, Carlos does not believe he is taking advantage of his suppliers; rather, he is providing them with an opportunity that would allow them to increase their efficiency and reach sales volumes that may otherwise be unattainable. However, existing literature also notes that some trade partnerships may appear cooperative when in reality, one trade partner simply has learned not to complain.[53] Thus, though Carlos's trade partner continues to pursue the relationship, it does not mean that it is not being exposed to opportunistic behaviour.

Although not in a relationship with a large buyer, several suppliers imagine that they would be subject to opportunistic behaviour if they were to begin dealing with the major supermarkets. Brian (seafood supplier) prefers to deal directly with small retailers and has no intention of trading with large buyers to move excess quantities. He believes that dealing with the major supermarkets would require him to sign a supply contract, which might allow the supermarkets to change the price agreement subsequently. Although the pressure on suppliers to maintain low prices is evident from discussion with major supermarket buyers and their trade partners, written contracts are rarely involved. Only one dyad in our research has a written contract, and even it is not legally binding but rather more like a memorandum of understanding.

The opportunistic behaviour of major supermarkets, as perceived by suppliers, is a hard reality for Ada (fruit supplier), whose products the supermarkets had rejected because they did not meet their quality specifications, though she believes they simply had over-ordered the product:

Going to Queensland, they're rejected before they actually get there. They'll accept [the value-added product] but the other ones don't pass. They're not good enough. They haven't even arrived to be inspected yet, so, you know, but that sort of stuff does happen and ... other times, they'll reject them once they've landed and we've had our agents from that particular market pick them up ... and they bring them back and there's nothing [wrong with them]; they're perfect, there's absolutely nothing wrong.

(Ada, fruit supplier)

Ada further notes that 'They try to push us down to our most competitive to a point where we can barely survive at times' – behaviour that in this case can only be viewed as opportunistic. However, such opportunistic behaviour is associated only with larger partners.

Conclusions

The lack of written contracts in this industry, combined with relationship-specific investments, leads many suppliers to feel quite vulnerable. This feeling gets exacerbated in many cases by issues of dependence, power, conflict and opportunism, which we consider further here. When a partnership is characterized by asymmetrical dependence, the supplier has grown reliant on the volume of sales available by dealing with the buyer.

Thus, the asymmetrical nature of the relationship gets exacerbated as the supplier becomes ever more reliant on the trade partner for sales volumes, whilst potentially sacrificing its own relationship with the end consumer.

Power is evident in most relationships, such that trade partners wish to gain access to the end consumer, and the trade partner closest to the end consumer tends to have the most power. In most cases, the buyer controls access to the consumer, though smaller suppliers could develop more direct methods of distribution and thus gain more control and reduce the dilution of consumer information by remaining closer to the points of sale. Such circumstances meant that suppliers often did not have to share power and enjoyed greater access to information. However, it also meant that access to greater volume sales was not possible.

Although in some relationships (that is, those involving suppliers that rely on the volume of sales promised by large buyers), power and dependence do not deter relationship-specific investments, the lack of strategic commitment to relationship development could result in the relationship being not entirely collaborative. Both parties retain their own agendas and pursue their own strategies separately. Both functional and dysfunctional conflict also emerge in trade dyads and, in the worst cases, ended in relationship dissolution. When dysfunctional conflict is apparent, a positive relationship is not. Such conflict results in one or both trade partners retracting from the relationship and simultaneously withdrawing their resource commitments. Opportunistic behaviour by the buyer also affects the development of positive relationships, such that suppliers that feel open to opportunistic behaviour appear reluctant to invest in the relationship with their trade partner.

MANAGERIAL RECOMMENDATIONS

Complexities are an ever-present feature of dyadic relationships between suppliers and their buyers. We provide evidence of the increasing power of buyers within the Australian food industry, a concern for many in the sector who seek a more equitable balance. As with the UK context,[54] many suppliers feel dwarfed by the size and resultant power of their retail trade partners and have little option but to acquiesce to the requirements of the supermarkets or choose distribution methods that do not include dealing directly with large retailers. Large suppliers, though not as weak as smaller suppliers in relation to their retail trade partners, still face a scenario of dependence because they require large volumes to sustain their business. Such large volumes require a relationship with at least one of the major retailers.

For managers, being aware of these relationship complexities is important and may allow them to minimize the hindrances to their trade relationships. For example, recognizing conflict in a particular relationship might allow it to evolve from a dysfunctional situation to one that is functional, possibly strengthening the relationship between trade partners. A small supplier that recognizes its dependence on a larger counterpart may reduce this inequity by seeking out alternative partners and/or new product lines. Retail buyers, especially those in larger organizations, also would do well to recognize the pitfalls inherent in relationships characterized by these complexities. Seeking to lessen situations of asymmetrical dependence, power, conflict and opportunistic behaviour may allow more collaborative relationships to develop, ultimately leading to a potential sustainable competitive advantage. Such collaborative behaviour could be reinforced through the enactment of written contracts, however unpalatable this option may seem to some buyers.

Policymakers might consider legislation that recognizes the complex nature of relationships in these sectors and takes into account the power of buyers, especially retailers, within the Australian marketplace. However, in addition to the primary concerns about the price received by the supplier for the product, there should perhaps be wider consideration of the impact of power on the supplier's strategic direction. As suppliers become tempted (or forced) to abandon their own strategies to retain partnerships with major retailers, the Australian marketplace comes at risk of further enhancing the power of the retailers. The dependence of suppliers on major retailers, coupled with the potential for power issues and opportunistic behaviour that suppliers may be subjected to, may require the consideration and assistance of federal and/or state governments. An industry watchdog might be established that would monitor the behaviour of those involved in the industry. In addition, new legislation could seek to curb potentially opportunistic behaviour by supermarkets, while also protecting the rights of their supply trade partners.

FUTURE RESEARCH AVENUES

This dyadic approach provides rich data on which we draw. In terms of further research, we suggest that this methodology might be extended into other national and supply chain contexts. Moreover, research that investigates agri-food chains with intermediaries would be welcome. Such investigations could include, for example, how adding more chain members affects perceived relationship complexities.

References

1. Day, G. S. (2000), 'Managing market relationships', *Journal of the Academy of Marketing Science*, Vol. 28, No. 1, pp. 24–30; Wagner, S. M. & Boutellier, R. (2002), 'Capabilities for managing a portfolio of supplier relationships', *Business Horizons*, Vol. 45, No. 6, p. 79.
2. Day, op. cit.
3. Kasabov, E. (2007), 'Towards a contingent, empirically validated, and power cognisant relationship marketing', *European Journal of Marketing*, Vol. 41, Nos. 1/2, pp. 94–120.
4. Hunt, S. D. (1994), 'On rethinking marketing: our discipline, our practice, our methods', *European Journal of Marketing*, Vol. 28, No. 3, pp. 13–25.
5. Kinsey, J. D. (2001), 'The new food economy: consumers, farms, pharms and science', *American Journal of Agricultural Economics*, Vol. 83, No. 5, pp. 1113–1130. See p. 1113.
6. Ibid.; Kinsey, J. D., & Senauer, B. (1996), 'Consumer trends and changing food retailing formats', *American Journal of Agricultural Economics*, Vol. 78, No. 5, pp. 1187–1191; Ould, M. (2006), *'Challenges and opportunities facing the Australian food industry'*, paper presented at the ABARE Outlook Conference, <http://www.abareconomics.com/interactive/outlook06/outlook/>, accessed 17 August 2006.
7. Treadgold, A. (1996), 'Food retailing in Australia – three retailers, three strategies', *International Journal of Retail and Distribution Management*, Vol. 24, No. 8, pp. 6–16.
8. IBISWorld (2006), 'Supermarkets and other grocery (except convenience) stores in Australia', <http://www.ibisworld.com.au>, accessed 1 August 2006.

9. Ibid.
10. Harvey, M. (2000), 'Innovation and competition in the UK supermarkets', *Supply Chain Management: An International Journal*, Vol. 5, No. 1, pp. 15–21.
11. Grant, H. (1995), 'The challenge of operating in the new Europe', *British Food Journal*, Vol. 97, No. 6, pp. 32–35.
12. Coviello, N. E., Brodie, R. J., & Munro, H. J. (2000), 'An investigation of marketing practice by firm size', *Journal of Business Venturing*, Vol. 15, No. 5, pp. 523-545.
13. Ould, op. cit.
14. Pritchard, B., Burch, D., & Lawrence, G. (2007), 'Neither 'family' nor 'corporate' farming: Australian tomato growers as farm family entrepreneurs', *Journal of Rural Studies*, Vol. 23, No. 1, pp. 75–87.
15. Hughes, D. & Merton, I. (1996), '"Partnership in produce": the J Sainsbury approach to managing the fresh produce supply chain', *Supply Chain Management: An International Journal*, Vol. 1, No. 2, pp. 4–6.
16. Jacenko, A. & Gunasekera, D. (2005), 'Australia's retail food sector: some preliminary observations', <http://pandora.nla.gov.au/pan/30281/20060323/abare/PC13141.pdf>, accessed 15 August 2006.
17. Hornibrook, S. A. & Fearne, A. (2001), 'Managing perceived risk: a multi-tier case study of a UK retail beef supply chain', *Journal on Chain and Network Science*, Vol. 1, No. 2, pp. 87–100.
18. Parker, M. M., Bridson, K., & Evans, J. (2006), 'Motivations for developing direct trade relationships', *International Journal of Retail and Distribution Management*, Vol. 32, No. 4, pp. 121–134.
19. Dapiran, G. P. & Hogarth-Scott, S. (2003), 'Are co-operation and trust being confused with power? An analysis of food retailing in Australia and the UK', *International Journal of Retail and Distribution Management*, Vol. 31, No. 5, pp. 256–267.
20. Hogarth-Scott, S. & Parkinson, S.T. (1993), 'Retailer-supplier relationships in the food channel: a supplier perspective', *International Journal of Retail and Distribution Management*, Vol. 21, No. 8, pp. 11–19.
21. Wilson, D. T. (1989), '*Creating and managing buyer-seller relationships*', Working Paper, The Institute for the Study of Business Markets, Pennsylvania State University, University Park, PA.
22. Zhou, N., Zhuang, G., & Yip, L. S. (2007), 'Perceptual difference of dependence and its impact on conflict in marketing channels in China: an empirical study with two-sided data', *Industrial Marketing Management*, Vol. 36, No. 3, pp. 309–321.
23. Hatton, M. J. & Matthews, B. P. (1996), 'Relationship marketing in the NHS: will it bring the buyers and suppliers together again?', *Marketing Intelligence and Planning*, Vol. 14, No. 2, pp. 41–47.
24. Ganesan, S. (1994), 'Determinants of long-term orientation in buyer-seller relationships', *Journal of Marketing*, Vol. 58, No. 2, pp. 1–19.
25. Ibid.
26. Hogarth-Scott & Parkinson, op. cit.
27. Batt, P. (2001), '*Building trust in the fresh produce industry*', Proceedings of the Australian and New Zealand Marketing Academy, <http://smib.vuw.ac.nz:8081/WWW/ANZMAC2001/home.htm>, accessed 10 August 2007.
28. Joshi, A. W. & Arnold, S. J. (1998), 'How relational norms affect compliance in industrial buying', *Journal of Business Research*, Vol. 41, No. 2, pp. 105–114.
29. Dapiran & Hogarth-Scott, op. cit.

30. Hingley, M. K. (2005), 'Power imbalanced relationships: cases from UK fresh food supply', *International Journal of Retail and Distribution Management*, Vol. 33, No. 8, pp. 551–569.
31. Dapiran & Hogarth-Scott, op. cit.
32. Fearne, A., Duffy, R., & Hornibrook, S. (2005), 'Justice in UK supermarket buyer-supplier relationships: an empirical analysis', *International Journal of Retail and Distribution Management*, Vol. 33, No. 8, pp. 570–582.
33. Kumar, N. (1996), 'The power of trust in manufacturer-retailer relationships', *Harvard Business Review*, Vol. 74, No. 6, pp. 92–106.
34. Morgan, R. M. & Hunt, S. D. (1994), 'The commitment-trust theory of relationship marketing', *Journal of Marketing*, Vol. 58, No. 3, pp. 20–36. See p. 35.
35. Kumar, op. cit.
36. Anderson, J. C., Håkansson, H. & Johanson, J. (1994), 'Dyadic business relationships within a business network context', *Journal of Marketing*, Vol. 58, No. 4, pp. 1–15. See p. 2.
37. Kumar, op. cit.
38. Balabanis, G. (1998), 'Antecedents of cooperation, conflict and relationship longevity in an international trade intermediary's supply chain', *Journal of Global Marketing*, Vol. 12, No. 2, pp. 25–46; Palmer, A. (2002), 'The role of selfishness in buyer-seller relationships', *Marketing Intelligence and Planning*, Vol. 20, No. 1, pp. 22–27.
39. Dwyer, F. R., Schurr, P. H., & Oh, S. (1987), 'Developing buyer-seller relationships', *Journal of Marketing*, Vol. 51, No. 2, pp. 11–27; Morgan & Hunt, op. cit.
40. Fontenot, R. J. & Wilson, E. J. (1997), 'Relational exchange: a review of selected models for a prediction matrix of relationship activities', *Journal of Business Research*, Vol. 39, No. 1, pp. 5–12.
41. Morgan & Hunt, op. cit.
42. Rosenbloom, B. (1973), 'Conflict and channel efficiency: some conceptual models for the decision maker', *Journal of Marketing*, Vol. 37, No. 3, pp. 26–30.
43. Duarte, M. & Davies, G. (2003), 'Testing the conflict-performance assumption in business-to-business relationships', *Industrial Marketing Management*, Vol. 32, No. 2, pp. 91–99.
44. Zhou et al., op. cit.
45. Ibid.
46. Murphy, P. E., Laczniak, G. R., & Wood, G. (2007), 'An ethical basis for relationship marketing: a virtue ethics perspective', *European Journal of Marketing*, Vol. 41, Nos. 1/2, pp. 37–57.
47. Williamson, O. E. (1985), *The Economic Institutions of Capitalism: Firms, Markets, Relational Contracting*. New York: The Free Press. See p. 47.
48. Provan, K. G. (1993), 'Embeddedness, interdependence and opportunism in organizational supplier-buyer networks', *Journal of Management*, Vol. 19, No. 4, pp. 841–856.
49. Gulati, R., Lawrence, P. R. & Puranam, P. (2005), 'Adaptation in vertical relationships: beyond incentive conflict', *Strategic Management Journal*, Vol. 26, No. 5, pp. 415–440.
50. Lindgreen, A. (2001), 'A framework for studying relationship marketing dyads', *Qualitative Market Research: An International Journal*, Vol. 4, No. 2, pp. 75–87. See p. 78.
51. Ibid.
52. Morgan & Hunt, op. cit.
53. Palmer, op. cit.
54. Fearne et al., op. cit.

Appendix 3 Characteristics of Interviewees

Interviewee characteristics and industry	Name (pseudonym)	State	Job title	Full-time employees in company
Dairy	Debbie	TAS	Owner	3
Trade Services	Alison	TAS	Manager	20
Dairy	Kylie	QLD	Director	2
Retail	Winona	QLD	Owner	5
Dairy	Delia	QLD	Owner	2
Retail	Kent	QLD	Manager	7
Nuts	Sam	QLD	Owner	3
Retail	Paulo	QLD	Owner	1
Fruit	George	SA	MD	2–5
Retail	Jasmine	VIC	Owner	2.5
Dairy	Jemima	TAS	MD	15
Retail	Pedro	VIC	Buyer	1000+
Dairy	Louise	TAS	Manager	8
Retail	Larry	VIC	Owner	50–60
Dairy	Sandy	SA	Co-Owner	5
Retail	Don	SA	Manager	5
Fruit	Heather	SA	Owner	1
Retail	Rochelle	SA	Manager	3
Fruit	Warner	SA	MD	4.5
Retail	Emily	SA	Co-Owner	2.5
Vegetable	Terry	SA	GM	120
Retail	Carlos	SA	Marketing Manager	1000+
Dairy	Miranda	SA	Owner	2
Retail	Jordan	SA	Owner	6
Vegetable	Kenya	TAS	Marketing Manager	5
Retail	Valerie	TAS	Owner	2
Seafood	Brian	TAS	Manager	1
Retail	Sally	TAS	Owner	3
Vegetable	Nick	TAS	Farm Manager	4

Interviewee characteristics and industry	Name (pseudonym)	State	Job title	Full-time employees in company
Trade Services	Sahara	TAS	Owner	5
Vegetable	Mel	TAS	Director	4
Trade Services	Spike	TAS	Owner	3
Vegetable	Maurice	TAS	MD	20
Retail	Sean	TAS	Manager	15
Fruit	Ada	VIC	Owner	8
Retail	Salma	TAS	Manager	4
Meat	James	TAS	Co-Owner	2
Trade Services	Percy	TAS	Owner	2
Fruit	Abi	TAS	Owner	2
Retail	Aileen	TAS	Owner	1
Vegetable	Hannah	VIC	Marketing Manager	20
Trade Services	Tom	TAS	Owner	7
Fruit	George	NSW	Owner	1
Retail	Kathy	NSW	Retail Officer	1.5
Beef	Romilda	QLD	Marketing Manager	1.5
Retail	Spencer	VIC	Buyer	1000+
Fruit	Faith	QLD	Marketing Manager	150
Retail	Maddox	NSW	GM	1000+
Fruit	Alannah	VIC	Marketing Manager	5
Retail	Mabel	VIC	GM	18

18 Are Supermarkets Poor-friendly? Debates and Evidence from Vietnam

BY PAULE MOUSTIER,* MURIEL FIGUIÉ,† DAO THE ANH‡
AND NGUYEN THI TAN LOC§

Keywords

supermarkets, Vietnam, poverty, street vendors, farmer organizations, quality.

Abstract

Are supermarkets poor-friendly? Scientific literature offers a controversial answer to this question, and this study compares the viewpoints of different authors for the case of Vietnam. Surveys of the access of poor consumers, traders and farmers to different food retailing points in the country's main cities show that supermarkets are not adapted to the specific constraints of poor consumers. Street vending and informal markets generate more employment than supermarkets, especially for the poor. Poor farmers have no direct access to supermarkets because of the high volume, stringent payments and quality requirements they impose. Recommendations aimed at maintaining the diversity of retail trade and supporting supermarket access for poor producers emerge from this analysis.

* Dr Paule Moustier, CIRAD, UMR MOISA, Montpellier, F-34398 France. E-mail: paule.moustier@cirad.fr. Telephone: + 33 844 719 73 20.

† Dr Muriel Figuié, CIRAD, UMR MOISA, Montpellier, F-34398 France. E-mail: muriel.figuie@cirad.fr. Telephone: + 33 467 617 586.

‡ Dr Dao The Anh, Vietnam Academy of Agricultural Sciences, Food Crop Research Institute, Centre for Agrarian Systems Research and Development (CASRAD), Thanh tri, Hanoi, Vietnam. E-mail: daotheanh@fpt.vn. Telephone: + 844 346 508 62.

§ Mr Nguyen Thi Tan Loc, Fruit and Vegetable Research Institute, Trau Quy, Gia Lam, Hanoi, Vietnam. E-mail: thiloc@netnam.vn. Telephone:+ 844 876 7997.
The authors are all members of the MALICA consortium (Markets and Agriculture Linkages for Cities in Asia) and collaborators with the ADB/DFID, Making Markets Better for the Poor project, which funded this research.

Introduction

This chapter investigates the potential benefits and risks of the development of supermarkets relative to other forms of food distribution, especially for the poor population of Vietnam. The poor here include consumers and traders as well as farmer producers. The first section of this chapter explores the rapid development of supermarkets throughout the world and particularly in Asia and Vietnam. This development has been accompanied by marketing innovations and the concentration of capital among selected investors. Issues relating to the distribution of value generated by supermarket development, especially in Vietnam where poverty alleviation is a major policy concern, are raised in the second section. The controversies spawned by these issues are then developed, including the positive approach applied to the development of supermarkets in Vietnam versus a more balanced approach investigating the potential exclusion of the poor due in particular to labour-intensive technologies and diversity in the types of food distribution. The relevance of this balanced approach is confirmed by empirical evidence gathered in Hanoi and Ho Chi Minh City regarding the access of poor consumers, traders and farmers to different food retailing points. Managerial recommendations suggest ways that each of these food retailing points could sharpen their competitive advantages relative to supermarkets, especially in terms of consumer proximity. This chapter also addresses supermarkets and ways to incorporate small-scale suppliers. The conclusion provides an examination of the intrinsic value that the Vietnamese situation demonstrates and the need to preserve diversity and consumer proximity in food distribution, which has now declined in many other countries.

Background

Although Vietnam continues to receive praise for its success in poverty alleviation, poverty and unemployment remain major concerns for the government and donors. In 2004, the poverty rate was estimated at around 20 per cent (compared with 58 per cent in 1993). Vietnam would like to achieve the status of a middle-income country by 2010 and increase its present gross domestic product of 600 (US) dollars to a level in excess of 1000 (US) dollars per year.[1] Poverty in Vietnam is mostly rural; the rural poverty rate is 25 per cent compared with 4 per cent in cities. Yet urban poverty often gets underestimated, because most migrants are not registered and do not benefit from social services. Cities also are growing at a rate of 3 per cent per year, though urbanization remains limited in comparison with other Asian countries (for example, 25 per cent in 2002 compared with an average of 36 per cent for Southeast Asia). Reducing rural and urban poverty is one of the four pillars – termed social inclusion – of Vietnam's 2006–2010 Socio-Economic Development Plan.

Food distribution is a key factor for the social inclusion of the poor, because it creates small-scale business activities and affects the access of the poor to food commodities.[2] Although not as rapid as in other countries of Asia, such as Thailand, the development of supermarkets in Vietnam is progressing at a steady pace. In late 2001, Vietnam had a total of 70 supermarkets, 32 in Hanoi and 38 in Ho Chi Minh City, whereas there had been none before 1990.[3] Consumers also are expressing growing concern for the quality – and more importantly safety – of food products.[4] This demand has encouraged the development of supermarkets as the point of sale for food products and is promoting new retailing enterprises operating through market stalls or shops, for which efforts toward visual quality (for example, attractive presentation or packaging) and communication about product safety represent major promotional tools.

The rapid development of supermarkets in both developed and developing countries has been extensively reported on in the last decade, particularly by Reardon and Berdegué in Latin America[5] and through a recent workshop organized by FAO in Malaysia with regard to the Asian context.[6] In Asia, the first supermarkets emerged in the 1990s, and Malaysia is reportedly the most advanced in terms of supermarket development. China has seen a staggering rise in supermarket development, with annual growth rates of 40 per cent for supermarket outlets and 80 per cent in the value of sales.[7]

In Vietnam, the marketing of fruits and vegetables is still characterized by a diversity of distribution chains that include formal and informal markets, street vendors, shops and supermarkets. Formal markets are planned by the state (may be totally or partially roofed), and each has a management board. In these markets, retailers pay monthly rent for their stall, plus taxes. Informal markets are open air and not subject to state planning. Street vendors usually are mobile vendors selling from baskets, bicycles or motorbikes who move from one place to another. Street vendors also may sell as groups at certain times in the day. Stores are defined as a shopping area of less than 500 m², with walls and roofs. They are commonly set up on the ground floor of residences. According to the Vietnamese Ministry of Trade, supermarkets are diversified retail sale units occupying an area of more than 500m² and characterized by self-service and various infrastructures, including parking areas.

The government is promoting the expansion of supermarket distribution and had plans to eliminate all informal trade at the time of this research work. The examples reported from other countries raise doubts about the sustainability of forms of food distribution other than supermarkets. In Latin America for example, where poverty affects 40 per cent of the population, supermarkets experienced rapid growth, initially in major cities (1980s), followed by small cities and towns (late 1990s). Supermarkets were initially established in wealthier neighbourhoods and then middle-class areas, before finally reaching working-class areas. Their share in supplying consumer demand increased from 10 to 20 per cent by 1990 and then to 50–60 per cent in 2000, a growth pattern that took 50 years in the United States. In Guatemala, the poorest country in Latin America, supermarkets increased in number from 66 in 1994 to 128 in 2002 and sell 35 per cent of the total quantity of food sold in retail outlets (up from 15 per cent in 1994). Since the 1990s, Asian countries such as Singapore, China and Thailand have had increased access to supermarkets, the diversity of which reflect the living standards of the customer base. In Thailand and Taiwan, large-volume distribution represented 20 per cent of fresh product purchases in 2000.[8] Three main factors explain the growth of large-volume distribution worldwide: (1) urbanization; (2) growth in incomes, which has both direct and indirect effects (for example, purchase of refrigerators, means of transportation); and (3) increasing proportions of working women.

Issues

The development of supermarkets goes hand in hand with value-adding activities: 'Modern' wholesale and retail firms are characterized by investments in shelf presentation (packaging, storage), advertising and selection of suppliers based on quality and regularity criteria. These investments, together with the economies of scale generated by volume distribution, add much value to the business of food retailing, compared with more traditional retail outlets that sell uniform, undifferentiated products to consumers.[9] Thanks to economies of scale, they also have the potential to cut distribution costs and offer more affordable products to consumers. Although at present, most supermarkets still sell at higher prices than marketplaces in developing countries, the situation may change as supermarkets expand, as was the case in Latin America in the 1990s. The challenge is how to ensure that the value added by these new enterprises can be distributed effectively to the poorest people rather than primarily bringing profit to those who are able to invest in this demanding business and excluding the poor who cannot compete with them. What are the effects of new distribution chains on the poor, who are both consumers and suppliers of food products? How can the modernization of distribution better fit the demands for food and income for the poor? Which, among present and alternative distribution chains, should be promoted and supported with more positive effects for the poor?

These issues are all the more acute in developing countries, where poverty reduction, employment generation and affordable food are major concerns of many governments and donors. Currently, the planned rapid increase in supermarkets and the elimination of provisional markets and street vendors appear in the strategy of the Vietnam Ministry of Trade's Domestic Trade Department from the present until 2020, based on the grounds of 'modernization' and 'civilization'.[10]

Controversies

The debate about rapid supermarket development throughout the world can be expressed as follows: It is a positive approach to food modernization, an innovation that benefits all actors in the chain, or it is a negative approach that injects foreign capital into the economy to the detriment of local actors. In Vietnam, the first position usually appears in documents and in the policy statements of public officials and international experts. The idea of modernizing food distribution as a positive innovation has been convincingly summarized by Hagen.[11] One of the main features of retailer innovation is self-service, which goes hand-in-hand with pre-packaging and thus protects products from damage by handling. Mass distribution enables economies of scale and market power, themselves drivers of cost reductions. According to Hagen, most of these innovations have a positive effect on cost reductions, and if the private sector is slow to adopt them, it is mostly because of a lack of external impetus.

Yet this characterization may be a naïve assessment of innovations. Innovations are rarely neutral, relative to the factors of production; they are generally biased in favour of capital.[12] Most of the innovations listed by Hagen require substantial capital investment. Some of them are labour-saving, such as self-service, mass volume distribution and scanning cash registers, whereas others transfer labour from traders to consumers (for example, self-service), from traders to employees (cash registers) or from traders (and consumers) to farm enterprises (pre-packaging, processed food). Because food retailing has very high labour costs relative to profits, modern retailers such as Wal-Mart have paid much attention to managing them and increasing labour productivity.[13] The problem with labour-saving and scale-biased innovations is that they have a greater negative impact on employment of the poor and may be less suitable to a country like Vietnam, where labour is in excess supply, than is the case with capital-saving or neutral innovations (without massive credit programmes focusing on the poor).

A major challenge for poverty reduction may be the need to develop capital-saving and scale-neutral innovations. The flexibility of street vending may enable access to consumers in a decentralized way, which could be regarded as an innovation that reduces costs (time, money) for the consumers compared with the more centralized distribution patterns of supermarkets. In Europe and the United States, the development of modern distribution system has coincided with a greater concentration of power in the hands of a few multinationals (for example, Wal-Mart), less favourable working conditions, lower salaries for employees and the creation of 'food deserts' (areas where it is difficult to purchase food if consumers do not have their own means of transportation).[14] Numerous advocates of 'alternative distribution food chains' claim that citizens should be able to access local, neighbourhood, small-scale retail points – run directly by farmers if possible – rather than being limited to mass-scale monopolistic distribution.[15] Supermarkets vary with regard to social objectives (balancing ethical standards versus competitive pricing); there should at least be the promotion of 'responsible' supermarkets. As for the impact of supermarkets on food prices, the situation is highly variable according to the stage of supermarket penetration and the nature of the products they sell.[16]

Reports from other countries show that supermarkets have brought about several changes that challenge small-scale farmers. Food quality standards developed by supermarkets compensate for the absence or inadequacy of public standards. They also serve as marketing tools, enabling supermarkets to compete with the informal sector by claiming superior product attributes.[17] In addition, large volume requirements, daily delivery obligations, requests for deferred payments and the need for bank accounts all result in the exclusion of small-scale farmers.[18] The characteristics of small-scale family agriculture, with its diverse farming systems and practices that result in disparity and a lack of uniformity in agricultural produce, complicate matters for supermarkets that have exacting requirements and standards. This typical situation marks Vietnam, where tens of millions of farmers cultivate less than one hectare per household.[19] Thus, intermediaries such as wholesalers or farmer organizations can play important and needed roles by connecting farmers and supermarkets, providing economies of scale and offering specialized skills in product assembly, grading and the transfer of information between buyers and sellers.

We advocate a balanced attitude when analyzing the impact of innovations on all actors in the chain of supermarket development, one based on stringent assessments of the impact of different types of distribution with variables pertaining to of capital and labour resources for consumers, traders and farmers. These issues are not commonly dealt with in existing literature, which highlights the value of this research presented for the case of Vietnam, which investigates the impact of supermarkets and alternative distribution chains on the poor, who include consumers, traders and farmers.

Methododology

To test our hypotheses, we collected data about the access enjoyed by both farmers and consumers to the various distribution points and the reasons that determine their choice of retail outlets, with a focus on poor households. We also used secondary and census data to estimate the impact on employment (induced by centralization) of the various distribution points. The data collection methods are summarized in Table 18.1 and Table 18.2.

Table 18.1 Nature of data collection by issue

Issue	Nature of information	Collection method
Trends in the nature of distribution points	Changes in retail points; policies on food distribution	Documents from/interviews with Vietnamese Department of Trade and other government levels
Consumer access	Where and what they buy and why they buy at certain places	Surveys of poor households: 110 in Hanoi, 52 in Ho Chi Minh City (+65 non-poor)
	Price differences between supermarkets, markets and street vendors	Comparison of 10 products in Hanoi and Ho Chi Minh City between supermarkets, shops, street vendors and markets (3 points/type, 3 vendors/type, randomly selected)
Supplier access	4 value chains analyzed (vegetables in Hanoi and Ho Chi Minh City, lychee in Bac Giang, rice in Hai Hau)	
Impact on employment (retail)	Number of people employed by supermarkets, markets, shops and street vending	Census in 2 districts and extrapolation
Mapping of value chains	Nature and location of intermediaries	Cascade interviews from sample retailers to farmers
Organization of value chains	Relationships (horizontal + vertical)	In-depth interviews among a sample group of traders (retailers, wholesalers, collectors) and farmers
Performance of value chains	Constraints and opportunities in production and marketing	
	Distribution of costs and benefits	

CONSUMER SURVEYS

Our results combine quantitative and qualitative data for a simple descriptive analysis. Surveys on the access of poor consumers to different retailing points were performed in Hanoi (the capital of Vietnam) and Ho Chi Minh City (the country's largest city), together with a comparison of prices among different points of sale in these cities.

The poor are rarely considered consumers. Most studies conducted in Vietnam pertaining to purchasing habits attempt to assess the potential for expanding modern distribution systems in Vietnam. They thus focus on middle- or high-income consumers. This study is the first to analyze the purchasing habits of poor consumers, an approach with some specific difficulties and limitations (for example, surveying a sufficient number of people to obtain a representative sample). We choose to combine the data on declared practices with observed practices through a follow-up study of 107 poor families. This approach succeeds because the focus of the study is mostly a description of the constraints on consumer food purchases and because it sidesteps the specific limitations and difficulties of an impoverished sample population.

The chosen survey area, Quynh Mai district, is a poor district in Hanoi,[21] populated by government factory workers and their families (their official monthly income at the time of the survey was 40 US dollars for at least a 48-hour week). In practical terms, the families earn a low but regular income, which reflects the increasing number of factory workers in Vietnam. Although this sample group may be quite different from a group of unregistered families who have settled in Hanoi and are unemployed or without a regular income (and

Table 18.2 Sample interviews of farmers and traders

Sample	Commodities			
	Lychee (to Hanoi)	Vegetables (to Hanoi)	Vegetables (to HCMC)	Rice (to Hanoi)
Supermarket managers or purchasers (1)	13	13	8	19
Wholesalers (2)	3	4	4	6 Hanoi wholesalers 3 food companies 20 Hai Hau wholesalers
Market retailers (3)	6	8	6	10
Shop vendors (2)	6	11		10
Street vendors (4)	(all fruit)	30		10
Collectors (2)	3	5	4	13
Farmers	North of Vietnam: 70 in Yen The and 10 in Luc Ngan (randomly chosen from list given by local authorities), The head and five members of lychee association of Thanh Ha; 30 Thanh Ha farmers outside the association, 30 members of the association.	North of Vietnam Moc Chau: 32 randomly chosen from list given by collectors. The head of Moc Chau farmer's association and five members of the association. Soc Son: 4 farmers in the groups supplying supermarkets, 12 farmers outside the groups supplying supermarkets. Dong Anh: the head of farmer association (Van Noi) supplying one supermarket.	South of Vietnam Cu Chi district: 2 heads of farmer organizations, 5 members of the organization, 5 outside the organization, Lam Dong Province: 3 heads of farmer organizations, 120 farmers, including one third members of organizations.	North of Vietnam, Hai Hau district: 44 farmers in 2 communes (Hai Phong, Hai Toan) randomly chosen from list given by local authorities including 24 non-members of the organization and 20 members. Head of rice farmers' association.

(1) This number accounts for more than 80 per cent of the total number of supermarkets selling the selected products in the selected cities; the choice of supermarkets is representative of their diversity in terms of scale and location.

(2) This number accounts for more than 30 per cent of the total number of traders selling the selected products in the selected cities; they were chosen to be representative of the diversity of traders in terms of scale and location.

(3) The study also used the results from secondary studies of the organization of traditional fruit and vegetable markets in Hanoi that provide data about the source and nature of mediators in commodity chains, based on a representative sample of traders.[20]

(4) Street vendors were interviewed in two districts: one medium to high income and one low to medium income (randomly chosen in all areas of these districts).

who perhaps still depend on their village of origin for food provision), it nevertheless draws a comprehensive picture of the poorer strata of society. Similarly, the Ho Chi Minh City sample group, though small in absolute size (52 poor households and 62 non-poor household surveyed on a single occasion), provides interesting points of comparison.

CASE STUDIES OF VALUE CHAINS

Four value chains were considered for case studies:

1. Lychee to Hanoi from Yen The and Luc Ngan districts in Bac Giang province in the north.
2. Vegetables supplied to Hanoi from Soc Son (in suburban Hanoi) and Moc Chau (a secondary town in the northwestern mountains, in Son La province).
3. Vegetables supplied to Ho Chi Minh City from peri-urban areas (Cu Chi district), as well as from the Duc Trong and Don Duong districts in Lam Dong province.
4. Flavoured rice from Nam Dinh Province (Hai Hau district) in the North of Vietnam.

The choice of the case studies was based on the involvement of the poor (albeit in small numbers) in production and trade, as well as the involvement of supermarkets and other quality chains in the marketing process.

To identify different supply and distribution chains, we use representative samples of traders. To quantify the access of farmers to the chains, we gathered data from interviews with commune and district leaders, as well as in-depth interviews with a sample of poor and non-poor farmers in selected villages and with the heads of eight selected farmer organizations supplying supermarkets. The villages from which the farmers were selected are representative of the diversity of farmer profiles (even though the sample size used is too small to draw definitive scientific conclusions). We gathered information from the interviews with the provincial authorities and a representative sample of farmers from these villages, determined by random selection from a list supplied by the local authorities. In-depth interviews of stakeholders along the chains helped us investigate three key interlinked conditions: the horizontal and vertical coordination that links the poor to the markets; the distribution of costs and benefits between the farmers and the traders along the chains (incomes have been estimated by adding monetary incomes and the monetary value of self-consumption, if any); and the respective advantages and drawbacks involved in supplying different types of outlets, as perceived by the stakeholders.

POVERTY STANDARDS

Prior to commencing the study, we established a baseline poverty level. The national standard of poverty, 100 000 VND[¶]/month in rural areas and 150 000 VND in urban areas (before 2006), then 200 000 and 260 000, respectively (after 2006), is defined at the overall national level. In reality, each province has its own standard for poverty. In its 2004 Vietnam Development Report on Poverty, the Vietnam Consultative Group acknowledges that the national standard is useful for time-based comparisons of poverty rates but is inadequate for practical purposes to conduct surveys on poverty or for aid

---

¶   1 USD = 16 000 VND in 2005.

allocation. This study therefore uses the standards established by the local administrations. For the survey of poor consumers in Hanoi and Ho Chi Minh City, we adopt the standard of the Women's Union, an organization responsible for allocating support funds to poor households, which also guided us toward an ideal target group for surveying. The following definitions of poverty provide the baseline for this study: 80 000 VND/month (5 USD) in rural areas of northern Vietnam, 130 000 VND/month (8 USD) in the peri-urban districts of Hanoi, 250 000 VND/month (16 USD) in Lam Dong province; 300 000 VND/month (19 USD) in Hanoi and 500 000 VND/month (31 USD) in Ho Chi Minh City. Poverty criteria in Vietnam are based on the cost of obtaining a basket of food and non-food products in the respective locality, including expenditures for access to basic services (health, education and transportation), as well as self-consumption of foodstuffs that may be grown and not sold. Thus, though it is a financial definition, it also reflects the difficulties that households experience in accessing primary services.

Empirical Evidence

The empirical data collected in Vietnam confirm the importance of maintaining a diversity of existing food distribution systems for the poor. Despite the advantages of supermarkets and other distribution value chains, the poor still have limited involvement as consumers, traders and farmers.

LIMITED ACCESS FOR POOR CONSUMERS

Viewed from the perspective of poor consumers, more than 60 per cent of households surveyed (110) in Hanoi have never shopped in a supermarket, only 2.7 per cent shop in such stores regularly (a few times a month to a few times a week) and 95 per cent purchase their food from mobile vendors or informal market traders on the street more than once a week. In Ho Chi Minh City, 33 per cent of poor households (52) have never shopped in a supermarket, and only 38.5 per cent shop there regularly. Among poor households in Ho Chi Minh City, 60 per cent visit formal markets to purchase food on a weekly basis, 40 per cent buy from street vendors, 42 per cent purchase from shops and 13 per cent buy from supermarkets. These figures contrast with those for non-poor households, among which only 2 per cent (out of 65) have never shopped in a supermarket and as many as 81.2 per cent shop there regularly. The consumers who do not shop in supermarkets cite prices, time and distance as the main reasons. Comparing prices for 10 products among samples of street vendors, market retailers and supermarkets, we find an average difference of 20 per cent in Hanoi between street vendors and retailers versus supermarkets. Furthermore, poor consumers have positive opinions of their primary food outlets (generally formal markets or informal markets such as street vendors and shops). In particular, poor consumers have specific and opposing perceptions of supermarkets and street vendors: The former offer good quality in terms of food safety but also command high prices and time commitments, whereas the latter offer lower quality for a lower price and are more convenient in terms of time and freshness.

Street vending is perceived by trade authorities at the national, city and district levels as having various negative impacts, as reflected by the legislation, including (1) traffic congestion, (2) poor food safety, (3) attraction of illegal migrants to Hanoi and (4) a bad image for the city. Yet these alleged problems are difficult to assess. Buying from supermarkets implies using a motorbike or car, whereas many consumers can access street food on foot. Food safety also has various dimensions, including the amount of chemical residues, such as fertilizers and pesticides, in the products bought. Because the sources of supply used by both street vendors and fixed formal market retailers are similar, that is, mostly night wholesale markets, the food safety of commodities should be similar. This assumption is confirmed by the quick test analyses of pesticide residues carried out by the Vietnam Fruit and Vegetable Research Institute (FAVRI) in 2004 and 2005. Of a total of 25 samples from street vendors and 23 from fixed formal retailers, only one offending case was found in the sample from street vendors (0.4 per cent) and 2 in the market sample (8 per cent).[22] A higher difference in pesticide residues was observed between products sourced from ordinary markets or street stalls and 'safe vegetable' stalls, shops and supermarkets than between informal and formal trade. Excess pesticide residues were also tracked on one sample collected from one supermarket by FAVRI. Another dimension is the problem of waste collection, which may be aggravated by street vendors selling on the tarmac, but this point has not been assessed.

Regarding the bad image of the city, this problem was never mentioned by consumers interviewed in Hanoi. The image of street vendors actually appears as an attractive feature in various Vietnam tourism promotion campaigns.

LIMITED INVOLVEMENT OF THE POOR AS TRADERS

With regard to the poor as traders, according to the survey, supermarkets create less employment in a geographical area and per unit of volume sold than do markets and street vending. The supermarket share of total employment in the retail trade in Hanoi is estimated at approximately 6 per cent (directly) or 11 per cent (if we include indirect employment). Using the limits of Hanoi defined before 2004, street vending accounts for around 32 per cent of retail quantities traded and 37 per cent of employment created by the vegetable retail trade, whereas supermarkets represent 1.3 per cent and 0.6 per cent, respectively; shops provide 9 per cent for both measures; and retail markets account for 58 per cent and 53 per cent, respectively. The figures are even higher for lychee street vending. In addition, we estimate that the retail sale of one ton of vegetables per day provides jobs to 13 street vendors, whereas big supermarkets provide employment to only 4 employees to handle the same volume. Whereas street vending and informal markets employ mainly the poor, with a required investment limited to 400 000 VND, entry to formal markets is constrained in terms of the investment required (around 12 000 000 VND). It may be difficult for the poor to be hired by supermarkets because of their low education. In our sample of 60 street vendors, 18 per cent are poor according to the 2005 threshold of poverty in Hanoi (500 000 VND per month). Other forms of food distribution do not employ the poor. Most street vendors (89 per cent) are part-time farmers who come from rural areas on the periphery of Hanoi. They cannot generate enough income from their farms to feed their families, so street vending provides their main source of income, which supplements their home-grown food and the income generated by the farm. The remaining street vendors (11 per cent) are Hanoi residents of limited income, such as retired women. For these women, street vending is a means of subsistence. In contrast with the rural background of the street vendors, the fixed market retailers interviewed by Van Wijk et al. generally have backgrounds as either industrial workers or small traders.[23]

LIMITED INVOLVEMENT OF POOR FARMERS

As for poor farmers, though they are potential suppliers to supermarkets, they often lack direct access because of the strict requirements of the supermarket in terms of safety, quantity and provision of invoices. Especially in the case of vegetables, supermarkets want to work with suppliers who can display quality control certificates (even if out of date) and can deliver vegetables and/or other products daily. The bulk of vegetables supplied to Hanoi supermarkets now originate from 'leading safe cooperatives' outside the city (for example, Ba Chu, Dao Duc) or semi-public companies (for example, Hadico, Bao Ha). 'Safe vegetables' refer to those vegetables produced in areas where farmers have received training in low chemical production. Safe vegetable cooperatives consist of voluntary associations of farmers who have neighborhood and/or family relationships and a higher-than-average financial capacity and land size. The cooperatives also have small vans to transport vegetables to the supermarkets and can collect enough vegetables to meet the supermarkets' requirements in terms of quality and diversity.

In our investigation of the value chains, we find poor farmers' participation in supermarket supply is limited to some members of farmer associations that supply supermarkets with specialty products, in addition to farmers supplying food companies with flavoured rice on an individual basis.[24] Of the 3000 farmers producing flavoured Hai Hau rice, approximately 20 per cent (600) are poor; 103 of them (3 per cent) have their products sold in supermarkets either through their farmer association (52) or through food companies. In the Moc Chau region, 102 farmers grow tomatoes, and 20 of this group sell to one cooperative that supplies safe vegetable shops. Among these farmers, four have contracts with the cooperative, and two are salaried by the cooperative. These six farmers are from the Thai ethnic minority. They have stepped out of poverty due to their involvement in the cooperative vegetable supply for the past 10 years. They were previously involved in rice and maize production for self-consumption, but the commitment of the cooperative to endorse the risks, in case of production losses, and the guaranteed purchase of all outputs by the cooperative convinced them to become involved in commercial tomato production. Signs of their gradual escape from poverty include extensions and improvements to their residences and investments in a motorbike.

In the Soc Son district of Hanoi province, 80 farmers belong to safe vegetable groups, including 20 farmers who supply supermarkets through a wholesale company, of whom six are poor. None of the vegetable farmers supplying Ho Chi Minh City supermarkets are poor.

Yet supermarkets can yield positive benefits for farmers. For example, the sale of commodities to retailers of high-quality products (including supermarkets and shops) can generate additional income for farmers, especially those who are organized in associations. For example, farmers in the Anh Dao cooperative who supply ordinary tomatoes from the Dalat area to Coopmart generate profits per kilo that are four times higher than those they might receive in traditional chains. In Hanoi, the Soc Son farmers who supply the Bao Ha Company, which in turn supplies supermarkets, receive 23 per cent higher profits than do the other Soc Son farmers. The main advantages of supermarkets for farmers stem from the stability in prices and quantities ordered. In Ho

Chi Minh City, vegetable prices paid by supermarkets can be 10–20 per cent higher than the prices paid by traditional chains. In both Ho Chi Minh City and Hanoi, supermarkets purchase consistent quantities of vegetables on a weekly basis at more stable prices than the traditional chains do. This stability can translate into yearly contracts with estimated quantities and prices that are negotiated more precisely each week. However, stability in quantities and prices varies among supermarket chains and is diminishing as a result of increasing supermarket competition and development. Moreover, the disadvantages of supermarkets, according to farmers, relate to their demands in terms of quality, diversity and delivery, as well as less favourable payment conditions (for example, a 15-day payment deadline is a minimum) and their possible opportunistic behaviour, with frequent changes of suppliers by one supermarket. These results are in line with the literature review, which suggests the exclusion of small-scale farmers from supplying supermarkets in Latin America and Asia.

Policy and Managerial Recommendations

The present diversity of retailing points in Vietnam should be maintained, because it fits the diversity of consumer purchasing power and allows small-scale traders to maintain their livelihoods. The different retailing points have comparative advantages for various consumers, which should be strengthened and the disadvantages reduced. Hence, the following recommendations are aimed at private enterprises in food commodity chains. Street vendors and market traders should bring to the fore their advantages, in terms of food freshness and low prices, but also improve their hygiene and develop relations with farmers to ensure food safety. Farmer groups should manage their own shops as a meaningful alternative to supermarkets; they could develop their perceived advantage in terms of food safety by combining internal and external food safety controls. Finally, supermarkets should target local farmer groups in their sourcing of food and provide them with support in terms of food safety development and control.

These recommendations aimed at private stakeholders should be backed up by public support, especially for players with less capital, namely, street vendors and small-scale farmers. The experiences of other countries, such as Korea, India and Singapore, provide examples of the successful integration of street vending in urban planning through the organization of street vendors and dialogue with authorities. Instead of outright prohibiting street vending and informal markets, authorities should support the 'formalization' of this sector. They could allow street vendors to operate in designated areas (off main streets). Decisions about relocating markets should occur only after consultation with the traders that may be affected.

Credit support should be provided to vendors who make an effort to upgrade their businesses, including street vendors and quality food shops managed by farmer groups. The organization of special farmers' markets as alternative distribution channels would enable consumers and farmers to benefit from proximity and the development of trust-based relationships.[25]

Greater support should be given to farmer associations that are involved in improving the quality of production, such as through the dissemination of success stories, the provision of advisory services (with a particular focus on technical training in the areas of physical quality and food safety), and better access to credit programmes. Another area for action is participatory food quality control. A widespread system could monitor food safety and impose sanctions in cases of non-compliance, provide laboratories and certification bodies with accreditation standards and encourage participatory guarantee systems (PGS) in which farmer associations, consumer groups and supermarkets form sustainable partnerships. Farmers also could benefit from capacity-building assistance related to forming contractual arrangements with supermarkets, especially through broader awareness of successful examples, access to training in farmers' rights and responsibilities in supplier contracts, and the development of codes of good practices for supermarkets. Finally, farmer groups should receive support in undertaking the administrative steps required to become registered and issue invoices.

Ultimately, consumers should be better informed about the social impact of supermarket development. Public authorities may have little power over the development of supermarkets, given the huge attraction they exert upon consumers. Yet consumers also may not be sufficiently aware of the indirect effects of supermarket development on the network of alternative food distributors. Because their individual purchasing behaviour has broad macro-economic consequences in terms of employment and social welfare, consumers are crucial stakeholders who should be made aware of the consequences of their choice and stand up in defence of their rights in terms of access to diverse food distribution. The present turmoil created by the increased rigidity of municipal laws related to street vending in Vietnam (that is, 62 streets on which street vending is prohibited) has pushed the administration to postpone enforcement of the new laws. Yet these reactions are mostly limited to intellectuals in the media, rather than coming from a consumer association that exists but still lacks financial and technical capacity.[26]

Conclusions

Vietnam is at a turning point with regard to the nature of food distribution. The present diversity in food distribution presents a unique picture and fits the disparity in the purchasing power of the population. Yet this balance is clearly jeopardized by the present administrative policies and the rapid development of supermarkets. Currently in Vietnam, supermarkets cannot be considered a poor-friendly distribution chain (especially in Hanoi). This finding reflects the capital-intensive rather than labour-intensive nature of the business and the rationale for a limited number of large-scale distribution centres that are not adapted to the transportation constraints of the poor segments of the population. Supermarkets may, however, offer income-generating opportunities for small-scale farmers who can form associations and guarantee product quality. Our study formulates some recommendations designed to help food distribution systems better fit the needs of the poor – be they farmers, consumers or traders. Some of these recommendations have already been acknowledged by public officials, who agree that they are worthwhile, especially the pilot action to allocate specific trading areas to street vendors who comply with rules of hygiene and food safety.

We also recommend further research that provides a more rigorous assessment of the impact of supermarkets on the price of food and farmers' incomes. Our study is limited in terms of time; time-series data could reveal whether, after a number of years in business, supermarkets succeed in cutting prices thanks to their economies of scale. Assessing the impact on farmers also requires time-series data. It would also be interesting to conduct a thorough comparison of farmers' incomes inside as opposed to outside supermarket chains, based on larger samples and econometric modelling. We share the view of Vorley et al.,[27] namely, that there is an urgent need for research into best practices in connecting small-scale producers with modern distribution channels, based on comparisons across countries and regions with varying degrees of market restructuring and policy environments.

References

1. World Bank (2007), *Vietnam-country partnership*, World Bank, Washington, p. 170. (accessed on http://www.worldbank.org).
2. Bhowmik, S.K. (2005), 'Street vendors in Asia: a review', *Economic and Political Weekly*, May 28–June 4, pp. 2256–2263.
3. Nguyen Thi Tan Loc (2003), 'The development of volume distribution' in P. Moustier et al. (eds), *Food Markets and Agricultural Development in Vietnam*, Hanoi: The Gioi Publishers, pp. 81–83.
4. Figuié, M., Bricas, N. Vu, P. N.T. & Nguyen, D. T. (2004), 'Hanoi consumers' point of view regarding food safety risks', *Vietnam Social Sciences*, Vol. 3, No. 101, pp. 63–72.
5. Reardon, T. & Berdegué, J.A. (2002), 'The rapid rise of supermarkets in Latin America: challenges and opportunities for development', *Development Policy Review*, Vo. 20, No. 4, pp. 371–388.
6. Shepherd, A. (2005), *The implications of supermarket development on horticulture and traditional marketing systems in Asia*, revised version of paper presented to FAO/AFMA/FAMA Regional Workshop on the Growth of Supermarkets as Retailers of Fresh Produce, Rome, FAO, p. 18 .
7. Zhang, X. Yang, J. & Fu, X. (2006), 'Vegetable supply chains of supermarkets in Sichuan, China and the implications for supply chain management', *Acta Hort. (ISHS)*, Vol. 699, pp. 507–516.
8. Cadilhon, J.-J., Fearne, A. P., Moustier, P. & Poole, N. D. (2003), 'Modelling vegetable systems in South-East Asia: phenomenological insights from Vietnam', *Supply Chain Management: An International Journal*, Vol. 8, pp. 427–411.
9. Goletti, F. (2004), *The Participation of the Poor in Agricultural Value Chains*, Agrifood Consulting International, Hanoi, p. 31.
10. Vietnam Ministry of Trade, Department of Domestic Market Policies (2006), *The Strategy for Domestic Trade Development for the Period 2010-2015 and Development Orientation to 2020*, Hanoi, Ministry of Trade, p. 30.
11. Hagen, J. M. (2002), *Causes and Consequences of Food Retailing Innovation in Developing Countries: Supermarkets in Vietnam*, New York, Cornell University, Department of Applied Economics and Management, p. 19.
12. Ellis, F. (1988), *Peasant Economics*. Cambridge University Press: Cambridge, p. 257.
13. Fox, T. & Vorley, B. (2004), *Stakeholder accountability in the UK supermarket sector*, Final report of the 'Race to the Top' project. IIED, London, p. 33.
14. Kinsey, J. D. (1998), *Concentration of Ownership in Food Retailing: A Review of the Evidence on Consumer Impact*, University of Minnesota, St. Paul.
15. Friedmann, H. (1994), 'Distance and durability: Shaky foundations of the world food economy', In Mc Michael (ed.), *The Global Restructuring of Agro-Food Systems*. Cornell University Press: Ithaca and London, pp. 258–277; Koc, M., Mac Rae, R., Mougeot, L.A. & Welsh, J. (1999), *For Hunger-Proof Cities: Sustainable Urban Food Systems*, Ottawa: CRDI.

16. Wrigley, N., Warm, D. & Margetts, B. (2003), 'Food retail access and diet: what the Leeds 'food deserts' study reveals', *British Retail Consortium Yearbook 2003,* London: The Stationery Office, Vol. 34, 228–230.

17. Ménard, C. & Valceschini, E. (2005), 'New institutions for governing the agri-food industry', European Review of Agricultural Economics, Vol. 32, No. 3; Reardon, T. & Timmer, C.P. (2005), *The supermarket revolution with Asian characteristics,* Paper presented at the International Conference of SEARCA, Makati City, Philippines, 10–11 November 2005.

18. Rondot P., Biénabe, E. & Collion, M.-H. (2004), 'Rural economic organisations and market restructuring: What challenges, what opportunities for smallholders?' *Regoverning Markets,* Phase 1: Review Workshop and International Seminar. Amsterdam, 14–19 November 2004.

19. Dao The Anh (2003), *Réformes socio-économiques et adaptation des choix d'activité des ménages ruraux dans le delta du fleuve Rouge au Vietnam [Socio-economic reforms and adaptation of activity choices for rural households in Vietnam's Red River Delta].* Montpellier, ENSAM, p. 300.

20. Moustier, P., Vagneron, I. & Bui Thi Thai (2004), 'Organisation et efficience des marchés de légumes approvisionnant Hanoi (Vietnam),' *Cahiers Agricultures,* Vol. 3, pp. 142–148; Van Wijk, M.S., Trahuu, C., Tru, N.A., Gia, B.T. & Hoi, P.V. (2006), 'The traditional vegetable retail marketing system of Hanoi and the possible impacts of supermarkets', *Acta Hort. (ISHS),* Vol. 699, pp. 465–477.

21. Parenteau, R. (ed.) (1997). *Habitat et environnement urbain au Vietnam [Habitat and urban environnement in Vietnam],* Paris: L'Harmattan.

22. Nguyen, Kim Chien & Moustier, P. (2006), *Vegetable Pesticide Residues in Selected Fields and Points of Sale in Hanoi, Hanoi,* FAVRI, CIRAD, p. 18.

23. Van Wijk et al., op. cit.

24. For details on these results, please see Moustier, P., Dao The Anh, Hoang Bang An, Vu Trong Binh, Figuié, M., Nguyen Thi Tan Loc, & Phan Thi Giac Tam (eds) (2006), *Supermarkets and the Poor in Vietnam,* Hanoi, Cartographic Publishing House. p. 324 <http://www.markets4poor.org>.

25. Kirwan, J. (2004), 'Alternative strategies in the UK agro-food system: interrogating the alterity of farmers' markets', *Sociologia Ruralis,* Vol. 44, No. 4, pp. 395–415; Braber den, K. (2006), *Developing Local Marketing Initiatives for Organic Products in Asia. A Guide for Small and Medium Enterprises.* A book written for IFOAM. Hanoi, ADDA (Agricultural Development Denmark Asia).

26. Diaz V. & Figuié M. (forthcoming), 'La protection des consommateurs au Vietnam [Consumer protection in Vietnam]', *Etudes Vietnamiennes.*

27. Vorley, B., Fearne A. & Ray D. (2007), 'Restructuring of agri-food systems and prospects', in Bill Vorley, Andrew Fearne, and Derek Ray (eds). *Regoverning Markets. A Place for Smale-Scale Producers in Modern Agrifood Chains?* Aldershot (UK), Burlington (USA): Gower, IIED.

19 An Appraisal of the Fair Trade System: Evidence from Small Producers in Emerging Countries

BY LUCIANA MARQUES VIEIRA* AND LUÍS KLUWE AGUIAR†

Keywords

fair trading, global supply chain, honey, market entry strategy.

Abstract

In this chapter, we propose an analysis of fair trading as a possible alternative route for the access of small producers to markets in developed countries. Honey producers in Brazil serve as a case for evaluating fair trading from different perspectives, namely, those of consumers, the global value chain and Fair Trade Labelling Organization (FLO) certification standards.

In this chapter, we will discuss fair trading as an alternative for small producers in light of globalization dynamics; provide an understanding of ethical demand; address chain governance issues; and, finally, assess how small producers' associations might use fair trading standards as value-additions, with a view toward market entry strategies.

Introduction

This study addresses the controversy regarding fair trading as an economic system and attempts to support farmers in developing countries whilst also supplying ethical‡ products to consumers and markets in developed countries. Fair trading has shaped

*   Dr Luciana Marques Vieira, Lecturer, Business School, UNISINOS, São Leopoldo/RS, Brazil. E-mail: lmvieira@unisinos.br. Tel. +55 51 35911100 ext. 1596.

†   Mr Luís Kluwe Aguiar, Senior Lecturer, School of Business, Royal Agricultural College, Stroud Road. Cirencester, GL7 6JS, UK. E-mail: luis.aguiar@rac.ac.uk. Telephone: + 44 1285 652 531.

‡   We consider 'ethical' those goods that are produced following a sustainable production system. Therefore, production methods take into account the preservation of the environment, peoples, local culture and traditions.

the way some smallholders in developing countries position themselves in the face of choices they must make about both production methods and market access. However, to evaluate current smallholder's situations and their capabilities regarding the choices they have to make to be sustainable, it is necessary first to understand the wider context, including the social and economic forces that shape current market performance and the opportunities that could be derived from it. In turn, to better appreciate this chapter, it is necessary to understand the requirement of fair trading certification and its likely impact in smallholder communities. Fair trading is a system that has been under the scrutiny and thus surrounded by controversy with regard to its economic function, its social fairness and the producer benefits.

The aim of this chapter is threefold: to characterize the fair trade (FT) market and ethical consumers in the United Kingdom; to present the role of third-party certifiers, such as the Fair Trade Labelling Organization (FLO-CERT), which provide clear rules, inspections and trustworthy certification to assure FT and organize and transfer technical and marketing knowledge from the consumer market to producers located in developing countries; and finally, to provide an empirical analysis of the main difficulties that small producers face in complying with FT regulations, according to evidence from honey producers in southern Brazil.

Discussion

Since the early 1980s, as a result of more open markets, ever-increasing economic globalization has dominated the way companies and consumers relate to the market. Globalization follows a neo-liberal economic agenda, a phenomenon that has led to market liberalization, deregulation and privatization of the means of production, which in turn have fostered the world's interconnectivity. As a result, products can be accessed from every corner of the world at low costs, and at the moment, international interdependency in the exchange of foodstuffs is a dominant feature.

In the process of globalization, large businesses have taken advantage of such opportunities and, in the search for decreasing unit costs of production and economies of scale, been able to source inputs or supply markets thousands of miles away. Globalization has provided for the internationalization of production processes, especially for large businesses for which goods could be assembled or foods manufactured at sites where the factors of production were more advantageous.

In the food chain, globalization can be illustrated well by the many divisions at different stages of production. Coffee serves as a good example. Coffee beans are produced in different countries, assembled by a handful of traders, roasted and blended by a small number of very large companies, and sometimes even sent back to markets in the countries where the beans were produced. In the internationalization of a production line, diverse countries provide various resources, such as labour and raw materials, that contribute to a final product. Globalization also provides a network of commodity exchanges, binding producers and consumers across the world under the dominance of large agri-food transnationals.[1]

Moreover, ethical products might be considered those that are organically and or locally produced, which reduces carbon emissions.

However, this globalization process also has been criticized as unfair. Businesses looking for increased economies of scale and declining unit costs operate in imperfect markets. The concentration of companies that sometimes dominate entire supply chains has implications for market transparency, price and information asymmetry. Concentration and domination also have implications for the power relationships in supply chains, which do not favour the more vulnerable members on the production side.

When globalization is analyzed from the perspective of the agri-food sector, a sudden increase in trade of all-seasonal, high-value and exotic foodstuffs, supplied to affluent populations through corporate sourcing arrangements, is characteristic.[2] Such a system creates discrepancies between the image and experience of globalization and the global reality, which is not equal for all.[3] As a result, globalization ultimately has pushed small farmers aside, due to their inability to engage in corporate-type supply contracts. Small farmers also lack easy access to credit, improved technology, information and markets.

GLOBALIZATION AND FAIR TRADING

The reality of globalization is often subject to criticism for its multiple devastating effects, resulting from its short-term, intensive, large-scale and low-cost processes, on the natural and human environment. Globalization demands more land for agricultural production at the expense of natural forests, monocrops, intensive usage of agrochemicals and the displacement of peasant farming communities. As a result of globalization, people face increasing concentrations of land, production control, marketing and, consequently, power in the hands of a few to the detriment and impoverishment of the many.[4]

Globalization is also an ever-expanding phenomenon, moved by pure short-term objectives.[5] Insatiable globalization requires increasing flows of information, technology improvements and exchanges. Many feel that feeding this ever-expanding and insatiable process serves as a guarantee against uncertainty.[6] In this sense, by enhancing connectivity through a network of relationships and interdependency, risk declines. However, such risk-avoidance behaviour, driven by the constant need to increase the scales of production, might also generate more uncertainty.

Nevertheless, as globalization intensifies and dependency on international supply increases, a system operating parallel to,[7] or perhaps counter to, the flow of globalization has thrived and developed, namely, fair trading. Whereas globalization is characteristic of a worldwide phenomenon in which suppliers and consumers seek food sourced on an international scale at ever-lower costs, disregarding how it has been produced, fair trading advocates more sustainable exchanges. In this sense, FT has attempted to transform the traditional *modus operandi* of international food chains by providing an alternative market for food products produced by small farmers in developing countries.[8]

Moreover, those who advocate FT as an alternative system for globalization perceive it as a countermeasure against the voracious system and the vagaries of the marketplace. Fair trading precepts propose that producers and consumers relate to the world and the market from an alternative and more sustainable perspective. The FT principles[§] seem logical enough, until we delve more deeply into its mechanics. Since the onset of FT

[§] The basic principles are (1) direct purchasing from farmers; (2) transparent and long-term trading partnership; (3) agreed minimum prices and (4) focus on development and technical assistance through the payment of an agreed social premium.[9]

in the 1970s, it has expanded by operating in the shadow of the dominant free-market economy. Despite the inherent contrast with the dominant paradigm, FT organizations have used conventional marketing practices and existing, traditional foodstuffs channels of distribution. Fair trading organizations, often antagonistic to the effects of globalization and neo-liberal practices, have demanded little alteration, if any, to accommodate moving fair-traded products through conventional marketing channels. As Barrat-Brown proposes,[10] fair trading should operate parallel to existing marketing channels. In reality, in our experience through our study of this and other fair trading arrangements, little evidence exists of any radical shifts from the dominant paradigm with regard to product moving, price negotiations or adhering to certification standards.[11]

AN OVERVIEW OF FT OVER TIME

Fair trading and ethical consumption is nothing new. Since the 1970s, increasing concerns about the unsustainable way the environment was being exploited have gained momentum amongst consumers. Fair trading, whilst still in its infancy, proposed to create alternative marketing channels for moving FT goods, usually craft artefacts. However, at the new millennium, a shift has occurred in FT practices. As more FT commodity foodstuffs, such as coffee, bananas and cocoa, were being traded, the need arose to move bulk products across great distances. Because they were dissimilar to craft objects, these commodity products had to rely on more established and conventional channels. The only difference in the fair trading exchanges thus was the removal of some intermediaries in the supply chain.

The actual operation of fair trading has raised some controversy. Those advocating neo-liberal FT criticize the new trading paradigm by challenging the system's fairness. For example, the price premium¶ that fair-trading farmer's associations receive might actually constitute an indirect subsidy, and the FT system is not truly open to all, which means that being part of a farmer's association would exclude some. Moreover, though gains for producers in the developing world may be attributed to the removal or by-passing of some intermediates, because FT typically entails a fragmented supply base, the presence of mediators might actually be beneficial.[13] In the case of coffee, for example, removing intermediaries would lock farmers into selling to a relatively small number of buyers.[14]

In the United Kingdom, FT products can be purchased from specialist retailers trading in alternative products. When these marketing and retailing practices were dominated by faith organizations, small businesses and cooperatives used alternative retail outlets and street market stalls to sell their products. Owners or managers of such retail outlets had strong ethical views regarding the world around them and the products they would stock. These pioneering retailers were also responsible for linking of FT production and consumption sites, thus establishing the supply channels. Moved by their beliefs, faith and moral values, the pioneers often did not consider quality a prime concern regarding the product offered. However, morality, altruism or charity clearly was not sufficient to sustain the market and drive specialist supply chains with international dimensions. On the consumption side, ethical consumers, not dissimilar to conventional ones, started to

---

¶   Producer's associations received a price premium as a result of engaging in a contract with a certifying body. The price paid to producers is based on the international price of the commodity in question, adding a premium that sometimes can reach up to 20 per cent. This price premium is paid by a cheque after 12 months. The sum has to be reverted back into the community.[12]

demand good quality FT products. Pioneering retailers then had to follow suit and satisfy consumers to ensure they would return to purchase more. Hence, guaranteeing repeat sales was key to business success.

In the 2000s, a shift occurred from an ideological orientation towards a market orientation among those responsible for FT initiatives. The belief was that it would be advantageous to all to make changes in the retailing structure. Thus, conventional retail outlets, such as major supermarket chains and other large specialist food retailers, should also sell FT products. This shift in philosophy was possible because of increased consumer awareness and latent demand for quality FT products. Subsequently, the conventional food retailing sector reacted rapidly and positively. In pursuing a growing consumer trend, large food retailers made it possible for FT to become mainstream. This development occurred in many Western markets, where the consumption of fair-traded goods is prominent. This shift also helped many food companies 'clean up their act'. As a result, many corporate social responsibility statements made by large food organizations now include references to some form of fair trading. This chapter does not address the extent to which such a turn in the marketing positioning of conventional companies represents a white – or 'green' – wash. We leave that analysis to other researchers and critics who have already expressed their concerns about corporate transparency and the match among a company's statements, its policies and its practices.

In the United Kingdom in the early 2000s, FT goods were no longer atypical or exclusive products controlled and moved by many fair trading certification bodies. Specialist food retailers such as Marks & Spencer and Waitrose not only carried an increasing range of FT goods but also embarked on their own FT initiatives.[15] Other larger food retailers followed in close pursuit after the Ethical Trading Initiative (ETI), began in the late 1990s.[16] The ETI works with an alliance of companies, nongovernmental organizations (NGOs) and trade unions to improve corporate codes of practice, including working conditions in the supply chains. Such an initiative quickly attracted the attention of large UK food retailers, because it was attempting to regulate and improve labour conditions at the sites of production. It therefore met international labour standards and fulfilled some of the FT criteria with regard to the well-being of the workforce, which counted heavily in ethical consumers' perceptions.[17]

At present, as the third largest market for FT products in Europe, after Germany and Holland, the UK offers some opportunities for FT product consumption in more differentiated formats.[18] FT products can be found in universities, governmental and private offices, schools, coffee shops, vending machines, service stations and fast-food chains, to name but a few. Whole towns claim to be FT. The distribution of FT products in the four largest UK food retailers is a testimonial to the popularization of fair trading. However, it also means that the initial niche market has finally broken into the mainstream.

In 2005, some £200 million FT goods sold in the UK.[19] This level reflects the direction that many corporations have taken with respect of their corporate and social responsibility strategies. Angela Webb, a BBC journalist, stated that in 2005, the FT market grew by 40 per cent and featured 1500 different lines of products, which in some cases made it responsible for 15 per cent of the total share of the market.[20] Tesco, currently the top UK food retailer, claims to have some 130 different FT lines in its outlets.[21] ASDA also mentions fair trading in its corporate responsibility statement and is a member of ETI. J Sainsbury's in turn claims itself proud to be the leader in sales of FT bananas and has recently created a Fair Trade Development Fund, responsible for supporting various initiatives in FT-producing countries.[22] The John Lewis Partnership, through its food retail arm Waitrose, has created the Waitrose Foundation to foster projects in South Africa, especially in citrus groves.[23] Marks & Spencer has launched Plan A, a 5-year plan aimed at addressing challenges around the world, including the safeguard of natural resources and ethical trading. Marks & Spencer, which carries only own-brand products, claims that its fair trading commitment benefits more than 1 million farmers and workers.[24] The list of examples goes on.

THE ETHICAL CONSUMER

To better understand how the market share of FT products has expanded, it is necessary to characterize who consumes fair-traded goods.

The FT paradigm embeds, among its principles, the notion of sustainability. In essence, sustainability is a new perspective on production and consumption, a 'green logic'.[25] The green logic centres on changing from an inherent short-term to a long-term view of how to use limited resources. Hence, it considers the extent to which there is a need for maximizing utility while satisfying present consumption of goods to achieve more long-term satisfaction.[26] Although on the one hand, globalization provides individual consumers with an infinite myriad of options (in this case, food options), on the other hand, fair trading relates to consumption with a more long-term perspective. As a result, fair trading should give consumers more conscientious choices, which in turn would be more collectively harmonious.

We propose an 'ethical logic' of consumption that expands Isaak's green logic. The green dimension can constrain understanding of the implications of what we attempt to propose in this chapter, because the green debate is strongly associated with organic or biological systems of production, which tend to focus narrowly on selecting products that do the least harm to the environment. In contrast, the fair trading system encompasses not only the care for the natural environment but also a human dimension. Moreover, the cultural and social values associated with the history and geography of production are also important.

Furthermore, this chapter does not attempt to debate ethics in depth but rather uses it as an argument to support the analysis of fair trading. Nevertheless, we acknowledge that acting more conscientiously when purchasing goods is only one side of the ethical consumption dimension; boycotting is the other.[27] Some call the conscientious purchase of goods 'positive ethical buying', whereas boycotting or anti-consumerism could be perceived as 'negative ethical buying'.[28]

Isaak investigates consumers and assumes that lifestyles translate into votes (consumption).[29] This concept fits in well with the discussion in this chapter regarding whether a preference for fair-traded goods is a conscious act of rejecting conventional ones. As part of an ethical logic, consumers may favour a more sustainable future.[30] This argument reinforces the idea that the impact of conscious choices applies not only to consumption, the last act in a chain of activities, but also to the production, processing and relationships among the actors in such an exchange.

The underlying issue for this chapter is that ethical consumption can be understood in terms of consumers making conscious and deliberate choices that either directly or indirectly impact their immediate or remote environment. Deliberate conscious choices are not fixed; rather such are affected by moral qualities, on an everyday basis.[31] The charitable notion of doing good for the immediate or remote environment, as consumers demonstrate it, depends on some attributes that are communicated to and sensitize the public. In this sense, FT marketing campaigns, consumer group practices and governmental policies constantly shape and influence individual choices. These campaigns send strong signals that inform not only consumers but also producers. In turn, contrary to proposing boycotts of products or brands, fair trading promotes the consumption of ethically produced goods according to an ethical logic. It does so by exploiting consumers' ability to change from one brand to another – in this case, from conventional to FT brands. On the production side, this change means shifting from conventional to alternative agricultural systems. However, we also recognize that fair trading is a consumer-driven system; without ethical consumption, there is little point in producing goods ethically.

Many attempts to identify and profile ethical consumers have emerged since Wagner's, in his research on ethical consumerism, attempted to identify a standard ethical consumer as typical of ABC, the ACORN market segmentation classification.[32] According to Wagner, ethical consumers in the late 1990s were affluent, high income, liberal people. Moreover, marketing research institutes tended to classify ethical consumers as typically female,[33] though gender dominance in ethical purchase decisions might not reflect reality. However, in the Western world, despite their increasing role in the workforce, women are still largely responsible for household shopping.

Such profiling says little about who the ethical consumer actually is. Harrison and colleagues compare traditional or conventional consumers with ethical ones.[34] They find that traditional purchase behaviour pertains to the fulfilment of consumption desires, based on cost, quality and the utility that a product can provide. An ethical consumer adds other criteria to the traditional purchase behaviour. Dickinson and Hollander also designate the ethical consumer's main motivation as an individual set of values and beliefs.[35] This notion is important in any attempt to understand ethical purchase behaviour, because ethical buyers are frequently interested in higher-dimension issues that transcend the self. Hence, a typical ethical purchase decision is concerned not only with the self and the personal effect and impact of choices but also with the other, or the collective being, whether it is close or remote. According to this notion, fair trading consumers can be characterized as those who respond positively to higher-level issues, such as the possibility of improvement in production systems, care for the environment and the sustainability of people's livelihoods in remote or developing countries. Values, beliefs and the collective** are as important as the intrinsic product, whether it be coffee, cocoa or bananas, in the ethical purchase behaviour.

We thus have addressed some controversial issues that are relevant to an understanding of fair trading. In particular, increased globalization and some of its less positive effects have paved the way for alternative marketing channels such as fair trading. Such a system has evolved and, in recent times, become a strong niche market for some foodstuffs, largely due to ethical consumption that has sustained and encouraged more producers to engage in FT activities. However, ethical consumption does not act in isolation, and consumer–producer relationships might not be sufficient to support a specialist supply

---

** In a world in which ideology is no longer the dominant philosophical engine, we argue that fair trading issues increasingly have filled gaps in consumers' mind.

chain. Therefore, it is necessary to understand how FT supply chains are organized and governed.

Because of its popularity, fair trading often has been suggested as the natural alternative for small producers to engage in activities that could provide them with some sort of sustainability, whether in the form of higher income, guaranteed outlets for products, greater efficiencies or access to information. Yet assessments of small producers' capabilities seldom occur before they engage in FT activities, which means they sometimes might be lured into schemes that might not be advantageous for them.

Therefore, we first address the role of governance. We also attempt to characterize fair-trading certification in light of its requirements. Next, we use small honey producer associations in southern Brazil as a case study to illustrate how fair trading criteria might be applied to ascertain whether these produces can comply with the requirements and the extent to which such compliance would be advantageous for them.

GOVERNANCE AND THE FLO

A global value chain (GVC) analysis provides an understanding of the governance structure and institutional framework within global production, as well as the spread of sourcing and manufacturing across developing countries. Gereffi differentiates between two types of chain configuration and governance structure: producer-driven and buyer-driven.[36] The first relates to chains in which large companies (usually transnationals) coordinate the whole supply chain, characterized by capital- and technology-intensive industries, such as automobiles and computers. The main strategy is to attain economies of scale for manufacturing. Conversely, buyer-driven chains focus on the domination of retail companies and brand-name merchandisers. They compete intensively against one another by introducing minor innovations of their products and packaging while also maintaining strict quality criteria and price levels.[37]

In chain governance, the key agent, or 'governor', delegates, manages and enforces the production process to ensure that everyone complies with the standards. Both buyer- and producer-driven systems may contrast, but they are not mutually exclusive.[38] Large companies usually play the role of the governor, creating and monitoring their own standards. These governors might be manufacturers, with technological and production information (producer-driven), or retailers and branded companies that concentrate on the possession and translation of market information.

Traditionally, the food industry has displayed the characteristics of a producer-driven chain, dominated by large processors such as Nestlé and Heinz. However, this pattern has been changing due to the concentration of retailing, which challenges the position of large processors. The governor of the chain, who is responsible for setting the standards, should have sufficient size and capacity to monitor the standards, whereas the supplier should have the capacity to invest to meet those standards. However, processors in developing countries have difficulties meeting the requirements of UK supermarkets with respect to, for example, food safety, care for the environment and labour standards.[39]

As new demands from leading companies increase in complexity, the original typology has received further development.[40] By identifying key determinants of the relationships, these developments highlight five forms of chain governance, as in Figure 19.1.

The key determinants are the complexity of transactions (CT), codifiability of information (CI) and capability of suppliers (CS). These key determinants drive different

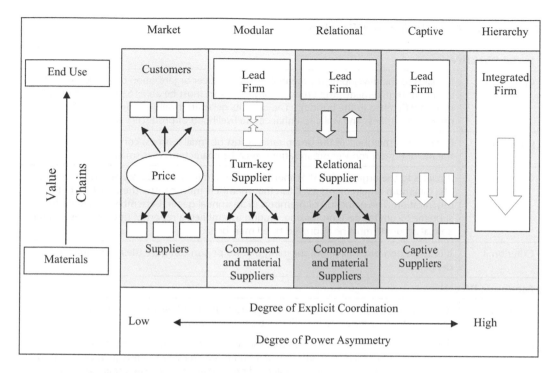

Figure 19.1 Global value chain dynamics[41]

kinds of inter-firm relationships, in which consumer culture and global scale regulations affect the configuration of the value chain.[42] Governance therefore can be exercised in different ways and through different agents along the length of the food supply process, depending on the standards (that is, codification of technical and marketing information). It is fundamental to identify the agent responsible for setting and monitoring the standards, particularly in a global food chain.

Market governance suggests a traditional view of price as the main requirement. In modular governance though, codification of information leads to production chain segmentation, with frequent transactions and relative supplier independence. Relational governance involves solidarity and cooperation with more balanced power between parties. Captive governance indicates a more clear coordination between processors and retailers, known as quasi-integration. Finally, hierarchy governance focuses on control of the whole process.

Raynolds et al consider this typology useful to understand FT relationships, which move from market, where price is the main requirement, to tighter relationships, where trust and knowledge transfer may evolve.[43] However, they also state that this system is co-coordinated not only by economic reasoning but also by the role of external agents and institutions.

In 1997, the FLO was founded.[44] Two overarching directives organize the structuring of the criteria regarding social and economical dimensions, as Table 19.1 shows.

To produce or manufacture products carrying the FT banner, producers' associations must adhere to a set of standards enforced by FLO with respect to economic, social and environmental directives that follow certain criteria and conventions that have been internationally recognized. These rules should facilitate the acceptance of products

Table 19.1 Criteria for certification

Social Directives	
Criterion 1	It relates to the development potential that a producer organization can have. As for the minimum requirements, the producer organizations must be able to prove that FT will make a difference to the business. The benefits generated will provide support for the growth of business and will also enhance the livelihood of producers and their families.
Criterion 2	It states that members of the organization must be small farmers; consequently, most of the members of producer organizations must be characterized as small farmers.
Criterion 3	It relates to the organization that should be democratic, participatory and transparent. The minimum requirements establish that members must control the structure of the organization. The gathering of members in an annual general assembly should be the supreme forum for decision making. During assemblies, a report of the activities and the annual accounts must be approved by all members.
Criterion 4	It is about non-discrimination, meaning that no person can have their participation denied.
Economic Directives	
Criterion 5	It is about the FT price premium. The organization should be able to manage the FT premium to the producers' benefit in a transparent way.
Criterion 6	It relates to the export capacity, so that the organization needs to possess some physical assets and be qualified to export. The minimum requirement to engage in exporting is a volume of 20 tonnes or the equivalent of one container load. The organization should have access to a telephone line, the Internet and a computer system and evidence of good administrative skills. The products traded must follow the current export quality standards, thus demonstrating that the organization has the ability to export successfully directly or, if necessary, indirectly through a partner. The contract established between seller and buyer must also provide a clear indication that the transaction is FT certified.
Criterion 7	It relates to economic growth. One of FT purposes is to increase the capacity of small farmers to work in groups aiming at the export market. Producers must develop their skills and capacities so as not to depend on other people who could behave opportunistically.
Environmental Directive	
Criterion 8	It relates to the environment and the way resources are managed, such as water, natural forests and other areas in the vicinity of the farm activity. The environment must be protected, including control against erosion and waste management. Environmental monitoring guidelines must be applied.

internationally. To be eligible for FLO certification, producers' organizations must comply with both general criteria (applicable to any product) and product-specific criteria. For each criterion, there is a minimum requirement for achieving certification. In this sense, fair trading could be characterized as a buyer-driven chain, governed by an external agent, namely, the certification body. Governance is exercised through complex relationships amongst actors, where information is codified and the capabilities constantly verified.

Understanding what fair-trading governance is has aided in the analysis of the likelihood of honey producers in southern Brazil, including whether they should pursue such certification.

Methodology

The research reported in this chapter is part of a study project focused on small honey producers in Brazil, financed by the Ministry of Science and Technology. The objective of the research is to ascertain whether beekeepers' associations comply with the criteria to obtain FLO certification, with a view towards developing a market entry for their products and especially aiming at the export market.

This study relied on descriptive and exploratory techniques, which provided further insights into this field of research.[45] The different techniques used included desk research of files, maps, reports and articles, as well as primary research. Key informants who demonstrated either special knowledge or experience in the subject area were selected to partake in rounds of semi-structured interviews. A combination of methods provided flexibility and interactivity for the quick completion of the task. It also allowed us to explore new issues that emerged during the data collection (that is, learning as we go). Nonetheless, this method also involves disadvantages, including limited validity, the dependence on our abilities to select an appropriate sample and the lack of quantitative information provided.

The different qualitative methods used in this research include:

- secondary data collection;
- participatory observations in meetings, seminars and events relating to the honey sector. Participatory observation took place in monthly meetings as part of a development project promoted by SEBRAE (an institution responsible for the development of small and medium-sized enterprises in Brazil);
- in-depth interviews, conducted with representatives of FTO, which is responsible for certification and auditing, based in São Paulo.

Regarding the selection of the focus region of the study, we choose Osório, located on the northeastern Atlantic coast of Rio Grande do Sul state. It is some 100 km from the state's capital, Porto Alegre. This region reflects several economic, social and environmental features that are appropriate for evaluating the FT criteria. Some 390 tonnes of honey is produced per year from the six associations selected in this region.

We also developed the case studies on the basis of the semi-structured interviews with beekeepers active in their associations, coupled with direct observations. A survey-type questionnaire based on the FLO criteria aided the semi-structured interviews with the leaders of the six selected beekeeper's associations. In addition, we conducted four interviews with members of those associations to validate some information.

Finally, we collected the data between November and December 2006 from the six honey-producing associations. In total, we interviewed 28 beekeepers using a semi-structured questionnaire. These interviews lasted approximately 1 hour each. All in-depth interviews were conducted and analyzed by each of the authors.

The last step in this study consisted of writing the case study. The reliability of the data increased because we developed clearly conceptualized constructs and used multiple indicators. To confirm validity, we discussed the findings with some key members of the associations and some of the respondents to ascertain their impressions. The use of multiple sources (in-depth interviews, annual reports, secondary data and direct observation) also helped improve the validity.

Results

As a result of the investigation, we developed Table 19.2, which provides a summary of the assessments of the six associations, according to FLO criteria.[46]

Data collected from the interviews revealed that small honey producers faced many difficulties when we applied the standard FT criteria for assessing their capabilities. The data also highlighted the main difficulties small farmers' associations confront in complying with and accessing FT marketing channels.

Although FT channels have provided opportunities for small farmers, the initial cost of certification can be prohibitively high. This cost has constrained the potential outreach of fair trading as an alternative. Because the FT label essentially is a private label, it operates parallel to other trade-linked certification mechanisms. To ensure that smallholders benefit from such a system, greater understanding of the adoption patterns, as well as the social-economic benefits of farmers' involvement in FT, is needed.

Moreover, the honey supply chain suffers some shortcomings, in that it typically operates at a regional level. The most difficult criterion to meet for the honey associations therefore is the export capacity. Honey producers are not even certified by the Brazilian Ministry of Agriculture to trade at an inter-state level. For the export dimension, this issue seems to be an even greater challenge.

The economic criterion also represents a main problem faced by the beekeepers interviewed. In terms of infrastructure, a lack of resources limits honey-processing installations and purchases of the necessary equipments. Some beekeepers increasingly have been partaking in other economic activities, off the farm, to generate enough income. This shift could be related to the effect of globalization, as discussed previously, because small farmers increasingly cannot generate enough income and therefore become marginalized.

Table 19.2　Fair trade criteria applied to associations

Summary of Associations							
Criteria		**Associations**					
		AAPO	**AAO**	**APITRA**	**APIMAQ**	**APA**	**AAFTV**
SOCIAL	1. Potential development	No	No	No	Partial	Yes	Yes
	2. Smallholders	Partial	No	Yes	Yes	Yes	Yes
	3. Democracy, participation and transparency	Yes	Partial	No	Partial	Yes	Yes
	4. Non-discrimination	Yes	Yes	Yes	Yes	Yes	Yes
ECONO	5. FT premium price	Yes	Partial	No	Partial	Yes	Yes
	6. Export capacity	Partial	Yes	No	Partial	No	No
	7. Economic growth	Yes	No	No	Yes	Yes	Yes
8. Environmental protection		Partial	Partial	Partial	Partial	Partial	Partial
Aptitude to obtain FT certification		No	No	No	Partial	Partial	Partial

Many farmers had to find work as temporary labourers, especially in the summer months, to be able to earn some savings to re-invest into their agriculture business. Furthermore, access to credit to invest in beekeeping activities and infrastructure was restricted.

The six associations only partially met the environmental criterion, or preservation of the environment. Although they were not using agrochemicals, no formal control existed regarding environmental practices, such as an internal control system.

Regarding discrimination, no overt perception of prejudice, be it racial, religious or gender, was observed. Therefore, though this criterion is not considered a problem in the region studied, it is important for FT networks to ascertain that no issues could jeopardize the possible international acceptance of the honey produced. Because it acts globally, FLO must be concerned about discrimination issues in many countries. Consumers' green or ethical logics play important roles in determining their acceptance of a product. If either environmental or discrimination criteria are not fulfilled, honey producers will face great challenges.

For the criterion of economic development, we attempted to verify the members' socio-economic condition, as well as the association's infrastructure and level of commitment to adhering to the certification rules. In the region studied, beekeeping was often a supplementary activity, used to complement the family's income. The interviews also indicated beekeeping as a hobby, a practice not uncommon in other parts of the world as well.[47]

Figure 19.2 attempts to show how the honey supply chain is configured. From suppliers of inputs to the market, the stages of honey processing can be clearly identified. The data collected also indicated reasonable institutional support available to the producers.

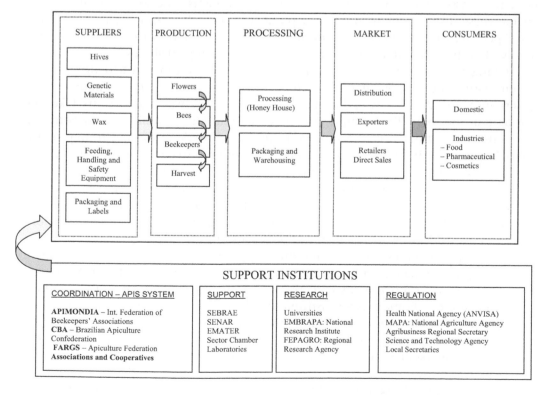

Figure 19.2 Generic honey chain in southern Brazil

From Figure 19.2, it is possible to identify the main segments that integrate the honey production chain:

- Inputs related to the production of honey, such as the acquisition of boxes to house the hives, wax matrix so that the bees can start producing honey, safety equipment (for example, overalls, gloves, masks, boots, fumigator), bee swarms, queens and an area to locate the beehives.
- Production carried out by the beekeeper, such as apiary installation, handling the hives, harvesting honeycombs and transporting them to the Honey House for processing.
- Agri-industrialization by the Honey House, such as centrifugation, filtering, decantation for 48 hours, stocking, classifying, packaging and labelling.
- Commercialization, which consists of two basic forms: when the sale is completed by the beekeeper at farmers' markets or by distribution to grocery stores, stands, bakeries and other small retailers, versus bulk quantities, such that the honey gets packaged in containers of 50, 100, 200 and 300 litres for export. This latter type of sale usually is performed through larger companies, which have sales warehouses or specialized companies that can offer the product on a large scale.

Of the six associations studied, only three could qualify for FT certification, according to Table 19.2: APIMAQ, APA and AAFTV. Nevertheless, to comply fully with FT criteria, even these associations would need to make adjustments. The FT certification would enable these associations to produce a product with high specificity. End consumers could perceive this specificity as a dedicated asset, related not only to the social attributes involved in the production but also to the product's quality attributes.

Having to adhere to FT certification criteria might mean certified honey uses a system with a specific governance structure. Thus, certification acts as a main driver of governance. Certification would require a change in the way transactions take place throughout the chain. For example, beekeepers and importers would rely on norms and sets of rules set out by the certification body. Importers would buy honey from producers following a set minimum price and premium. Importers also could provide assistance to producers through (1) advice and technical support of product development; (2) training; (3) supporting them during difficult social and economic moments; and (4) financing production starts or harvests. Such extra layers of care for small producers do not occur in conventional (free) market arrangements. Some observers consider such support a reflection of altruistic behaviour by the ethical consumer.

The complexity of the standards and thus the need to monitor and assure compliance requires tight governance. In the FT system, governance is exercised by FLO instead of an internal agent, as proposed by GVC theory.[48] Transactions between retailers and final consumers use certification as an instrument, which guarantees the consumer that the product has the desired social, environmental and quality attributes. Certified FT honey is an asset with high specificity, and FLO is responsible for organizing and transferring technical and marketing knowledge from the consumer market to producers located in developing countries.

The honey-producing associations studied currently find themselves in a pre-certification stage. Compared with more long-term, established fair trading associations, such as those for coffee,[49] these small farmers must go to great lengths to qualify to trade under the FT mark. To comply, honey producers have to fulfil all the basic steps that guarantee consumers fair trade satisfaction: higher ethical attainment, green ideology and sustainable practices. Previous coffee fair trading examples show that when small farmers realize that ethical consumers in developed countries can simply demand good quality products, issues such as environmental sustainability play a secondary role. In that case, fair trading could become an accessory.

The role of governance thus is once more relevant in controlling the FT supply chain, which in turn is not much different from what happens in conventional supply chains. Therefore, we might argue that fair trading practices are not dissimilar from those of conventional marketing channels.

Conclusions

As previously noted, the internationalization of fair trading supply chains is a recent development that requires further studies. The implications of FT compliance and governance for producers, intermediaries and consumers have not been fully explored. We might infer that the internationalization of fair trading is becoming a global phenomenon, yet production remains typical of developing countries, whereas consumption relates to developed countries.

A system that, at its onset, proposed to counterbalance unfair market imbalances for small farmers might be classed as buyer-driven, which is not dissimilar from conventional marketing systems. In the consuming countries, FT marketing is common among large food retailing formats that carry branded FT merchandises.

Fair trading as a proposition for adding value to honey products in Brazil does not clearly guarantee advantages. The exploratory nature of this study allows us to identify some difficulties that honey associations must address, especially with regard to organization and export capacity. These two features act as the primary obstacles to beekeepers qualifying for FT certification.

Furthermore, we argue that the risk of creating a high specificity asset may be too great, especially one that is not recognized in the Brazilian domestic market.[††] Without alternative markets, FT honey could become too dependent on international demand.

In addition, FLO certification criteria determine who is included in the FT system. Its role is typical of a chain governor, which makes FLO responsible for the coordination of information flows and tracking processes, which are key to the performance of the system. The studied associations did not provide enough capacity to embark on market entry strategies without assistance. The governance of FLO or other agents, such as retailers, could provide the much-needed collective organization and transfers of knowledge that enable technical upgrades.

Fair trading also has evolved from its original precepts and provides an alternative route to markets for agricultural products from small producers in different locations in developing countries. It is undeniable that the adoption of FT certification enables small producers to gain access to international market and add value to their products. In addition, FT has provided a cushion for small farmers against the adverse effects of

---

[††]   In Brazil, the market for fair trading is in its early stages.[50]

globalization in agri-food chains. But the extent to which FT can protect small farmers from market vagaries also is limited.

In this sense, FT certification acts as an entry barrier. To quality for FT certification, small honey produces must first set up an association to attain minimum export capacity and thus pay annual certification costs, among other requirements. As Figure 19.2 shows, the honey supply chain studied had not yet reached internationalization. The associations we studied were not ready to embark on exporting, because they had not gained enough knowledge to enter other markets at the domestic level in Brazil. However, non-economic factors also influence an FT system.[51] As it exercises its coordinating role in a honey FT chain, FLO-CERT also could provide the necessary conditions to overcome this limitation.

The FT certification also requires a change in the way transactions have been occurring throughout the chain. For example, currently, importers buy from beekeepers and are responsible for the payment of the minimum price. They also can give assistance to producers.

The complexity of transactions can be simplified through codification of information, transferred by FLO to suppliers. In the case of honey producers, suppliers have medium-level capabilities. Nevertheless, the three identified associations could be developed, with the help of institutions such as SEBRAE and the Ministry of Agrarian Development. In addition, we identified a major bottleneck for farmers that want to participate in FT initiatives, namely, the Brazilian Ministry of Agriculture. The ministry is the organ responsible for granting labels to food processing companies that qualifies them to engage in interstate trade and export.[‡‡]

Considering the GVC approach, the FT system would use a more relational form of governance, as seen in Figure 19.2. Such governance is exercised by FLO instead of by an internal agent, as proposed by GVC theory. In this case, certification acts as an instrument that guarantees consumers that the products they buy carry the desired social, environmental and quality attributes. Certification also facilitates transactions between retailers and end consumers. As such, the assets traded have high specificity. By taking responsibility for organizing and transferring technical and marketing knowledge from the consumer market to producers located in developing countries, FLO reinforces this specificity.

For small farmers in developing countries, understanding ethical consumers' consumption motivations, which ultimately may be about 'selling images of poor farmers' who make a living in degraded environments, can be difficult. However, small honey producers in Brazil may not need to go to such lengths provided they produce top-quality products. Yet understanding the ethical motivations of those in producing countries remains of extreme importance.

Moreover, the impact of ethical consumers' choices in developed countries is also felt in the producers' countries. High-level issues determine the demand that supports the expansion of a FT system with global ramifications. Thus far, ethical demand for FT products has sustained a growing market. What might happen though when the FT label that communicates stories about people and creates a message of trust, respect and partnership between producers and consumers is no longer sufficient to shape the

---

‡‡   The SIF label is granted by the Ministry of Agriculture to companies in the food sector and represents the Federal Inspection Service.

engagement of consumers in FT networks? We leave this thought-provoking question for further research projects.

References

1. Dolan, C. S. & Humphrey, J. (2000), 'Governance and trade in fresh vegetables: the impact of UK supermarkets on the African horticulture industry', *Journal of Development Studies*, Vol. 37, No. 2, pp. 147–176.
2. McMichael, P. (2001), 'The impact of globalisation, free trade and technology on food and nutrition in the new millennium', in *Proceedings of the Food and Nutrition Society*, Vol. 60, pp. 215–220.
3. Ibid.
4. Barrat-Brown, M. (1993), *Fair Trade, The Perfect Guide to the Maze of International Commerce*. Zed Books, London: Freidberg, S. (2003), 'Cleaning up down south', *Social and Cultural Geography*, Vol. 4, No. 1, pp. 27–43; Aguiar, L. K. (2005), *The sustainability of ethically produced coffee: the case of Brazil and Peru*, in Conference proceedings of the International Food and Agribusiness Management Association, June. Chicago, IL; Raynolds, L.T., Murray, D. & Wilkinson, J. (2007a), *Fair Trade: The Challenges of Transforming Globalization*. London: Routledge; Sejdaras, A. & Aguiar, L. (2007), *Recent developments in the apiculture sector in Albania: the insertion of beekeepers in the EU honey supply chain*, in Conference proceedings of the Second International Symposium on Enhancing the Performance of Supply Chains in the Transitional Economies, September. Hanoi, Vietnam.
5. Barrat-Brown, op. cit.
6. Ibid.
7. Ibid.
8. Lewin, B.; Giovannucci, D. & Varangis, P. (2004), *Coffee markets: new paradigms in global supply and demand*. Agriculture and Rural Development Discussion Paper 3. The World Bank Group. Available at http://lnweb18.worldbank.org/ESSD/ardext.nsf/11ByDocName/ExecutiveSummary-CoffeeMarkets (accessed 13 August 2004).
9. Raynolds et al., 2007a, op. cit.
10. Barrat-Brown, op. cit.
11. Vieira, L. & Aguiar, L. (2006), *Fair trade as product differentiation strategy and market access: an exploratory study of honey producers in southern Brazil*, In Conference proceedings of the International Symposium of Fresh Produce Supply Chain Management, December. Chiang Mai, Thailand.
12. Raynolds et al., 2007a, op. cit.
13. Zhener, D. (2002),'An economic assessment of fair trade in coffee', *Chazen Web Journal of International Business*. Columbia Business School.
14. Ibid.
15. Mark s& Spencer (2007),'Plan A'. Available at http://corporate.marksandspencer.com/howwedobusiness (accessed 3 September 2007); Waitrose (2007), 'The Waitrose Foundation'. Available at http://www.waitrose.co.uk (accessed 31 July 2007).
16. Ethical Trading Initiative (2007), 'Ethical trading initiative'. Available at http://www.ethicaltrade.org (accessed 12 June 2007).
17. Ibid.
18. The Cooperative Bank (2007), Ethical consumerism report.

19.   Ibid.
20.   BBC (2006), 'How fair is fair trade?' Available at http://news.bbc.co.uk/1/hi/business/4788662.stm (accessed 12 May 2007).
21.   Tesco (2007), 'Tesco sells 130 fairtrade lines'. Available at http://www.tescocorporate.com/plc/corporate_responsibility/resp_buying_selling/selling_responsibly/ (accessed 31 July 2007).
22.   J. Sainsbury's (2007), 'Fair development fund'. Available at http://www.j-sainsbury.co.uk/cr/?pageid=128 (accessed 31 July 2007).
23.   Waitrose, op. cit.
24.   Marks & Spencer, op. cit.
25.   Isaak, R. (1999),'Green logic: ecopreneurship, theory and ethics', *Teaching Business Ethics*, Vol. 3, No. 3.
26.   Ibid.
27.   Shaw, D., Newholm, T. & Dickinson, R. (2006), 'Consumption as voting: an exploration of consumer empowerment', *European Journal of Marketing*, Vol. 40, No. 9/10, pp. 1049–1067.
28.   Harrison, R., Newholm, T. & Shaw, D. (2005), *The Ethical Consumer*. Newbury Park, CA: Sage.
29.   Isaak, op. cit.
30.   Ibid.
31.   Barnett, C., Cloke, P., Clarke, N. & Malpass, A. (2005), 'Consuming ethics: articulating the subjects and spaces of ethical consumption', *Antipode*, Vol. 37, No. 1, pp. 23–45.
32.   Wagner, S.A. (1997), *Understanding Green Consumer Behaviour: a Qualitative Cognitive Approach*. Consumer Research and Policy, Routledge, London.
33.   Datamonitor (2002), *The Outlook for Organic Food and Drink*. London: Datamonitor Europe.
34.   Harrison et al., op. cit.
35.   Dickinson, R. & Hollander, S.C. (1991), 'Consumer votes', *Journal of Business Research*, Vol. 22, pp. 335–346.
36.   Gereffi, G. (1994), 'Organization of buyer-driven global commodity chains: how U.S retailers shape overseas production networks', in Gereffi, G. and Kzeniewicz M, (eds), *Commodity Chains and Global Capitalism*. New York: Praeger.
37.   Gereffi, G. (1999), 'International trade and industrial upgrading in the apparel commodity chain', *Journal of International Economics*, Vol. 48, No. 1, pp. 37–70; Kaplinsky, R. (2000), *Spreading the gains from globalisation: what can be learned from value chain analysis*, IDS Working Paper 110, p. 37; Dolan & Humphrey, op. cit.
38.   Gereffi, 1994, op. cit.
39.   Dolan & Humphrey, op. cit.
40.   Gereffi, G., Humphrey, J. & Sturgeon, T (2005), 'The governance of global value chains', *Review of International Political Economy*, Vol. 12, No. 1, pp. 1–27.
41.   Ibid.
42.   Ibid.
43.   Raynolds, L., Murray, D. & Wilkinson, J. (2007b), *Fair Trade in the Agriculture and Food Sector: Analytical Dimensions*. London: Routledge.
44.   Ibid.
45.   Yin, R. K. (2003), *Case Study Research: Design and Methods*, 3d edn. Newbury Park, CA: Sage.
46.   Fairtrade Labelling Organisation (2004), *Fairtrade Standards for Honey for Small Farmers' Organisations*. Available at http://www.fairtrade.net (accessed 20 December 2005).
47.   Sejdaras & Aguiar, op. cit.; Aguiar, op. cit.
48.   Gereffit, 1994, 1999, op. cit.
49.   Aguiar, op. cit.

50. Ibid.
51. Raynolds et al., 2007a, op. cit.

Index

Printed and bound by CPI Group (UK) Ltd, Croydon, CR0 4YY

21/10/2024

01777095-0010